Motion Planning for Humanoid Robots

Kensuke Harada · Eiichi Yoshida · Kazuhito Yokoi
Editors

Motion Planning for Humanoid Robots

Springer

Dr. Kensuke Harada
Dr. Eiichi Yoshida
Dr. Kazuhito Yokoi

National Institute of Advanced Industrial Science and Technology
Intelligent Systems Research Institute
Tsukuba Central 2
Umezono, 1-1-1
305-8568 Tsukuba-shi, Ibaraki
Japan

kensuke.harada@aist.go.jp
e.yoshida@aist.go.jp
kazuhito.yokoi@aist.go.jp

ISBN 978-1-4471-5705-2 ISBN 978-1-84996-220-9 (eBook)
DOI 10.1007/978-1-84996-220-9
Springer London Dordrecht Heidelberg New York

British Library Cataloguing in Publication Data
A catalogue record for this book is available from the British Library

Softcover re-print of the Hardcover 1st edition 2010
KineoWorks is a trademark of Kineo CAM Corporation, MINIPARC – Bat. 2, Rue de la Découverte, BP 57748, 31677 LABEGE CEDEX, FRANCE, (RCS TOULOUSE B 433 754 090), http://www.kineocam.com

Cover design: eStudioCalamar, Figueres/Berlin

Printed on acid-free paper

Springer is part of Springer Science+Business Media (www.springer.com)

Preface

Since one of the goals of robotics research is to realize the robot that can automatically work like a human, the research on the motion planning for a humanoid robot is considered to be essential and important. However, the motion planning for a humanoid robot can be quite difficult due to their complex kinematics, dynamics and environment. It is consequently one of the key research topics in humanoid robotics research and the last few years have witnessed considerable progress in the field. This book surveys the remarkable recent advancement in both the theoretical and the practical aspects of humanoid motion planning. Various motion planning frameworks are presented in this book, including one for skill coordination and learning, and one for manipulating and grasping tasks.

This book is composed of 10 chapters written by different robotics researchers. The editors organized workshop on motion planning for humanoid robots in conjunction with the IEEE-RAS International Conference on Humanoid Robots (Humanoids) on 29 November 2007 in Pittsburg, USA. The goal of this workshop was to provide an opportunity to discuss the recent advances in humanoid motion planning from both theoretical and experimental points of view. We were pleased to have 10 presentations by qualified researchers of humanoid motion planning. After the workshop, we asked Springer-Verlag to publish a book related to this workshop. The research workers, including the speaker of the workshop, agreed to each write a chapter for this book.

Several new advances are included in each chapter. Chestnutt looks at gait planning and locomotion for navigating humanoid robots through complicated and rough terrain. Sentis describes torque-level control realizing multi-contact behaviors. Gienger *et al.* reviews some elements for a movement control and planning architecture. Yoshida *et al.* describe the research results on planning whole-body locomotion, reaching, and manipulation. Vahrenkamp *et al.* present a motion planning framework for manipulating and grasping tasks. Escande *et al.* investigate the problem of planning sequences of contacts that support acyclic motion in a highly constrained environment. Harada presents motion planning based on a biped walking pattern generator. Stilman describes the autonomous manipulation of moving obstacles. Hauser *et al.* present a motion planner that enables a humanoid robot to push an object to a desired location on a cluttered table. Kallmann *et al.* present a motion planning framework for skill coordination and learning. The editors would like to express the sincere thanks to all the authors of this book for their excellent contributions.

Tsukuba, Japan,
November, 2009

Kensuke Harada
Eiichi Yoshida
Kazuhito Yokoi

Contents

List of Contributors

Joel Chestnutt
National Institute of Advanced Industrial Science and Technology (AIST), 2-3-26, Aomi, Koto-ku, Tokyo 135-0064, Japan, e-mail: joel@chestnutt.us

Luis Sentis
Department of Mechanical Engineering, The University of Texas at Austin, Austin, TX 78712, USA, e-mail: lsentis@austin.utexas.edu

Michael Gienger
Honda Research Institute Europe GmbH, Carl-Legien-Strasse 30, 63073 Offenbach, Germany, e-mail: michael.gienger@honda-ri.de

Marc Toussaint
Technical University of Berlin, Franklinstr. 28/29, 10587 Berlin, Germany, e-mail: mtoussai@cs.tu-berlin.de

Christian Goerick
Honda Research Institute Europe GmbH, Carl-Legien-Strasse 30, 63073 Offenbach, Germany, e-mail: christian.goerick@honda-ri.de

Eiichi Yoshida
National Institute of Advanced Industrial Science and Technology (AIST), 1-1-1 Umezono, Tsukuba 305-8568, Japan, e-mail: e.yoshida@aist.go.jp

Claudia Esteves
Facultad de Matematicas, Universidad de Guanajuato, Guanajuato, 36000 Gto., Mexico, e-mail: cesteves@cimat.mx

Oussama Kanoun
LAAS-CNRS, 7 avenue du Colonel Roche, F-31077 Toulouse, France, e-mail: okanoun@laas.fr

Mathieu Poirier
LAAS-CNRS, 7 avenue du Colonel Roche, F-31077 Toulouse, France, e-mail: mpoirier@laas.fr

Anthony Mallet
LAAS-CNRS, 7 avenue du Colonel Roche, F-31077 Toulouse, France, e-mail: mallet@laas.fr

Jean-Paul Laumond
LAAS-CNRS, 7 avenue du Colonel Roche, F-31077 Toulouse, France, e-mail: jpl@laas.fr

Kazuhito Yokoi
National Institute of Advanced Industrial Science and Technology (AIST), 1-1-1 Umezono, Tsukuba 305-8568, Japan, e-mail: kazuhito.yokoi@aist.go.jp

Nikolaus Vahrenkamp
University of Karlsruhe, Haid-und-Neu Str. 7, 76131 Karlsruhe, Germany, e-mail: vahrenkamp@ira.uka.de

Tamim Asfour
University of Karlsruhe, Haid-und-Neu Str. 7, 76131 Karlsruhe, Germany, e-mail: asfour@ira.uka.edu

Ruediger Dillmann
University of Karlsruhe, Haid-und-Neu Str. 7, 76131 Karlsruhe, Germany, e-mail: dillmann@ira.uka.edu

Adrien Escande
CEA-LIST, Fontenay-aux-Roses, France, e-mail: adrien.escande@cea.fr

Abderrahmane Kheddar
CNRS-UM2 LIRMM, Montpellier, France and CNRS-AIST JRL, UMI3218/CRT, Tsukuba, Japan, e-mail: kheddar@lirmm.fr

Kensuke Harada
National Institute of Advanced Industrial Science and Technology (AIST), 1-1-1 Umezono, Tsukuba 305-8568, Japan, e-mail: kensuke.harada@aist.go.jp

Mike Stilman
Georgia Institute of Technology, Atlanta, GA, USA, e-mail: mstilman@cc.gatech.edu

Kris Hauser
Computer Science School of Informatics and Computing, Indiana University at Bloomington, Bloomington, IN 47405, USA, e-mail: hauserk@indiana.edu

Victor Ng-Thow-Hing
Honda Research Institute USA, 800 California St., Suite 300, Mountain View, CA 94041, USA, e-mail: vng@honda-ri.com

Marcelo Kallmann
University of California, Merced; 5200 N. Lake Road, Merced CA 95343, USA,
e-mail: mkallmann@ucmerced.edu

Xiaoxi Jiang
University of California, Merced; 5200 N. Lake Road, Merced CA 95343, USA,
e-mail: janexip@ucmerced.edu

Chapter 1
Navigation and Gait Planning

Joel Chestnutt

Abstract Humanoid robots are wonderfully complex machines, with the potential to travel through the same formidable terrains that humans can traverse, and to perform the same tasks that humans can perform. Unfortunately, that complexity and potential results in challenging control problems and computationally difficult planning issues. In this chapter, we look at gait planning and locomotion for navigating humanoid robots through complicated and rough terrain. Humanoid robots have the ability to step over or onto obstacles in their path, and to move in an omni-directional manner, allowing the potential for increased versatility and agility when compared with wheeled robots. In this chapter we discuss some of the differences that arise when planning for legs instead of wheels, and some of the reductions and approximations that can be used to simplify the planning problem. We discuss how a safe sequence of footsteps can be chosen in rough terrain, and present planners which can efficiently generate collision-free, stable motions through complicated environments for real humanoid robots.

1.1 Introduction

Humanoid robots must constantly maintain balance using a small base of support which changes with every step the robot takes. Additionally, many different degrees of freedom must coordinate for even the simplest tasks. While we can apply whole-body motion planning methods to generate stable and safe motion [9], when applied to complex tasks and long sequences of motion these approaches become overwhelmed by the complexity of the problem. In this chapter we consider gait planning and locomotion, which cause very long and complex paths through the joint space of the robot. For this reason, we divide these large tasks into manageable

Joel Chestnutt
National Institute of Advanced Industrial Science and Technology (AIST), 2-3-26, Aomi, Koto-ku, Tokyo 135-0064, Japan, e-mail: joel@chestnutt.us

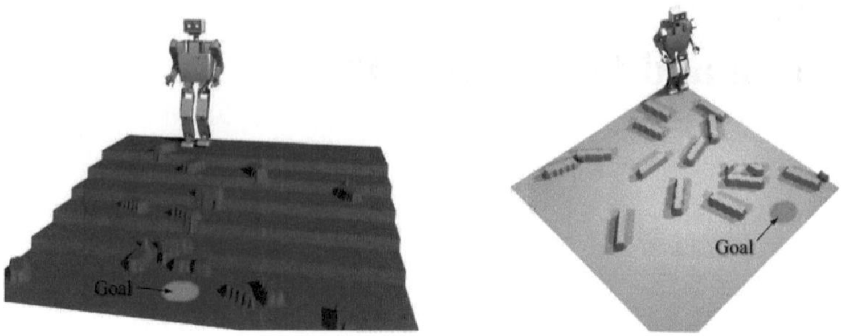

Figure 1.1 Example navigation problems a humanoid may face. In each case the robot must be able to reason about how it can safely step through the environment

chunks, and reduce the dimensionality of the problem to provide computationally-tracktable problems to tackle. For legged robots, locomotion is a largely cyclical motion, stepping one foot after another. We can take advantage of this repetitious nature of legged locomotion to solve navigation problems efficiently, while still utilizing the abilities that the legs give to the humanoid robot for traversing rough and complicated terrain.

1.1.1 Navigation Planning

Navigation for mobile robots is often a low-dimensional problem. For wheeled robots, the configuration space of the robot is usually represented as $SE(2)$, and the metrics to rate costs in the world are often only 2D. Due to this low dimensionality, algorithms such as A* can be used to provide optimal solutions for wheeled robots and still provide real-time performance. For real-world situations, the D* family of algorithms [12] provide inexpensive replanning to moving robots reacting to new sensor data.

To further increase performance, or for planning in higher dimensions (such as including velocities), a coarse–fine or hierarchical approach is used in many systems, in which a coarse global plan is first computed, and a finer-grained planner or controller follows the global path, while adjusting for the dynamics of the vehicle and newly-sensed obstacles. An example of this approach for a cart-like robot is shown in Figure 1.2. By layering planners and controllers in this way, the low-level control and planners only need to consider the near future, trusting to the higher-level planners in the hierarchy that the rest of the path will be executable. By shortening the planners' horizons, the individual planners can perform much quicker. The planners then replan as the robot begins to execute the plan to fill in details along the remaining path.

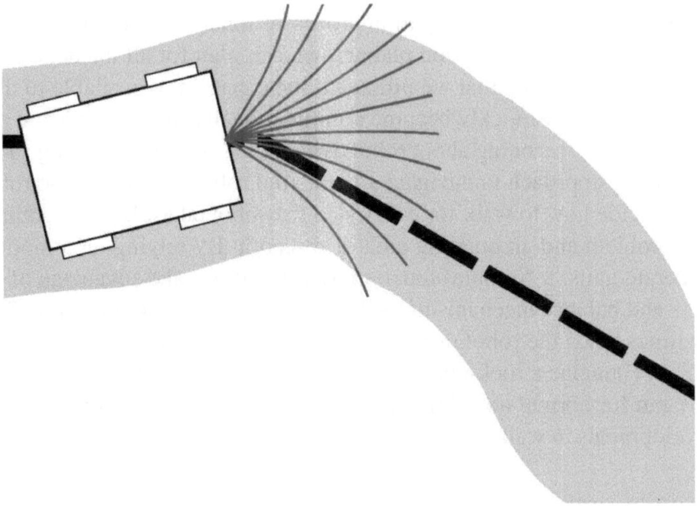

Figure 1.2 An example of coarse–fine planning for a mobile robot, where the coarse planning provides a rough path, and the fine planning is performed by a set of trajectories accounting for vehicle dynamics

1.1.2 Navigation and Legs

Despite the established research into navigation strategies for mobile robots, there are some significant differences between humanoid robots and wheeled robots. Unlike wheeled robots, which can simply turn their wheels to move through the world, for a humanoid to travel from one location to another involves constant coordination of many degrees of freedom to provide balance and support.

Another significant difference for legged locomotion is the form of interaction with the environment. Wheeled robots follow a continuous path through the environment, with mostly continuous change in contact with the ground (the exception being wheels leaving or re-contacting the ground). In contrast, while legged robots travel a continuous path through their state space, the contact with the environment changes discretely with every step taken. The dynamics of a legged robot are a hybrid dynamic system, containing discrete regions of continuous dynamics which change whenever a foot touches down or lifts off.

It is still possible to directly use navigation planners developed for wheeled robots for humanoid robots [8, 11]. For example, imagine bounding a humanoid robot by a cylinder, and planning a path for that cylinder through the environment so that it avoids all obstacles. This example amounts to "steering" the robot through the environment, and while effective and efficient in simple situations, this approach ignores the fact that the robot has legs, and with those legs the ability to step over obstacles, or to climb stairs, or to do the many things which make legged robots interesting.

To use those abilities, the navigation processes must consider them and plan for them. By using a full-body motion planner, we can plan for all the degrees of freedom of the robot, ensuring that we utilize all the capabilities available to the robot. However, this can very quickly become computationally intractable when planning long sequences and reasoning about robot dynamics. Additionally, using a full-body motion planning approach to navigate a humanoid robot means that the robot must essentially decide how to walk for every step the robot takes. Biped walking is still a difficult problem and an ongoing area of research. By relying on a motion planner to generate gaits, it becomes much more difficult to take advantage of walking controllers and balance mechanisms as they are developed and improved. There are certainly times when the robot may need to fall back to this kind of low-level planning [1, 2, 6] (imagine a rock-climbing situation, where each motion must be made carefully), but for normal navigation the robot can use an efficient gait utilizing the latest developments in walking and balance algorithms.

1.2 Dimensionality Reductions

Because the number of degrees of freedom of humanoid robots is too high to allow full-body motion planning for a large motion sequence, we must reduce the problem to a manageable size, while retaining as much of the abilities of the robot as possible.

In order to apply this planning approach we need to make an important decision, namely, what are these actions that the robot can take? Are they trajectories, control policies, or something else? This choice of what the actions consist of and how they are chosen has serious consequences for both how difficult the planning problem becomes, as well as how much of the robot's capabilities are used in the resulting plan. As stated earlier, the full motion planning problem for all the degrees of freedom of a legged robot can quickly become intractable. However, we can reduce this problem to planning in a lower-dimensional subspace in an efficient way by looking at the structure of the problem. After performing our planning in this low-dimensional space, we can turn our plan back into a full-body motion for the robot, allowing the robot to move to the goal.

In choosing how to reduce our search space, we have several goals for our final action representation:

- *Re-use of existing control.* A great deal of research has been performed on the subject of balance, walking, and running for legged robots. Ideally, our planning strategies can make use of the controllers that have resulted from that research, rather than requiring the planner to discover how to walk or run from scratch for each situation.
- *Planning simplicity.* The fewer dimensions in our search space, the easier it is for a planner to explore. Simplifying the planning space has a large impact on real-time performance on a robot in real environments. For humanoid robots in challenging, changing environments, quick planning allows for reaction to new sensor data and execution errors before they become problems.

- *Capability usage.* While we want to reduce the dimensionality of our problem and simplify the actions for our planner, another goal is to make sure that we do not sacrifice the unique abilities of legged robots in the process. We strive to find the right balance between utilizing as many capabilities of the robot as possible while at the same time not overwhelming the planner with too many details of the locomotion process.
- *Executability.* In addition to having enough freedom in the planner to use the full capabilities of the robot, we need to have enough information in the plan to ensure that we are not exceeding the robot's capabilities. Depending on the details of the robot and its controllers, some dimensions may be safely ignored. However, it is important that we do not reduce our problem to the point where the planned action sequences exceed the abilities of the robot and controller.

To plan for stances for a particular legged system, we will need to provide several descriptions of robot capabilities. Each of these capabilities will be based on both the physical limits of the robot as well as the limitations of any underlying gait controllers that the humanoid will use. For a humanoid system we must define the following:

- The *state space* through which the planner will search. This is a projection of the full state of the robot, which will be based around stance configuration.
- The *action model* which describes the connectivity of our states. In planning for stances, our path will not be continuous through our state space, so we must explicitly define connectivity.
- The *state-action evaluation function* which scores locations in the environment, and the actions the robot can take, providing a description of the types of terrain the robot and its controller can safely traverse. The function also is the objective function we wish to minimize, defining the optimal path.

Note that these are requirements for performing optimal discrete planning in any domain. We will be using the idea of planning for stances to collapse the motion planning problem down to a manageable state space, with the action model encapsulating the complexity of the true robot motion.

1.3 Contact Forces and Hybrid Dynamics

As mentioned earlier, biped locomotion is a system of hybrid dynamics, where the discrete changes in the dynamics model happen upon the touchdown and liftoff of the feet. As the robot travels through its environment, its contact configuration with the terrain changes, and the reaction forces it can generate from those contacts change with each contact configuration. We can classify each of the states of the robot by the support contact configuration, or a *stance*, the set of contact points between all parts of the robot and the environment. In a particular stance, there is a particular contact constraint being satisfied, but the robot still has freedom to move

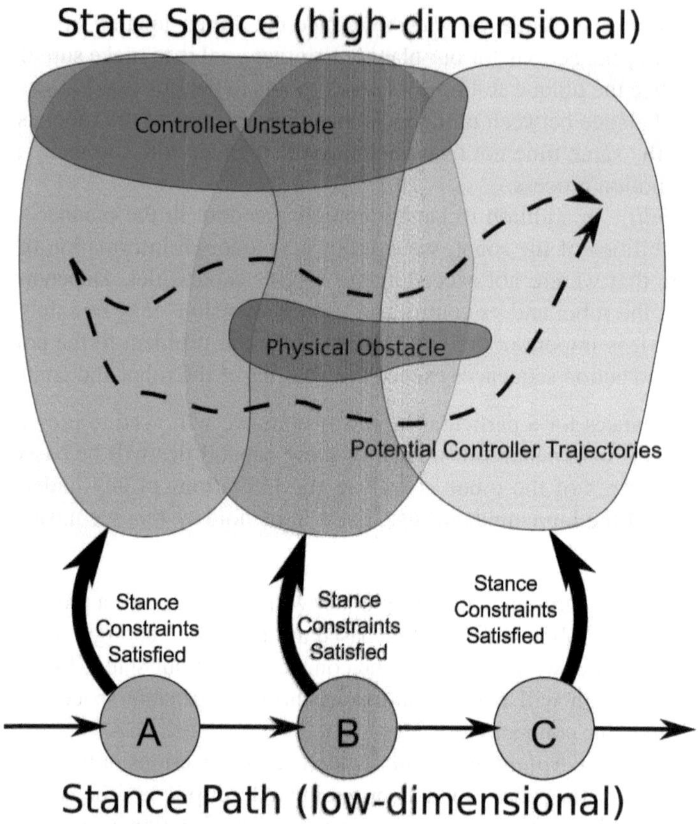

Figure 1.3 Simplifying the planning process by planning for support contact (stances), not full-body motions. Once the sequence of stances is found, a safe trajectory can be generated by the humanoid's locomotion controller through the regions of the state space where those stance contact constraints are satisfied

in the null space of that constraint. Any locomotion of the robot will move through a sequence of these stances, as the support contact configuration changes. For a walking biped, any possible path involves switching back and forth between various single and double support stances, the double support stance being the intersecting region of two adjacent single support stances.

This classification provides a natural way to break up the problem into discrete chunks, in order to build our action set. By planning a path first as a sequence of constrained surfaces through which the motion must pass, we can significantly reduce the dimensionality of the search and simplify the checking of environment constraints, without the need to sacrifice the robot's capabilities. For the walking biped example, every path will be made up of a sequence of double support phases, connected by paths through a single support phase.

This breakdown also provides a very natural level of abstraction for behaviors and controllers. The behavior or controller can handle the unconstrained degrees of freedom of the robot to maintain balance and walk efficiently, while the planner plans in the space of changing support constraints. In this way, the planner only needs to consider the set of contact points between a foot (or other part of the robot) and the environment (*footholds* and *handholds*) and how those contacts relate to the requirements of the locomotion controller. Thus the current stance is made up of the union of the current set of footholds. The humanoid can connect two stances by operating in the null space of the stance contact constraints to generate footstep motions.

In this way, our action set becomes the set of possible footsteps the robot can make. Collision checking in the planner becomes a matter of evaluating footholds and stances at the border between actions, as well as the motion of the connecting footsteps. The planning process with this action model breaks the full motion planning problem into a planning problem in the reduced dimensionality of relevant stances, and then the problem of generating footstep motions, paths through those constraint null spaces.

To re-use existing control strategies with this action set, we can use one of many previously developed locomotion controllers to solve the problem of connecting various stances. For a walking humanoid, we can use the stances from the planner as input to a walking controller which then generates a dynamically stable full-body motion taking the robot from one foothold to the next.

1.4 Stance Connectivity

One of our goals was to ensure the executability of generated plans. Some limits come from the physical specifications of the robot, such as leg geometry limiting step length and height, or join velocity limits restricting the step cycle. In addition to these physical limits, any locomotion controller will add extra constraints to the valid robot states within which the controller is stable, depending on the particular capabilities of the controller or behavior. The use of particular controllers can also limit the capabilities beyond the physical hardware limits. For example, a walking controller may not be able to handle large step lengths, or stances involving the knees or hands, or airborne states. The planner's problem is then to find a sequence of stances which take the robot from the start to the goal, such that the stances and the constraint null spaces connecting those stances lie within the capabilities of the available controllers. The resulting path of stances will jump from point to point in the space of stances, not move in a continuous path. Thus, we need ways to determine the connectivity of those points with regard to the humanoid system.

The dimensionality reductions possible are determined by the minimal information needed to determine the connectivity of stances for a given robot and controller. For example, some controllers limit the body velocity enough that the connectivity of adjacent stances is independent of walking speed. Thus, for these systems, the planning state space can be reduced to just the space of stances, significantly reducing the dimensionality of the space the planner must explore.

To ensure that our sequence of stances can be followed, we need to know two important pieces of information based on the robot and controllers. First, we need to know which stances are valid. This involves evaluating the terrain to determine if it is indeed a place to which the robot and controller can step. Second, we need to know how the potential trajectories of the robot affect the connectivity of the valid configurations. This is determined by both the kinematic reach of the robot, but more importantly by the limitations of the underlying controller which will be tasked with moving from one support state to another. The connectivity problem can be phrased as the question: given two support states and the environment, can the controller safely move from one to the other?

This second piece of information is the action model of the robot. It describes what actions are possible, and encodes the capabilities of the robot and its controller, allowing the robot to generate plans which will be executable by the underlying system.

One way to create this action model is to use a motion planner to connect up individual stances by planning from one to the next. This provides a very accurate model of the robot's capabilities, but once again puts us in the situation of having the robot re-determine how to walk and balance at every step of the plan. While this can be used as an efficient way to break up the full-body motion planning problem in very difficult environments [7], it is still very computationally expensive for real-time operation. Without exactly planning full motion of the robot at every step, and to quickly determine whether a step is feasible for our locomotion controller, we instead approximate the robot's true set of actions with a model operating on our low-dimensional space.

Even in this low-dimensional space, we can develop good approximations of the capabilities of the robot and its controller. The model is very tied to the particular controller used during locomotion, and the exact limits of what the controller is capable of may be unknown, even to the controller's designers. As an example of this difficulty, refer back to Figure 1.3. In that illustration, one region of the state space is clearly known to be unstable to the controller. However, it is often much more difficult to tell in the presence of arbitrary physical obstacles whether or not the controller can still generate stable and safe trajectories, and exactly how much it can be perturbed before it will fail.

1.5 Terrain Evaluation

Due to the fact that humanoid robots have only two legs, they spend a significant amount of time in single support configurations. Even in double support phases, the overall base of support may be small. As a result, the contact of the feet with the terrain takes on paramount importance. The robot must be able to generate required reaction forces to maintain balance as well as to perform its next footstep. More importantly, the contact of the feet with the terrain must not violate any assumptions that the locomotion controller makes about a particular foothold. This requirement

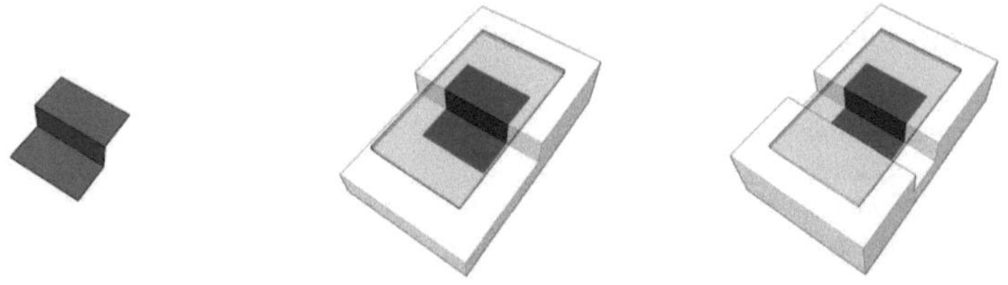

(a) (b) (c)

Figure 1.4 Evaluating less area than an entire footstep can lead to ambiguous situations: (a) a small patch of terrain appears to be unsafe. (b) In one situation with that terrain patch, the foot is unsupported. (c) In another situation, the terrain supports the foot well

can be difficult to satisfy, as it needs an in-depth understanding of the detail of the locomotion control. However, the more capable and robust the locomotion control of the humanoid robot, the less exacting the planning process needs to be to protect the robot from moving to an unsafe location.

Evaluating the terrain to determine safety and stability is very dependent on both the physical robot used and the control scheme used to balance and locomote, and must be adjusted individually for every robot and controller. Additionally, a truly accurate evaluation of the terrain must consider the entire contact of the foot (or hand) with the ground. Figure 1.4 shows one ambiguous case that arises from analyzing only a small feature of the terrain. While that particular feature looks unstable by itself, the surrounding terrain is necessary to either confirm that instability or to recognize a safe foothold.

1.6 A Simple Example

To provide a more concrete illustration of this idea of navigation planning based on footstep choice, this section describes a simple footstep planner for a humanoid robot, which can efficiently plan safe navigation paths through complicated environments.

The planner takes as input a map representing the terrain to plan over, an initial state, a goal set, and an action model consisting of a discrete set of possible footsteps that can be taken. This set represents a fixed sampling of the full space of capabilities of the robot. If a path is found, the planner returns the solution as an ordered list of the footsteps that should be taken to reach the goal. Because we will be using planning through a low-dimensional state space, we can plan using an A* search over the possible footsteps of the robot to find the optimal sequence. The algorithm for this planning is given in Algorithm 1.1.

Algorithm 1.1 PLANPATH($x_{init}, \mathscr{G}, e, \mathscr{A}$)

QueueInsert(x_{init}, 0, 0, NULL);
while *running_time* $< t_{max}$ **do**
    $x_{best} \leftarrow$ QueueExtractMin();
    if $\neg$ *Closed*(x_{best}) **then**
        AddClosed(x_{best});
        if $x_{best} \in \mathscr{G}$) **then**
            return PathTo(x_{best});
        end
        foreach $a \in \mathscr{A}$ **do**
            $x_{next} \leftarrow T(x_{best}, a, e)$;
            $c_l \leftarrow$ LocationCost(x_{next}, e);
            $c_s \leftarrow$ StepCost(x_{best}, x_{next}, a, e);
            $c_e \leftarrow$ ExpectedCost(x_{next}, e);
            QueueInsert($x_{next}, x_{best}.cost + c_l + c_s, c_e, x_{best}$);
        end
    end
end

1.6.1 Environment Representation

The terrain map is represented by a grid of cells. Each cell is represented by

$$(x, y, h, i) \in \Re^3 \times \{0, 1\},$$

where (x, y) is its location in the grid, h is the height of that cell, and i is an information value used to encode extra information about the validity of the terrain that may not be apparent from its shape. Together, the cells create a height map representing the shape of the terrain the planner must overcome. The information values provide extra knowledge of the terrain, for places which appear to be safe to step on but should be treated as obstacles. This representation is easily generated from sensor data or other available representations of the terrain. It provides a simple way to represent many different kinds of environments, with the restriction that it cannot model certain details, such as overhangs or the areas underneath tables the way a full 3D representation can.

1.6.2 The State Space

For evaluating footholds, we model the biped as its footprint shape at each step. We represent the state of the robot by

$$(x, y, \theta, s) \in SE(2) \times \{L, R\},$$

where x and y are the coordinates of the center of the rectangle in a fixed world coordinate system, θ is the angle between the x-axis of the world coordinate system and the forward direction of the foot, and s denotes the support foot (left or right). While the true configuration of the foot is in $SE(3)$, the height, roll, and pitch of the foot are determined by the contact with the ground and the shape of the terrain at that location. Notice that there is no joint information, and no velocities or body position included in this state representation. In this case our state representation is merely the foothold the robot will stand on during its footstep. As mentioned earlier, the path will jump between different points in this space, so the start and goal do not have to be in the same connected component of the (x, y, θ) space.

1.6.3 The Action Model

The actions of the robot in this model are the footsteps which the robot can make, represented by

$$(x, y, \theta, c, h_{high}, h_{low}, h_{obst}) \in \Re^2 \times [-\pi, pi) \times \Re^4,$$

where x and y represent a relative displacement in the robot's current coordinate frame, θ represents the relative orientation change of the robot, and c represents the cost of making the step. h_{high} and h_{low} represent the allowable relative height change the action can make from one step to the next, and h_{obst} represents the maximum relative height of an obstacle that this action can step over. The set of actions the planner uses are sampled from the range of footstep location for which the robot's controller is capable of generating trajectories. The action set is constructed in such a way as to ensure that all states generated by applying the actions are reachable for the robot. The parameters h_{high}, h_{low}, and h_{obst} are used to verify that connectivity in the presence of the terrain. Four examples of footstep action sets that we used in our experiments are illustrated in Figure 1.5. The actions are grouped into sets for the left and right feet, the two sets being mirror images of each other.

1.6.4 The State–Action Evaluation Function

The planner evaluates all actions from a state, generating three costs for each one. First is the *location cost* $L(x)$, which evaluates the action's destination state as a potential foothold. This cost uses a variety of metrics to quickly compute how viable a location is for stepping onto. Second is a *step cost* $S(x_{next}, a, x_{current})$, which computes the cost of reaching the state x_{next} by taking the action $a \in \mathscr{A}$ from the current state $x_{current}$. This cost describes the connectivity of x_{next} and $x_{current}$. This cost includes the action's associated cost, a penalty for height changes, as well as an obstacle clearance check of the terrain between the foot's last position and the

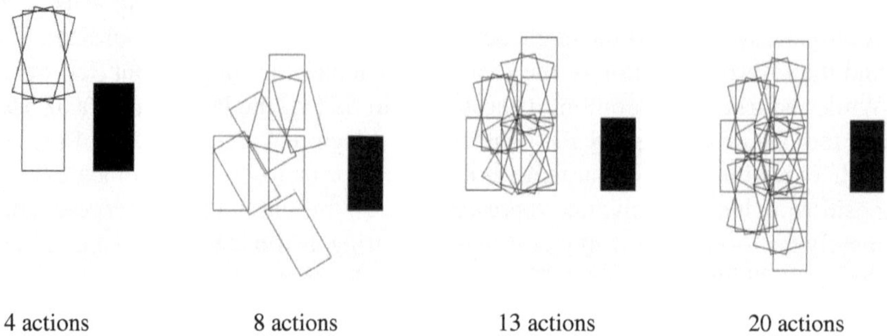

Figure 1.5 Footstep action sets. The actions displayed are only those for the left foot (relative to the right foot shown in black)

new foothold. Finally, the third cost is a heuristic which estimates the *remaining cost* to reach the goal state. This remaining cost can be computed in several ways, for example using the Euclidean distance to the goal, or the result of a traditional mobile robot planner. Heuristics are discussed in more detail in Section 1.7. These three costs are then used by the planner to determine the cost of each node in an A* search. Note that in this case the location cost is independent of the action or current state of the biped, and thus can be precomputed for all locations if desired, rather than computed for the needed states at run-time.

1.6.4.1 Location Metrics

To evaluate a location's cost, we would like to know exactly how the foot would come to rest on the surface, which parts of the foot would be supported, and be able to build a support polygon of the foot based on which parts of the foot are touching the ground and evaluate how stable that support polygon is. Unfortunately, this is very expensive to compute exactly. Instead, we use a set of metrics which can be quickly computed and serve to approximate this ideal location cost. To be useful, a metric should be quick to compute, invariant to the resolution of the heightmap, and should eliminate or penalize an unwanted form of terrain while not eliminating or heavily penalizing good terrain.

To compute the metrics, we first determine the cells in the heightmap which will be underneath the foot for a particular step. This gives us a set of cells to use with each of the metrics defined below. Each of the metrics has both a weight and a cutoff value associated with it. If any metric passes its cutoff value, the location is considered invalid. Otherwise, the location's cost is the weighted sum of these metrics. Below are descriptions of the five metrics we have used for terrain evaluation. The weighted sum of these metrics is shown in Figure 1.8.

To evaluate locations in the terrain, we will manually design some metrics to add cost for features that could potentially cause instabilities for the robot. Based on the location and shape of the robot's foot, we can compute the set of cells, $\mathscr{C}$, of the heightmap which lie underneath or around the foot. Each metric then uses this set of cells to describe a particular feature of the terrain at that location.

The first metric we add is based on the angle of the terrain with respect to gravity. This captures both the increased frictional requirements of highly-sloped footholds, as well as the limits of ankle joint angles. This metric can be easily computed by fitting a plane $h_{fit}(x,y) = ax + by + c$ to the cells of the map which fall under the foot, and comparing the plane's normal to the gravity vector.

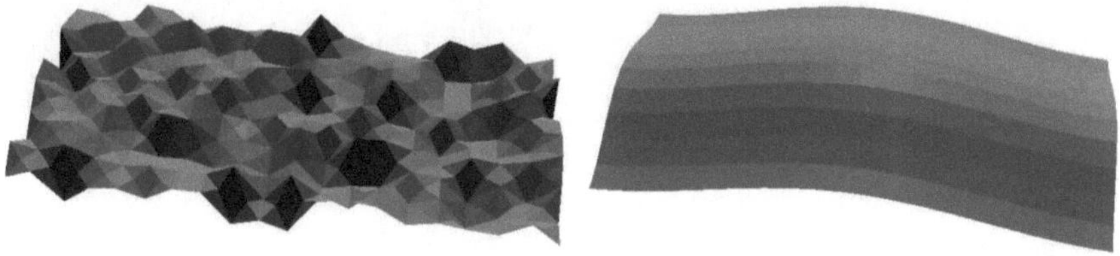

(a) (b)

Figure 1.6 (a) A rough foothold; and (b) a foothold that while not rough, is nonetheless unstable

In addition to being horizontal, the terrain should not be rough or bumpy, which can provide poor contact between the foot and the ground. To measure this we add a second heuristic: the "roughness" of a location. This can be estimated as the average deviation the surface from the fitted plane. This metric will catch many nasty features of the terrain, as well as having a high value when the fitted plane does not approximate the terrain well:

$$\frac{1}{N} \sum_{(x,y,h,i)\in\mathscr{C}} |h - h_{fit}(x,y)|. \tag{1.1}$$

However, even when the surface is not very rough, the location may not be stable. The peak of a mound may be close to flat, but still only supports the center of a foot that steps directly on top of that peak. To reject these kind of unstable locations, we add a metric to approximate the curvature of the foothold. While perfectly flat is desired, curving down at the edges is penalized more than curving up at the edges. One way to compute this is to fit a quadratic function to the cells. Any simple dome-shaped filter function will work, however, and in this example we use a cosine function:

$$\frac{1}{N} \sum_{(x,y,h,i)\in\mathscr{C}} \left([h - h_{fit}(x,y)]g(x_f,y_f)\right). \tag{1.2}$$

x_f and y_f are x and y in the foot's coordinate system. $g(x,y)$ is the dome-shaped filter. Restrictions on $g(x,y)$ are that it should be higher in the center than on the edges, and it should sum to zero over the area of the foot. An example filter is shown below:

$$g(x,y) = \cos\left(\frac{2\pi x}{w}\right) + \cos\left(\frac{2\pi y}{\ell}\right). \tag{1.3}$$

w and ℓ are the width and length of the foot.

(a) (b)

Figure 1.7 (a) Terrain with a bump which could tip the robot, and (b) a safe foothold surrounded by unsafe terrain — dangerous with execution error

Another terrain feature that can upset the balance of the robot is a bump in the terrain which is below the threshold of the "roughness" metric, but prevents the foot from lying flat. A marble or pen under the foot of the humanoid can disturb the walking controller fairly easily.

To estimate this cost we find the largest deviation above the fitted plane:

$$\max\{h - h_{fit}(x,y)\},\ (x,y,h,i) \in \mathscr{C}. \tag{1.4}$$

One final issue that concerns us is the cells around the location we wish to step to. The target step location may be perfectly flat and inviting, but if it is surrounded by unsafe terrain, execution errors could easily cause the foot to land in an unsafe region. Terrain above or below our target step does not have a symmetric effect on our stability, however. Terrain below our target step (for example at the edge of a stair) would simply cause a small part of the foot edge to be unsupported in the case of an execution error. But a wall, or obstacle, or a even a small rise next to the target step can cause the robot to stub its foot or tip dangerously in the event of an execution error. To estimate this cost, we again find the largest deviation above the fitted plane, using the cells surrounding our target step (the width of this safety band is chosen based on the expected execution error of the robot and its walking controller).

The overall cost of terrain locations weighted by the above metrics is shown on an example terrain in Figure 1.8. Each cell shows the average of the cost of stepping to that location from several orientations. It is important to note that these metrics provide costs based solely on the *shape* of the terrain. Using only shape-based metrics, we do not reason about any physical properties of the terrain, such as softness and compliance, friction, or the ability of the location to support the weight of the robot. To reason about walking on ice, or grass, or other surfaces in which the available friction or shape will vary greatly based on physical properties, we would need to have information about those physical properties in our environment representation. Another limitation of the above metrics is that they are based only on the terrain location itself, and not on the action to perform in that location. A sharp turn or sudden acceleration may require a higher friction or prefer a certain slope to the terrain, or a particularly difficult step may have more stringent stability requirements. These kind of action-specific terrain costs are not captured by the above metrics.

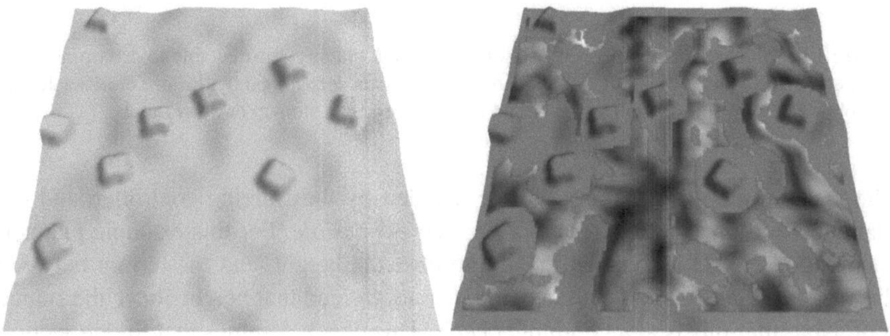

(a) (b)

Figure 1.8 (a) An example terrain, and (b) the weighted sum of the foothold metrics, applied to that terrain. The intensity corresponds to the cost, with some additional areas marked as unsteppable

1.6.4.2 Step Cost

The step cost computes the cost to the robot of making a particular footstep. In addition, it will determine if a particular footstep can be executed in the presence of obstacles (by returning infinite cost if it is unexecutable). If the step is executable, the cost of taking a step is given by

$$StepCost(x_{current}, x_{next}, a, e) = c_a + w_h \cdot |\Delta h(x_{next}, x_{current}, e)|. \tag{1.5}$$

c_a is the cost of the action a. $\Delta h(x_{next}, x_{current}, e)$ is the height change between the state x_{next} and the current state of the robot $x_{current}$ for the environment e. w_h is the penalty for height changes, chosen manually based on the user's desire to try to avoid paths that go up or down. If the height change is outside the action's allowable range, x_{next} is not reachable from $x_{current}$, and the step is discarded. An obstacle collision check is also done to determine if the foot can safely clear all the cells along the path from its previous location to its new location. Because this path may include many cells, quadtrees that encode the maximum height are used to quickly check for collisions. If collisions are found, once again x_{next} is not reachable from $x_{current}$, and the step is discarded. However, the collision check is an approximation.

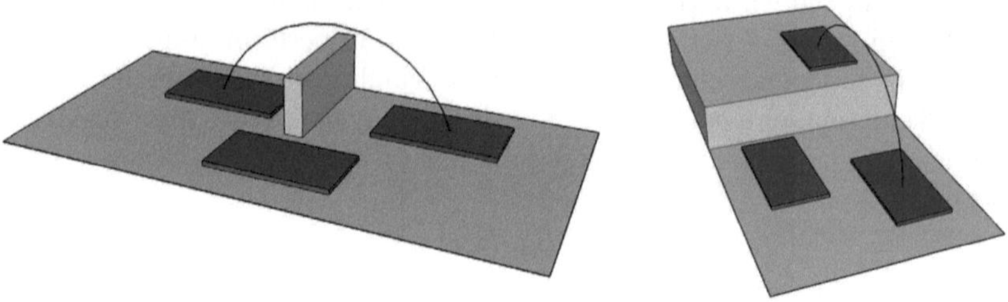

Figure 1.9 To determine the connectivity of stances, the planner must be able to determine if the robot can make a stepping motion connecting them. This is determined by the size of obstacle that the robot can step over, as well as the height that the robot can step up or down safely

The exact motion of the robot will be decided by the walking controller, and the planner simply needs to determine if a step is *possible*. For this example planner, each action has an associated maximum obstacle height, and the terrain between the start and end footholds in the plan must not exceed that height above the stance foot. Note that this is a very simple check. This check assumes that the projection of the foot onto the ground moves in a straight line, so motions which curve the foot around an obstacle are not considered. Furthermore, it assumes that a foot motion can be generated regardless of the shape of the intervening terrain, as long as it all remains below a certain threshold. A more accurate model of the kinds of stepping motions that the robot can make would allow for addition steps to be considered which are invalidated by the current collision check.

1.6.5 Using the Simple Planner

With these representations, metrics, and this action model we now have the necessary components to search through the space of footholds, and plan paths through many different types of terrains. Examples of the paths generated for some environments are shown in Figure 1.10. In each of these plans, the search process is evaluating the potential stepping motions the robot can make from each stance, allowing the final stepping motion to take advantage of the humanoid's capabilities. Thus the resulting paths can climb up and down stairs, step over or onto obstacles, and find stable footholds in rough and non-flat terrain.

A visualization of the planning process is shown if Figure 1.11. The tree is displayed for every ten nodes that are expanded. In this example, informed heuristics are used to guide the search (described in Section 1.7), and as a result the search does not need to branch out into the rest of the environment to find its path.

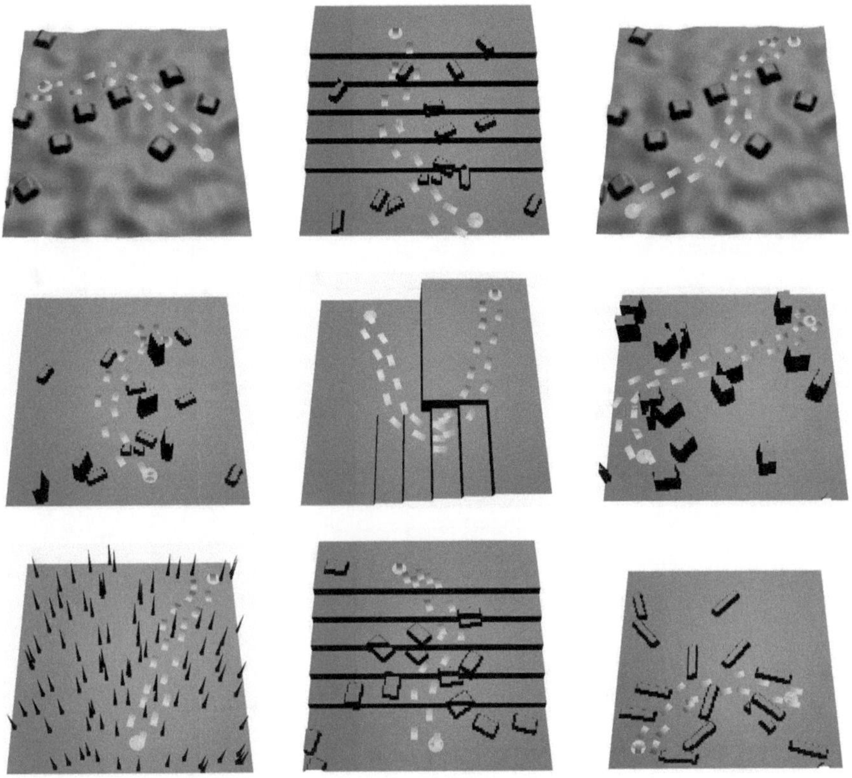

Figure 1.10 Example plans generated by the simple biped planner

In order to successfully use this planner with real robots, we must tune the metrics and action set to match the capabilities of the biped and its walking controller. These values both define what the planner will allow the robot to step on, as well as defining the costs for what will be considered optimal for a path. For a large variety of environments, the planner was not very sensitive to the weights or cutoff values used in the metrics. The decision to rate a step as safe or unsafe is clear for the vast majority of steps that the robot will face, and a wide range of values were able to correctly classify safe and unsafe stepping locations. However, moving toward rougher terrains with more complex surfaces, the weights and features used must be more carefully tuned to match the robot's abilities for rough environments. Even with the simplifications made in this example planner, it can be used successfully to navigate a real humanoid robot through complicated environments. Figure 1.12 shows the HRP-2 humanoid robot approaching and climbing a set of stairs autonomously. Note that the robot has not been specifically programmed to climb this particular set of stairs with a preplanned motion, but is able to determine safe stepping locations and safe stepping motions which the walking controller then turns into a full-body, dynamically-stable walking motion.

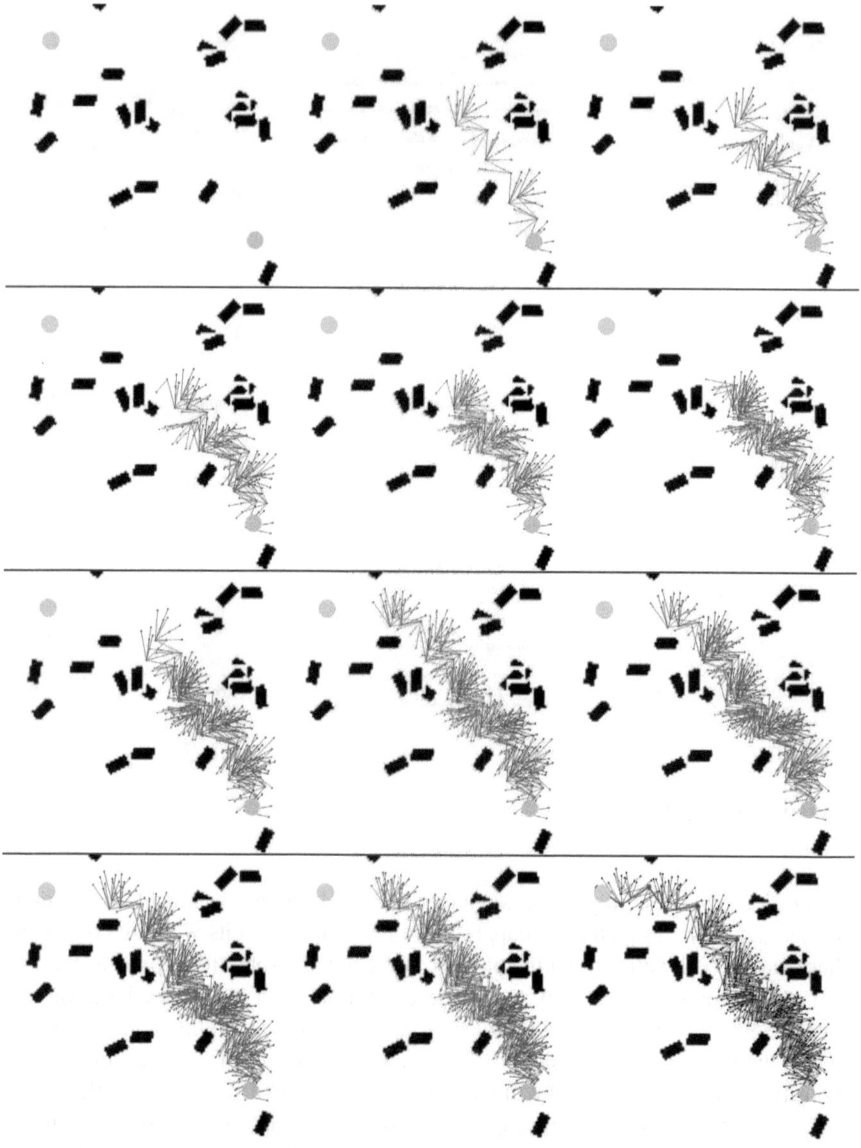

Figure 1.11 A visualization of the planning process generated by displaying the search tree after every ten node expansions. The planner is starting at the circle in the lower right of the terrain and planning to the goal in the upper left

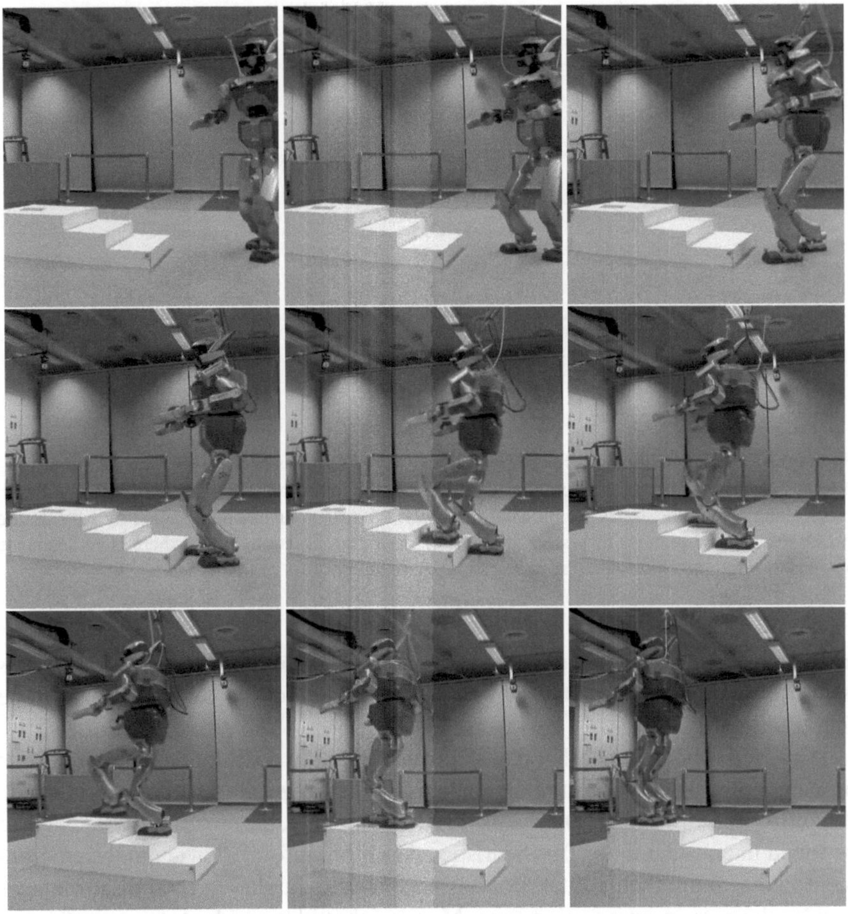

Figure 1.12 HRP-2 humanoid autonomously walking up a set of stairs

1.7 Estimated Cost Heuristic

The problem we are solving in humanoid navigation can be viewed as two subproblems: choosing a rough torso path, and choosing footsteps along that path. In the formulation described above, these two problems are combined and solved simultaneously. This is more computationally expensive than solving them separately, but ensures that the planner does not commit to a torso path that is unexecutable. In fact, choosing a torso path that is guaranteed to be executable and yet still utilizes the stepping abilities of the robot is an unexpectedly hard problem. Recall that from Figure 1.4, evaluating a small part of a terrain does not reliably determine if footholds which include that small part are safe or unstable. In addition, adding clutter to an environment can very quickly make estimating the correct torso path impractical. Figure 1.13 shows a few of the obstacle situations which can cause poor

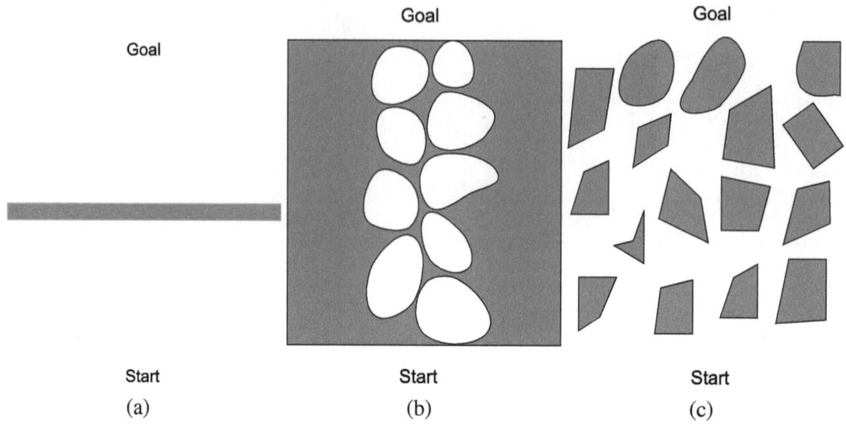

Figure 1.13 Difficult situations for computing overall torso paths or heuristic value estimates. (a) A torso path may avoid the horizontal obstruction, but the robot can step directly over it. (b) A path exists along "stepping stones" to the goal, however, the safe areas are surrounded by very unsafe terrain that a torso planner may steer around. (c) A similar situation, except there is no sequence of steps through the cluttered area, even though there are pockets of free space

torso paths or inaccurate heuristics. One way to make use of imperfect torso paths is to use them as heuristics to guide the footstep search, rather than a path to be followed exactly.

A mobile robot planner that plans outward from the goal state to the initial state provides a useful estimate of remaining cost, with the results stored in a grid which discretizes the workspace. During the footstep planning, the remaining cost can then be found in constant time. This heuristic takes more information about the environment into account than a Euclidean distance metric, but has several disadvantages besides the extra preprocessing time. Mobile robot planners look for a continuous path through the configuration space or workspace that connects the initial and goal states. Because the biped has the ability to step over obstacles, it does not require a continuous path through the workspace. The result of this difference is that the mobile robot planner can severely misjudge the cost of a location, as shown in a few example situations in Figure 1.13. In an environment with a long, low, thin obstacle, the mobile robot planner will provide lower cost to areas which send the biped the long way around instead of stepping over the obstacle, resulting in an overestimate. Also, it can underestimate when finding a path that seems that it may be clear enough, but where there are not actually alternating footholds the robot can step on. In general, the time complexity of A* search is an exponential function of the error in the heuristic used [10]. So while in many environments, this heuristic performs much better than a Euclidean distance heuristic, the worst case can be an arbitrarily large overestimate. An example environment using these heuristics is shown in Figure 1.14. For the Euclidean heuristic, the first half of the path has approximately the same value, while the heuristic from the mobile robot planner pushes the planner in the correct direction from the start. In this way, the search tree for the Euclidean

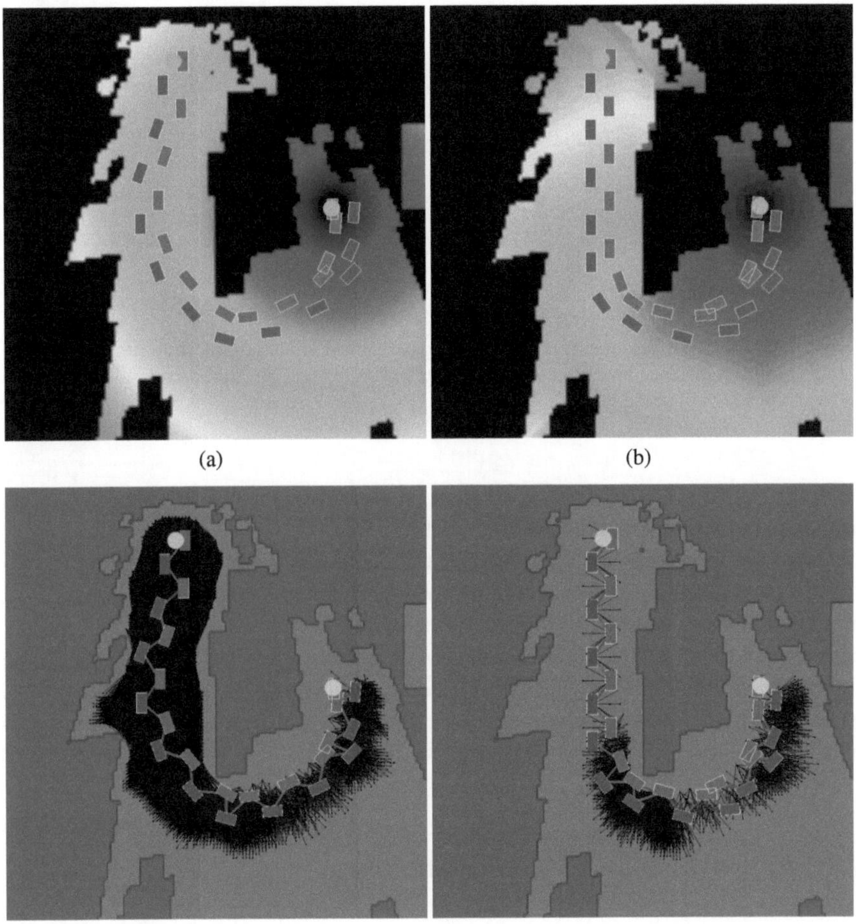

(a) (b) (c)

Figure 1.14 Examples of heuristics when planning to the circle on the right. (a) Euclidean distance heuristic. Notice that for the first half of the path, the heuristic does not provide useful information. (b) Heuristic from traditional mobile robot planner. (c) The search trees for both plans at the end of the search process

distance heuristic fills up the early part of the space, while the search tree for the reverse-planning heuristic moves immediately past that early section of the terrain. In this example, using the more informed heuristic allowed the planner to find a path over 200 times faster than when using a Euclidean distance heuristic, with only a 5% increase in path cost.

Because the time of arrival in the goal set is not known at the start of the planning process, it is difficult to use this reverse-planning heuristic in environments that vary with time. But there are some methods that we could use to still build a heuristic in these kinds of environments, such as choosing a range of likely arrival times and

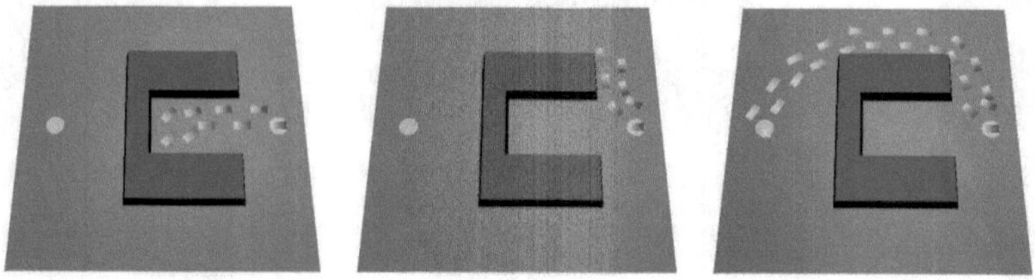

(a) (b) (c)

Figure 1.15 Planning from the right to the left around a local minimum: (a) time-limited planning with a Euclidean distance heuristic; (b) time-limited planning with the mobile robot heuristic; and (c) the complete path to the goal

planning backward from all of those times, or weighting the cost of a location in the reverse plan by the percentage of time that it is free.

Using the more informed heuristic on top of the low-level footstep planning results in the planning process becoming sped up in a large variety of terrains. Local minima no longer incur large processing penalties, because the heuristic guides the path around them. Given enough time, both planners return the path shown on the right in Figure 1.15. However, with a Euclidean heuristic the planner takes 117 s to find this path, but when combined with the more informed heuristic, it finds that path in only 560 ms total.

1.8 Limited-time and Tiered Planning

In the footstep planning examples shown so far, the planner always plans a full path to the goal. This is very different from how humans navigate through the world. As a human, you may know the next several steps you will take, but you certainly do not plan every step of a walk through the park in advance.

By choosing to plan at the level of footsteps, we limit the scope of our planning to essentially room-sized environments, within the reach of our sensing systems. Here we can apply the coarse–fine techniques that have been successful with wheeled robots to extend our planning abilities to larger environments and longer-range tasks. By using the footstep planner as the low-level planning, we can ensure that the robot is always executing safe motions to good footholds, and layering higher-level planners on top of it allows for long-range planning at reduced cost.

The heuristic described in Section 1.7 provides a coarse planner. It is integrated as a heuristic instead of a rough path, but it fulfills the same function. In addition to speeding the planning time, it increases usefulness of partial paths. When a partial path is returned due to time limitations in the planning process, a more informed heuristic makes the partial path more likely to be the beginning of the optimal path.

Figure 1.15 shows these advantages in the presence of a local minimum. In the left two examples, the total planning time was limited to 400 ms (the mobile robot planner finishes approximately 350 ms into the planning process). Even with only 50 ms to generate footsteps, the path found when using the mobile robot planner heuristic is a useful path to begin executing.

Above the level of the heuristic planner, we can extend the range with topological graphs for buildings and cities, and large-grid planners for outdoor environments, as are currently used with wheeled robots. These high-level planners are very fast, and do not need to re-plan very often, so they do not significantly increase planning overhead, but result in useful subgoals and guidance for the lower-level planners. This both shortens the length of the problems for the low-level planners, and guides the search process in an informed manner, resulting in 1–2 orders of magnitude speedup in planning times for large environments [3].

1.9 Adaptive Actions

A large drawback to the simple action model described this far is the fact that the actions the robot can take are limited to a very small subset of the robot's potential. By using a fixed set of actions, the planner can only solve problems for which that particular action set "fits." We can choose that action set so that it can still traverse a wide variety of terrains, but ultimately, certain terrains will be untraversable simply because the needed action was not a part of our model. To illustrate this problem imagine sets of stairs, where the particular desired step length for climbing them might not be in the action set. An example would be stepping stones across a river, where a very specific sequence of actions is required to negotiate the terrain. To remove this action set limitation, we need the action model to be capable of making any action that the robot and controller are capable of performing.

To improve planning performance, we wish to keep our branching factor small. Therefore, instead of adding a multitude of actions to our action model, we want to adapt our actions to the particular terrain we face during the search process.

The problem we are interested in is the choice of stepping actions used for expanding nodes in the A* search. Recall that an action a from the set of all actions of the robot, $\mathscr{A}$, describes a mapping from one robot state, $x \in \mathscr{X}$, to another, corresponding to a particular motion of the robot:

$$x' = Succ(x, a, e), \tag{1.6}$$

where e is the environment, and $x' \in \mathscr{X}$ is the new state of the robot after performing action a. In the simple example footstep planner, we have manually chosen a set of stepping actions which were applied during every node expansion. In constrained situations, the action needed to make progress from a particular state may not be present in that set of actions, although it is an action the robot and controller are capable of performing. So for that particular constrained situation, we need to have

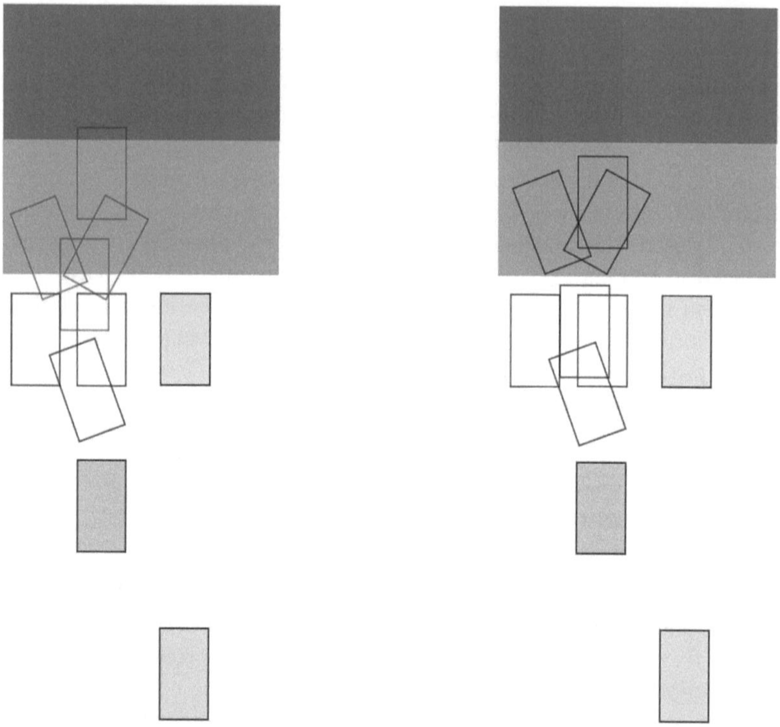

(a) (b)

Figure 1.16 Adapting actions to the terrain with stairs. (a) Several actions from a fixed set of action samples (the open rectangles) provide invalid steps (overlapping the stair edges), with no valid action from the set successfully climbing the stair. (b) The reference actions are adapted to find the closest fit in the terrain. With the result that the robot now has actions which can directly step up onto the stair

that appropriate action in our action set, but it is unnecessary at other times. If we have a desired branching factor, b, that we would like to maintain during the search process, the problem we wish to solve can be stated as follows: what are the best b actions to perform from state x in environment e?

In order to answer this question, we have to define what we mean by the "best" b actions. If we knew the value function for the navigation problem, the solution would be simple: perform the action that takes you to the best value in the value function. In that case, there would be no need to search at all, because the value function encodes within it the solution for every state. However, all that we have during the planning process is a cost function and a heuristic estimate of the true value function. Because we do not know the true value function, simply stepping to where the cost function and heuristic are a minimum will likely not result in a solution. So while choosing actions that are low in cost is important, we want to make sure that the actions we take allow us to reach a large number of other states, allowing us to explore the state space of the problem more efficiently. This reasoning provides the basis for one useful definition of the "best" b actions: the actions which will allow us to reach the largest amount of the state space from the b resulting states.

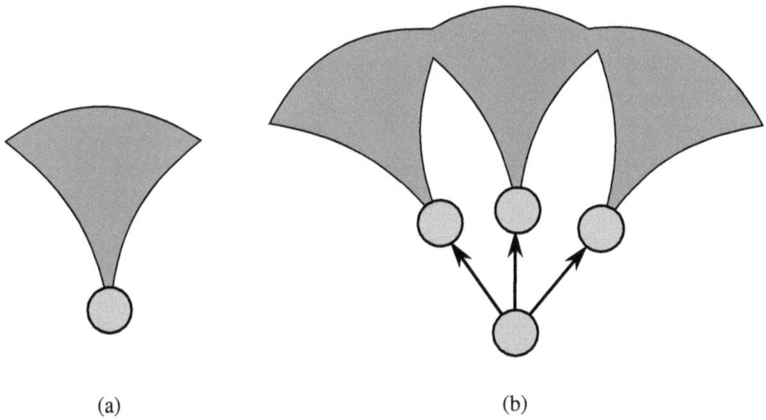

Figure 1.17 (a) The reachable region of a single state; and (b) the union of reachable states after applying a sampling of actions

The reachable space from a state, $\mathscr{R}(x)$, in an environment e can be given as the set

$$\mathscr{R}(x) = \{x' \in \mathscr{X} \mid \exists a \in \mathscr{A} \,.\, x' = Succ(x,a,e)\}. \tag{1.7}$$

Therefore, the reachable states from state x after performing the actions $a_1, a_2, ...a_b$ in our action set $\mathscr{S}$ are

$$\mathscr{R}(Succ(x,a_1,e)) \bigcup \mathscr{R}(Succ(x,a_2,e)) \bigcup ... \bigcup \mathscr{R}(Succ(x,a_b,e)), \tag{1.8}$$

as illustrated in Figure 1.17.

Thus we wish to find the action set $\mathscr{S}$, of cardinality b, which maximizes this region. Unfortunately, computing this region for all possible $\mathscr{S}$ is computationally intractable. Additionally, to pre-compute the necessary information for lookup would require computing the best $\mathscr{S}$ for all possible states in all possible terrains, which would be infeasible to store.

1.9.1 Adaptation Algorithm

Because it is infeasible to compute the truly optimal $\mathscr{S}$ for a given state and environment, we will settle for a suboptimal $\mathscr{S}$. To find this $\mathscr{S}$, we first start from a reference action set, $\mathcal{S}_r$, which has the good reachability property that we desire. On obstacle-free, flat ground, we can use this reference action set directly as our set of actions for node expansion. In the presence of rough terrain or obstacles, a state x' that an action in $\mathcal{S}_r$ maps to may be invalid. In this case we perform a local search around x' to find a new state, x'', and compute a new action to reach that state, a'. Pseudo-code for this approach is shown in Algorithm 1.2. Through this approach, we maintain our branching factor, replacing invalid actions with nearby valid ones. In addition, we retain the good reachability property that the reference action set had, although the reachable region will be slightly reduced by the modification of the result states of our action set.

Algorithm 1.2 ApplyActions ($\mathcal{S}_r$, x , e)

```
foreach a ∈ S_r do
    x' ← Succ(x, a, e)
    if VALIDSTATE(x', e) then
        AddToQueue(x', a)
    else
        x'' ← LOCALSEARCH(x', e)
        a' ← COMPUTEACTION(x, x'', e)
        AddToQueue(x'', a')
    end
end
```

In order to implement this approach, we need two extra pieces of information to model the robot and its walking controller. First, we need an explicit test of whether a state is in $\mathcal{R}(x)$. When using a fixed action set, we can be certain that our actions are drawn from the robot action space, $\mathcal{A}$. However, if we are allowing the resulting state to be adapted to the terrain, the local search needs to remain within the reachable region, so that an action exists which can move the robot to that state. The second new requirement is an inverse of the action set mapping, such that we can find the action which will move us from one state to another within its reachable range. Previously, we only required a forward mapping of the robot's actions. This inverse mapping is what allows us to search in the state space to adapt to the terrain.

This local search can be achieved by sequentially testing nearby states until a valid step is found, or by following gradients in the cost function to find a low-cost step. This local search to determine local actions results in the ability to find paths in constrained environments quickly, as well as finding lower-cost paths and lower planning times [5]. The algorithm described in Algorithm 1.2 only adjusts actions on an individual basis, however. This is a design choice to reduce the time required to expand a node in the search, which is usually performed thousands of time for each plan. In a situation where more computational resources are available, or time limits not as severe, a local optimization over the action set as a whole would be able to provide more applicable action sets for each state of the search.

1.10 Robot and Environment Dynamics

In the algorithms described above, the state of the robot used in the planner only contained the support configuration of the robot. This is reasonable on some humanoids due to the walking controller used, which can find feasible body and leg motions for any sequence of footsteps requested of it, as long as each new footstep is in an easily reachable region relative to the stance foot. However, in many cases, the connectivity of two points in different stances is dependent on the dynamics of the robot, and stance information alone is insufficient to determine connectivity. There are two main ways that we can deal with this difficulty. First, we can continue with the same state representation, and then require that all dynamic actions return the robot to some safe state before they complete. This approach may be appropriate in a situation that has a large action space which fits nicely with a low-dimensional state space as well and a few specialized actions which need extra state information to determine their validity. In this case, those few special actions can be padded with special setup and ending movements which allow these extended actions to fit within the original low-dimensional state space. As an example, imagine a walking biped which also can make long jumps. The jumping motions introduce large velocities into the robot's motion, which the walking controller may not handle correctly. By adding a recovery at the end of the jump as a part of the action, the planner can use this jumping motion and still return to a safe state with low velocities at the end which fits within the planner's low-dimensional state space.

However, if we wish to string many dynamic actions together, without returning to a particular state in between, we must move our search into a higher-dimensional state space which can describe the dynamics which we care about. Note that while we need this larger state space, it is still very likely to be much lower-dimensional than the configuration space of the humanoid. The exact number of dimensions needed will depend on the details of the particular walking or running controller in use by the humanoid, but only adding the expected torso velocities into the state space captures many of the important dynamics of running and walking robots. With an estimate of torso velocities for each state in the plan, the range of controllers and abilities which can be included in the planning process increase, while still providing efficient planning that can be used in a real-time manner [4].

1.11 Summary

Walking and running with humanoid robots remains a challenging problem, to improve on robustness, efficiency, and the range of terrains and surfaces which can be stably traversed. However, when planning navigation paths for legged robots, many of the complicated details of legged locomotion can be delegated to a walking or running controller. Planning for these controllers involves knowing and understanding their limitations. While these limitations are extremely controller-specific, they generally are dependent upon the support contact with the environment, the physical reach of the robot, and at times the velocity of the center of mass of the robot or its legs. Thus the limitations of the robot and its controller can be approximated by a model operating in a comparatively low-dimensional space, providing efficient, real-time, safe solutions for navigation for these complicated machines in real-world environments.

References

[1] Bretl T (2006) Motion planning of multi-limbed robots subject to equilibrium constraints: the free-climbing robot problem. Int J Robotics Res. 25(4):317–342

[2] Bretl T, Rock S, Latombe JC (2003) Motion planning for a three-limbed climbing robot in vertical natural terrain. In: proceedings of IEEE international conference on robotics and automation, Taipei, Taiwan

[3] Chestnutt J, Kuffner J (2004) A tiered planning strategy for biped navigation. In: proceedings of IEEE-RAS international conference on humanoid robots, Santa Monica, California

[4] Chestnutt J, Lau M, Cheng G, Kuffner J, Hodgins J, Kanade T (2005) Footstep planning for the Honda ASIMO humanoid. In: proceedings of IEEE international conference on robotics and automation, Barcelona, Spain

[5] Chestnutt J, Nishiwaki K, Kuffner J, Kagami S (2007) An adaptive action model for legged navigation planning. In: proceedings of IEEE-RAS international conference on humanoid robots, Pittsburgh, PA

[6] Hauser K, Bretl T, Latombe JC (2005) Learning-assisted multi-step planning. In: proceedings of IEEE international conference on robotics and automation, Barcelona, Spain

[7] Hauser K, Bretl T, Latombe JC (2005) Non-gaited humanoid locomotion planning. In: proceedings of IEEE-RAS international conference on humanoid robots, Tsukuba, Japan

[8] Kuffner J (1998) Goal-directed navigation for animated characters using real-time path planning and control. In: proceedings of CAPTECH: workshop on modelling and motion capture techniques for virtual environments, pp 171–186

[9] Kuffner J, Nishiwaki K, Kagami S, Inaba M, Inoue H (2001) Motion planning for humanoid robots under obstacle and dynamic balance constraints. In: proceedings of IEEE international conference on robotics and automation

[10] Pearl J (1984) Heuristics. Addison-Wesley, Reading, Ma.

[11] Pettre J, Laumond JP, Simeon T (2003) A 2-stages locomotion planner for digital actors. In: proceedings of SIGGRAPH symposium on computer animation

[12] Stentz A (1995) The focussed D* algorithm for real-time replanning. In: proceedings of international joint conference on artificial intelligence

Chapter 2
Compliant Control of Whole-body Multi-contact Behaviors in Humanoid Robots

Luis Sentis

2.1 Introduction

In 2008, the US National Intelligence Council published a report listing six disruptive technologies by the year 2025. It included subjects in biology, energy, robotics, and information technology. On the subject of robotics, it reports: "Robots have the potential to replace humans in a variety of applications with far-reaching implications. [...]. The development and implementation of robots for elder-care applications, and the development of human-augmentation technologies, mean that robots could be working alongside humans in looking after and rehabilitating people". In this spirit, we explore here a methodology to enable greater maneuverability and interactivity of robots in human environments (Figure 2.1).

As the expectation of humanoid robots to operate as human helpers and social companions grows, the need to improve their motor skills becomes increasingly important. Equipped with highly dexterous anthropomorphic systems, sensors for environmental awareness, and powerful computers, humanoids should be capable of handling many basic chores that humans do. Yet they cannot, not to the level that would make them practical.

The goal of this chapter is to summarize our collaborative work in the area of compliant control of humanoid robots with focus on the design and execution of advanced whole-body multi-contact skills. In particular, we address the important subject of unified whole-body force and motion behaviors under geometrical and multi-contact constraints, and the synthesis of whole-body skills using action primitives. In essence, I summarize here the following contributions: (1) whole-body task and postural control published in [29]; (2) prioritized torque control under geometric constraints published in [53]; (3) synthesis of whole-body behaviors published in [53]; and modeling and control of multi-contact behaviors published in [54].

Department of Mechanical Engineering, The University of Texas at Austin, Austin, TX 78712, USA, e-mail: lsentis@austin.utexas.edu

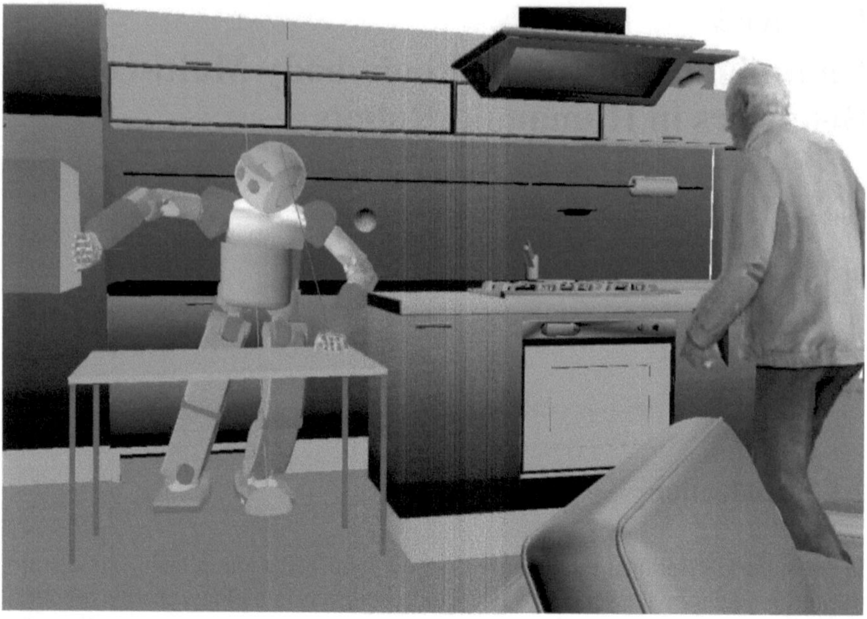

Figure 2.1 Service application. This snapshot taken from a simulated control scenario, depicts the operation of mobile robots in human environments. In the near future, tedious chores such as tidying up the house or cooking might be performed by skilled robots

Humanoids are difficult systems to model and control because they are highly dimensional and underactuated, and because they operate under contact and geometrical constraints. The work summarized here, describes efforts towards a unified torque-level control methodology that simultaneously optimizes all required processes to achieve advanced compliant manipulation and maneuvering behaviors under the physical constraints imposed by human environments.

The strength of this methodology relies on the underlying whole-body kinematic and dynamic models, which exploit the high dexterity of the humanoid's mechanism and account for the underactuated modes to create high-end skills without relying on high-dimensional motion planning computations. Operating at the torque control level enables the intuitive implementation of force, motion, and compliant multi-contact behaviors while removing the need to generate inverse kinematic trajectories. Dynamic models are incorporated into the framework to enhance physical performance, allowing controllers to use lower feedback gains to achieve whole-body compliant contact interactions.

A hierarchical organization of the controller imposes priorities between the control objectives according to their relative importance in the execution chain. Priorities are used as a mechanism to temporarily override non-critical criteria in order to fulfill critical constraints. A supervisory layer monitors the feasibility of the control objectives and provides feedback to high-level planners to modify the control policy or reason about the current action.

In the last two years we have co-developed an embedded software architecture that implements the methods described here. This architecture currently operates the upper body of a research version of the Honda Asimo humanoid robot which uses torques as control inputs [30]. Although, the architecture is not yet published it is based on collaborative work initiated in [28] and [46]. To handle the computational complexity of the dynamic models, we have based our design on a multi-process and multi-rate architecture where several servo processes running at fast rates control all interactive operations while another process running at low rates updates the models. Performance benchmarking has been conducted for simulated robotic structures, achieving the necessary speeds to govern high-dimensional humanoid robots.

Section 2.2 focuses on the description of whole-body kinematic and dynamic models under contact constraints and describes recent work on modeling and control of multi-contact behaviors. Section 2.3 summarizes contributions to prioritized torque-level control in the form of an architecture that unifies motion and force behaviors while fulfilling geometric and contact constraints. Section 2.4 describes several simulated examples involving whole-body multi-contact and manipulation behaviors.

2.2 Modeling Humanoids Under Multi-contact Constraints

One of the fundamental differences between humanoid robots and fixed robotic manipulators is their ability to maneuver in their environment using whole-body multi-contact interactions. Gravity forces push the humanoid's body against the ground, providing a supporting platform to move in the terrain. The robot's behavior is therefore determined not only by the displacements of its articulated joints but also by the maneuvers of its body and by the interactions between closed loops formed by the bodies in contact.

This section describes (1) kinematic models that take into account underactuated degrees of freedom and multi-contact constraints, (2) whole-body kinematic and dynamic models of task objectives, and (3) mechanical models of the internal forces and moments taking place between contact closed loops. These models will be used in the next section to construct compliant torque controllers for whole-body force and motion interactions under geometrical and contact constraints.

Contact interactions in robots have been addressed since the early 1980s with work on dynamics and force control in the context of robotic manipulation [24, 49]. Cooperative distributed manipulation became important to enable the handling of big or heavy objects [1]. To describe the behavior of the object independently of the manipulators, an augmented object model was proposed based on dynamically consistent models [27]. Research began to focus on modeling multi-grasp behaviors and the associated internal forces acting between manipulators [44]. Using a closed-chain mechanism called the virtual linkage model, decoupled object behavior and accurate dynamic control of internal forces was addressed [63]. Mobile robotic platforms equipped with robotic manipulators were developed [19] and multi-grasp manipulation was implemented using efficient operational space algorithms [7]. Related work on constrained kinematic and dynamic models include [14, 42, 61, 65].

The work described in this section is closely connected to locomotion. It is therefore important to review modern developments on this field of research. Dynamic legged locomotion has been a center of attention since the 1960s [15]. The Zero Moment Point criterion (ZMP) was developed to evaluate center of mass (CoM) acceleration boundaries in coplanar multi-contact situations [60]. Implementations of simple dynamic control algorithms for multi-legged running robots followed [48]. ZMP methods for humanoid robots where pioneered around the same times [57, 58] and later perfected as part of the Honda humanoid program [18]. To enable generalized multi-contact locomotion behaviors, extensions to the ZMP dynamic evaluation criterion were developed [17]. Compliant multi-contact behaviors using optimal distribution of contact forces have been recently explored [20]. Finding CoM static placements given frictional constraints for multi-legged systems was tackled in [2, 9]. The field of legged locomotion is vast and continues to broaden with pioneering contributions in areas such as hybrid non-linear control [59, 62], biomimetic coupled oscillators [5, 21], and passive dynamic walking [10, 41].

2.2.1 Kinematic and Dynamic Models

The kinematic and dynamic behavior of a humanoid robot interacting with its environment is determined not only by the robot's movement but also by its interactions with the environment through contact forces. To characterize these complex interactions, we represent humanoids as free floating articulated systems pushed against the ground by gravity forces and in contact with ground surfaces using their limbs or body parts to gain maneuvering support.

In Figure 2.2 we show a kinematic representation of a humanoid robot supported by its four limbs. The mobility of the robot with respect to the environment is determined by the relative position and orientation of a root base located on its body with respect to an inertial frame of reference, *i.e.*,

$$x_b = \begin{pmatrix} x_{b(p)} \\ x_{b(r)} \end{pmatrix} \quad \varepsilon\, \mathscr{R}^6, \tag{2.1}$$

where $x_{b(p)}$ and $x_{b(r)}$ represent, respectively, the relative position and rotation of the base with respect to the inertial frame of reference.

Computationally, we describe whole-body kinematics by assigning revolute and prismatic joints to the robot's free floating base, including a spherical joint representing the base orientation and three prismatic joints representing the base Cartesian position. We use efficient algorithms to compute branching kinematic variables using the algorithms described in [7, 13]. The humanoid system is therefore treated as a branching articulated system with n actuated joints and 6 underactuated DOFs.

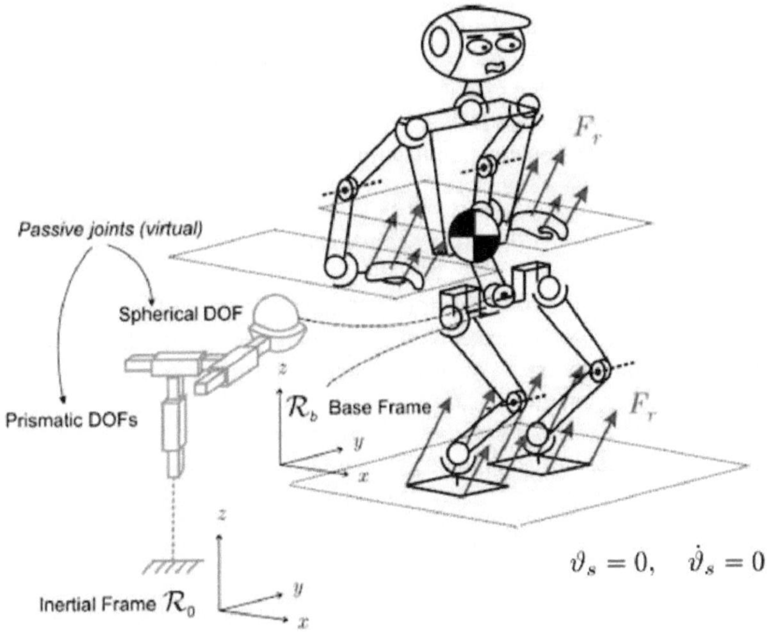

Figure 2.2 Kinematic representation of a humanoid robot. The free moving base is represented as a virtual spherical joint in series with three prismatic virtual joints. Reaction forces appear at the contact points due to gravity forces pushing the body against the ground. Contact constraints are expressed as rigid constraints with zero velocities and accelerations at the supporting bodies

Definition 2.1 (Robot Generalized Coordinates). The robot position and orientation in space and the position of its articulated joints can be fully described by the set

$$\{x_b, q\} = \{x_{b(p)}, x_{b(r)}, q_1, q_2, \ldots, q_n\} \quad \varepsilon \mathscr{R}^{6+n}, \tag{2.2}$$

where the vector x_b represents the coordinates of the base link and the $n \times 1$ vector q represents actuated joint positions.

Using Euler–Lagrangian formalism we can derive the following equation of motion describing the system's generalized dynamics under actuator torques and external contact forces:

$$A \begin{pmatrix} \dot{\vartheta}_b \\ \ddot{q} \end{pmatrix} + b + g + J_s^T F_r = U^T \Gamma, \tag{2.3}$$

where A is the $(n+6) \times (n+6)$ inertia matrix, b and g are $(n+6) \times 1$ vectors of Coriolis/centrifugal and gravity forces respectively,

$$U = \begin{pmatrix} 0_{n\times 6} & I_{n\times n} \end{pmatrix} \ \varepsilon \mathscr{R}^{n\times(n+6)} \tag{2.4}$$

is a selection matrix selecting actuated DOFs, and Γ is the $n \times 1$ vector of actuation torques. As such, U determines the underactuated/actuated degrees of freedom of the robot plus its free floating base by assigning zero torques to the base. Moreover,

J_s is the cumulative Jacobian of all supporting bodies in contact with the ground and F_r is the associated vector of reaction forces and moments.

As shown in Equation 2.3, reaction forces against the ground translate into forces acting on passive and actuated DOFs. With the premise that the contact bodies lay flat against the ground, a simple contact model can be defined where no relative movement occurs between contact surfaces and the supporting surfaces [64], *i.e.*,

$$\vartheta_s = 0_{6n_s \times 1}, \quad \dot{\vartheta}_s = 0_{6n_s \times 1}. \tag{2.5}$$

Here $\vartheta_{s(i)}$ is the 6×1 vector of linear and angular velocities of the ith contact support. Note that more sophisticated contact models should be considered in case that the contact interactions take place against deformable surfaces, *e.g.*, a stiffness/damper mechanical model.

Solving Equation 2.3 for the above set of constraints leads to a model-based estimate of reaction forces and moments in terms of actuation torques, *i.e.*,

$$F_r = \overline{J}_s^T U^T \Gamma - \overline{J}_s^T (b+g) + \Lambda_s \dot{J}_s \begin{pmatrix} \vartheta_b \\ \dot{q} \end{pmatrix}, \tag{2.6}$$

where Λ_s is the apparent inertia matrix projected in the space defined by the supporting bodies, J_s is the associated Jacobian matrix, and $\overline{J}_s$ is its dynamically consistent generalized inverse [25].

Plugging the above Equation 2.3 we obtain the following dynamic model under contact constraints

$$A \begin{pmatrix} \dot{\vartheta}_b \\ \ddot{q} \end{pmatrix} + N_s^T (b+g) + J_s^T \Lambda_s \dot{J}_s \begin{pmatrix} \vartheta_b \\ \dot{q} \end{pmatrix} = (U N_s)^T \Gamma, \tag{2.7}$$

where N_s is the dynamically consistent null space of J_s. Notice that the projection N_s in the above equation ensures that the applied torques fulfill the contact constraints of Equation 2.5.

Because the robot is constrained by the supporting ground, the position of its base can be derived from the position of the actuated joints alone. In turn, the coordinates of arbitrary task descriptors can be obtained from actuated joint positions only. Let us study this dependency more closely. The velocity constraint on the supporting extremities given in Equation 2.5 means that base and joint velocities lie in the algebraic null space of the support Jacobian. This dependency leads to the following expression:

$$\begin{pmatrix} \vartheta_b^* \\ \dot{q}^* \end{pmatrix} \triangleq N_s \begin{pmatrix} \vartheta_b \\ \dot{q} \end{pmatrix}, \tag{2.8}$$

where ϑ_b and $\dot{q}$ are arbitrary vectors of base and joint velocities, and ϑ_b^* and $\dot{q}^*$ are the corresponding constrained variables. Moreover, constrained joint velocities alone can be obtained by multiplying the above equation by the actuation matrix U, *i.e.*,

$$\dot{q}^* = U N_s \begin{pmatrix} \vartheta_b \\ \dot{q} \end{pmatrix}. \tag{2.9}$$

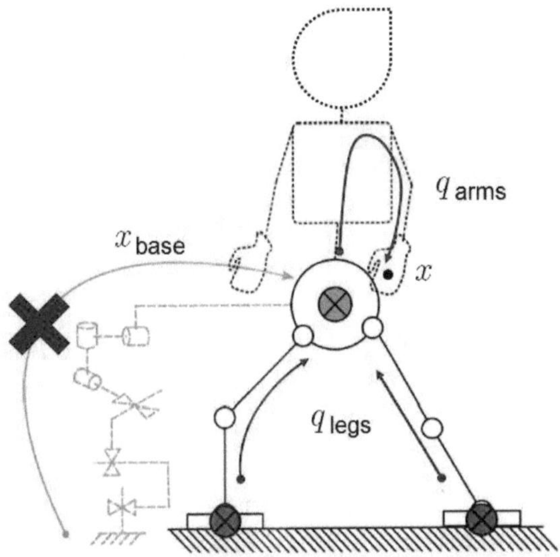

Figure 2.3 Dependency between base and joint displacements. Due to the effect of supporting contact constraints, the position of the end effector, x, can be expressed using joint positions alone, q, *i.e.*, disregarding the position of the base, x_{base}

When the robot is in single support stance, the matrix UN_s is full rank and therefore $\dot{q}^*$ can take arbitrary values. However, when the robot is in multi-contact stance, the presence of closed loops causes UN_s to be singular and as a result there are inter-dependencies between the constrained joint velocities $\dot{q}^*$. Solving the above equation yields the equality

$$\begin{pmatrix} \vartheta_b^* \\ \dot{q}^* \end{pmatrix} = \overline{UN_s}\,\dot{q}^*, \tag{2.10}$$

where $\overline{UN_s}$ is the dynamically consistent generalized inverse of UN_s. To verify that the above expansion is correct it must produce zero velocities at the support points, *i.e.*,

$$\vartheta_s = J_s \begin{pmatrix} \vartheta_b^* \\ \dot{q}^* \end{pmatrix} = J_s\overline{UN_S}\,\dot{q}^* = 0. \tag{2.11}$$

This is fulfilled when $\overline{UN_s}$ is the dynamically consistent generalized inverse of UN_s as defined in [52], *i.e.*,

$$\overline{UN_S} \triangleq A^{-1}(UN_S)^T \left(UN_sA^{-1}(UN_s)^T\right)^+, \tag{2.12}$$

where $(.)^+$ is the Moore–Penrose pseudo-inverse operator. To verify the correctness of Equation 2.11 we use the equalities $A^{-1}N_s^T = N_sA^{-1}$ and $J_sN_s = 0$ which are demonstrated in [52].

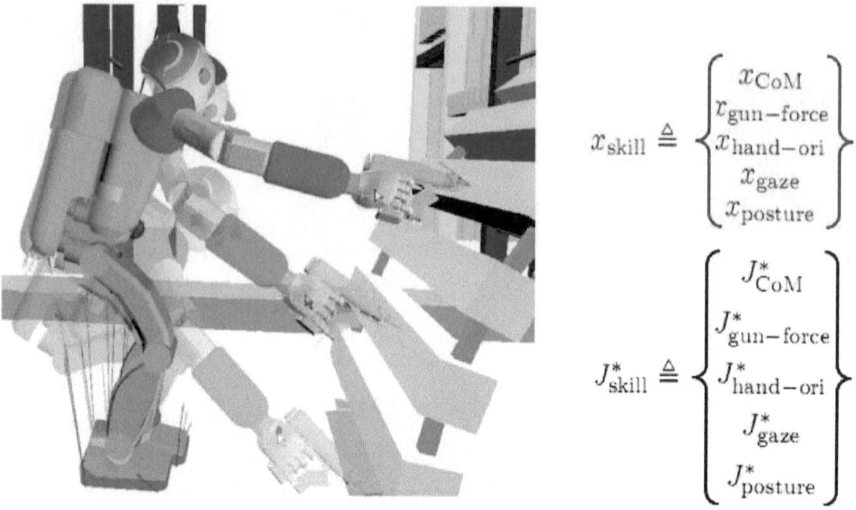

Figure 2.4 Whole-body manipulation behavior. We show here overlapping pictures from a simulated example consisting of a behavior of placing screws on a stair by means of a screwgun. To achieve this complex behavior we simultaneously optimize multiple task objectives shown in the right hand side if the image. Each objective is represented by a coordinate vector and a Jacobian transformation to express its relationship with respect to joint displacements

2.2.2 Task Kinematics and Dynamics Under Supporting Constraints

To execute complex actions, a humanoid must simultaneously coordinate the behavior of multiple task objectives. We consider the vector of task descriptors (*e.g.*, see Figure 2.4):

$$x_{\text{skill}} \triangleq \begin{pmatrix} x_1 \\ x_2 \\ \vdots \\ x_{n_t} \end{pmatrix}, \tag{2.13}$$

where each x_k describes the coordinates of a desired task descriptor, *e.g.*, the Cartesian position and orientation of a point in the robot's body, and n_t is the number of task descriptors. A general definition of task descriptors is any linear or non-linear transformation of the generalized coordinates defined in Definition 2.1.

When using (2.10), task velocities can be expressed in terms of joint velocities alone, *i.e.*,

$$\dot{x}_k = J_k \begin{pmatrix} \vartheta_b^* \\ \dot{q}^* \end{pmatrix} = J_k \, \overline{UN_s} \, \dot{q}^*. \tag{2.14}$$

Definition 2.2 (Support consistent reduced Jacobian). The task operator $J_k\overline{UN_s}$ in the above equation, behaves as a constrained Jacobian transformation mapping joint velocities into task velocities and we refer to it using the symbol

$$J_k^* \triangleq J_k \overline{UN_s}. \tag{2.15}$$

This expression is motivated by the dependency of base velocities on joint velocities, as shown in Figure 2.3.

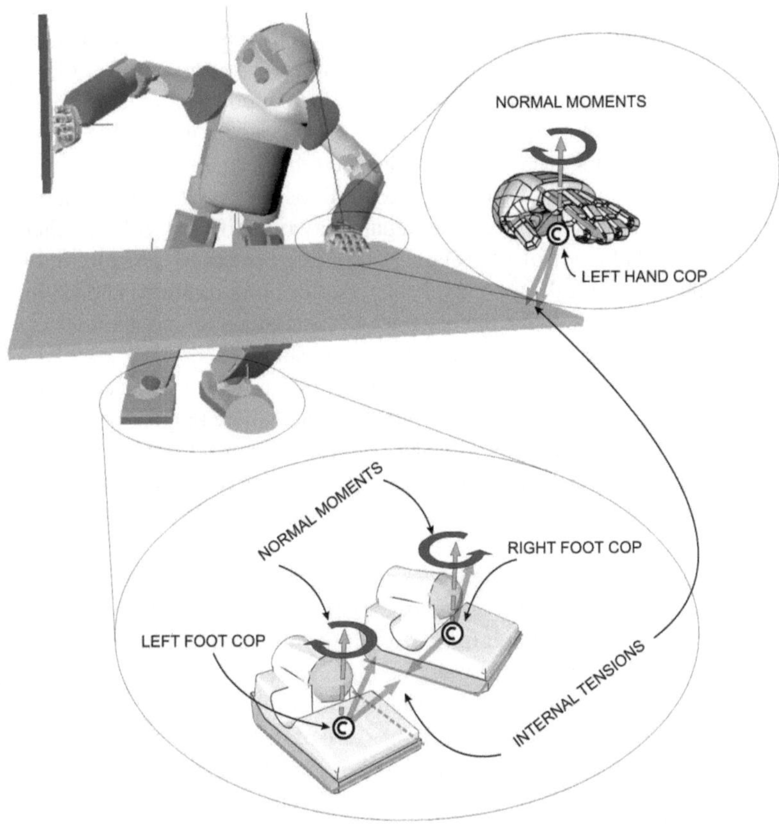

Figure 2.5 Decomposition of internal forces and moments. We decompose internal forces and moments into contact centers of pressure, internal tensions, and normal moments. Contact centers of pressure allow us to control contact rotational behavior while internal tensions and normal moments allow us to control the behavior of contact points with respect to surface friction properties.

2.2.3 Modeling of Contact Centers of Pressure, Internal Forces, and CoM Behavior

We consider whole-body contact scenarios where multiple extremities of the robot are in stable contact against flat surfaces (see Figure 2.5). In this case, every contact imposes six constraints on the robot's mobility. We assume that each extremity in contact has enough joints with respect to the base link to enable the independent control of its position and orientation. This condition translates to the existence of six or more independent mechanical joints between the base link and the extremity in contact. We consider contact scenarios involving an arbitrary number of supporting extremities, represented by the number n_s. Flat supporting contacts impose $6n_s$ constraints on the robot's motion, where 6 of these constraints provide the support to maneuver the robot's center of mass and the other $6(n_s-1)$ describe the internal forces and moments acting on the closed loops that exist between supporting extremities [63]. Internal forces and moments play two different roles in characterizing the contact behavior of the robot: (1) contact centers of pressure define the behavior of the contact links with respect to edge or point rotations and (2) internal tensions and moments describe the behavior of the contact links with respect to the friction characteristics associated with the contact surfaces.

For n_s links in contact we associate n_s contact CoPs. Each contact center of pressure is defined as the 2D point on the contact surface where tangential moments are equal to zero. Therefore, $2n_s$ coordinates describe all contact pressure points. Figure 2.5 illustrates a multi-contact scenario involving three supporting contacts on the robot's feet and left hand and an operational task designed to interact with the robot's right hand.

We focus on the analysis of the forces and moments taking place on a particular contact body, as shown in Figure 2.6. Based on [60], we abstract the influence of the robot's body above the kth supporting extremity by the inertial and gravity force f_{s_k} and the inertial and gravity moment m_{s_k} acting on a sensor point S_k. For simplicity, we assume the sensor is located at the mechanical joint of the contact extremity. Here, P_k represents the contact center of pressure of the kth contact extremity, f_{r_k} is the vector of reaction forces acting on P_k, and m_{r_k} is the vector of reaction moments acting on P_k. The frame $\{O\}$ represents an inertial frame of reference located outside of the robot, and the frame $\{S_k\}$ represents a frame of reference located at the sensor point. All force quantities are described with respect to the sensor frame. Assuming the supporting extremities are in static contact, the center of pressure for the kth contact link can be written as

$$P_{kx} = S_{kx} - \frac{f_{r_{kx}}}{f_{r_{kz}}}\left(S_{kz} - P_{kz}\right) - \frac{m_{sy}}{f_{r_{kz}}}, \tag{2.16}$$

$$P_{ky} = S_{ky} - \frac{f_{r_{ky}}}{f_{r_{kz}}}\left(S_{kz} - P_{kz}\right) + \frac{m_{sx}}{f_{r_{kz}}}. \tag{2.17}$$

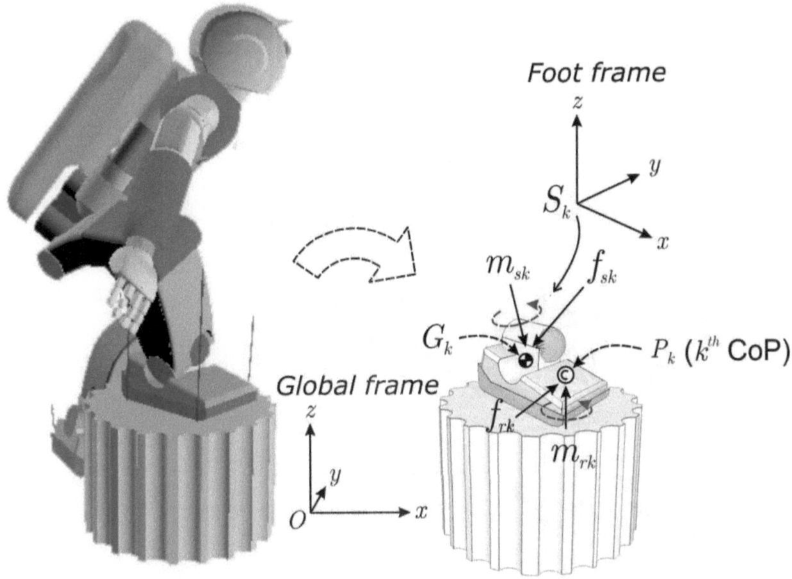

Figure 2.6 Forces acting on a supporting body (*e.g.*, the right foot). Establishing the balance of moments on each contact body allows us to determine the position of contact centers of pressure

Here, x and y refer to the tangential directions with respect to the local surface frames. The same analysis applies to the other extremities in contact, defining the n_s contact centers of pressure.

Characterizing contact CoPs is an important step towards developing contact models that describe intuitively the contact state of the robot. Based on these models, in the next section we describe methods that efficiently control the internal force state of the robot. In particular, we present control methods that enable the manipulation of contact CoPs to desired positions on the contact surfaces. By manipulating contact CoPs away from contact edges we ensure that contact surfaces stay flat against the supporting surfaces avoiding undesired contact rotations. Additionally, controlling contact CoPs will result in compliant contact behaviors since they imply neutralizing tangential moments exerted by contact surfaces. The various properties of contact CoPs make them an effective abstraction for the control and analysis of contact rotational behaviors.

The analysis of contact CoPs and CoM behaviors will be exploited next to develop a virtual linkage model that characterizes the interaction and maneuvers of humanoids in unstructured multi-contact environments.

In [54], a new instance of the virtual linkage model [63] to describe the complex contact relationships formed between contact closed loops was presented. The virtual linkage model is a parallel multi-articulated mechanical model connecting closed loops between contact extremities using virtual prismatic and spherical joints. It was first used to describe the relationship between resultant and internal forces of a shared object between multiple manipulators. In the case of humanoid

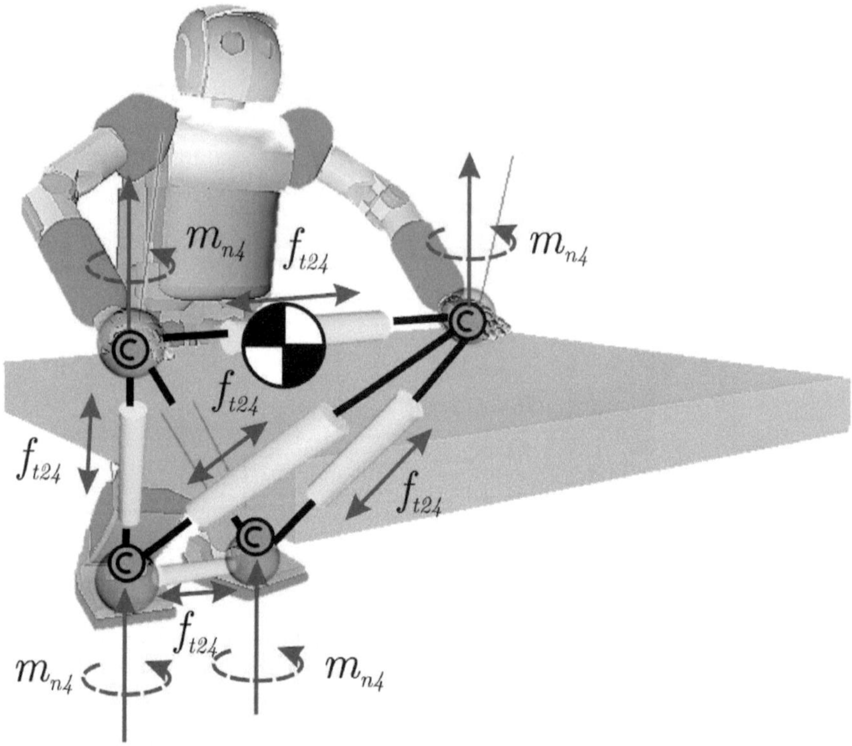

Figure 2.7 Virtual linkage model for humanoid robots. We define a virtual linkage model anchored through contact CoPs. It enables the characterization of internal tensions and moments against contact surfaces. The virtual linkage model abstracts the behavior of internal and CoM forces with respect to reaction forces. These characteristics make the virtual linkage model a powerful tool for the analysis and efficient control of CoM maneuvers and contact interactions

robots, we use the virtual linkage model to describe the behavior of internal forces and moments as well as the behavior of forces and moments acting at the robot's CoM.

To define the mechanical structure associated with the virtual linkage model, we choose anchoring points located at the contact CoP positions previously defined. Therefore, contact CoPs act as the nodes connecting the linkages. To prevent supporting extremities from rotating along contact edges, our approach is to place contact CoPs as close as possible to the geometric centers of the extremities in contact. Hence, unwanted transitions to edge or point contacts will be avoided in the case that contact disturbances occur.

Note that placing contact CoPs at the centers of the links is not a necessary constraint. They can be placed at any position below the links in contact, but away from contact vertices and edges to prevent rotations. Therefore, in this paper we only consider flat surface contact interactions. Extensions to corner and edge contacts could also be explored using a similar methodology We associate a virtual linkage model connecting all contact centers of pressure. In the scenario shown in

Figure 2.7 each body in contact introduces a tension with respect to its neighboring nodes as well as normal and tangential moments. For contacts with $n_s > 2$ we can independently specify $3(n_s - 2)$ tensions, n_s normal moments, and $2n_s$ tangential moments describing the centers of pressure. Any additional link in contact introduces three new tensions with respect to its neighbors and three more moments. No more than three tensions between neighboring nodes can be independently specified for a given contact. Internal tensions and normal moments characterize the behavior of contact bodies with respect to the friction cones and rotational friction properties of the surfaces in contact.

In [54], we derived the following expression describing the virtual linkage model:

$$\begin{pmatrix} F_{\text{com}} \\ \hline F_{\text{int}} \end{pmatrix} = G F_r \tag{2.18}$$

where F_{com} is the 6D vector of inertial and gravitational forces and moments at the robot's center of mass and F_{int} is the vector of internal tensions between closed loops and tangential and normal moments to the contact surfaces,

$$F_r \triangleq \begin{pmatrix} f_{r1} \\ \vdots \\ f_{rn_s} \\ \hline m_{r1} \\ \vdots \\ m_{rn_s} \end{pmatrix} \varepsilon \, \mathscr{R}^{6n_s}, \tag{2.19}$$

is the cumulative vector of reaction forces and moments projected onto the contact CoPs and expressed in global frame, and

$$G \triangleq \begin{pmatrix} W_{\text{com}} \\ \hline W_{\text{int}} \end{pmatrix} \varepsilon \, \mathscr{R}^{6n_s \times 6n_s}, \tag{2.20}$$

is the grasp/contact matrix defined in [54]. The upper part of the grasp/contact matrix defines the CoM behavior as

$$W_{\text{com}} \triangleq \left(\begin{array}{c|c} [I]_{3\times3} & [0]_{3\times3} \\ \hline \widehat{P}_{\text{com}} & [I]_{3\times3} \end{array} \right)^{-1} \left(\begin{array}{ccc|ccc} [I]_{3\times3} & \cdots & [I]_{3\times3} & [0]_{3\times3} & \cdots & [0]_{3\times3} \\ \hline \widehat{P}_1 & \cdots & \widehat{P}_n & [I]_{3\times3} & \cdots & [I]_{3\times3} \end{array} \right), \tag{2.21}$$

where, $\widehat{P}_{\text{com}}$ is the cross product operator associated with the position of the CoM with respect to the global frame, $\widehat{P}_i$ is the cross product operator associated with the

ith center of pressure point, and $[I]_{3\times3}$ and $[0]_{3\times3}$ are the 3×3 identity and zero matrices respectively. The lower part of G describes the internal force behavior, *i.e.*, the tensions between contact nodes and the surface moments, and its detailed expression can be found in [54].

In the next section, we will use these virtual linkage models to develop controllers that can govern internal forces and CoM behavior. We will also study the application of the matrix G to find solutions of CoM and internal force behaviors that comply with rotational and frictional contact constraints.

2.2.4 Friction Boundaries for Planning CoM and Internal Force Behaviors

The control of center of mass and internal force behavior is direclty related to the friction properties of the supporting surfaces. We analyze here this relationship and the effect on contact stability with respect to surface friction boundaries. The torque command that we will see in Equation 2.49 entails controlling both the robot's center of mass and the internal force behaviors. As discussed in [54], center of mass behavior is always one of the task objectives involved in the torque reference of a humanoid robot.

The trajectory and values of CoM and internal force behavior cannot take arbitrary values. They need to be chosen to fulfill contact and frictional constraints at the supporting surfaces. Ensuring that reaction forces remain within friction boundaries is needed to prevent robot contact extremities from sliding and rotating with respect to the environment.

To facilitate the analysis of friction behaviors with respect to surface properties we rotate reaction forces and moments, which are normally expressed in global frame as shown in Equation 2.6, to align with the frames attached to the contact surfaces, so their z components are parallel to surface normals, *i.e.*,

$$F_{\text{surf}} \triangleq R_{\text{surf}}\, F_r. \tag{2.22}$$

Here F_{surf} represents the rotated forces and moments and R_{surf} is a $6n_s \times 6n_s$ rotation matrix that aligns z components to surface normals. Using the above transformation, Equation 2.18 can be written as

$$\begin{pmatrix} F_{\text{com}} \\ \hline F_{\text{int}} \end{pmatrix} = G R_{\text{surf}}^{-1} F_{\text{surf}}, \tag{2.23}$$

where

$$F_{\text{com}} \triangleq \begin{pmatrix} f_{\text{com}} \\ m_{\text{com}} \end{pmatrix}, \quad F_{\text{int}} \triangleq \begin{pmatrix} f_{\text{int}} \\ m_{\text{int}} \end{pmatrix}, \tag{2.24}$$

are the forces and moments associated with the robot's center of mass and internal force behaviors. and R_{surf}^{-1} is an inverse cumulative rotation. The above expression can be used to estimate the surface forces due to center of mass and internal force commands, *i.e.*,

$$F_{\text{surf}} = R_{\text{surf}} G^{-1} \begin{pmatrix} F_{\text{com,ref}} \\ F_{\text{int,ref}} \end{pmatrix}, \tag{2.25}$$

where $F_{\text{com,ref}}$ and $F_{\text{int,ref}}$ are reference values induced by CoM and internal force behavior policies.

Using the above expression we can analyze the feasibility of whole-body maneuvers and force behaviors with respect to frictional properties of the contact surfaces. To evaluate the feasibility of these behaviors, we consider the following indexes associated with linear frictional cones and rotational frictional ratios for the kth contact extremity

$$\alpha_{\text{surf},k} \triangleq \begin{pmatrix} \tan^{-1}\left(\frac{f_{\text{surf},kz}}{f_{\text{surf},kx}}\right) \\ \tan^{-1}\left(\frac{f_{\text{surf},kz}}{f_{\text{surf},ky}}\right) \end{pmatrix}, \tag{2.26}$$

$$\beta_{\text{surf},k} \triangleq \frac{m_{\text{surf},kz}}{f_{\text{surf},kz}}, \tag{2.27}$$

where $\alpha_{\text{surf},k}$ is the vector of angles measured between the surface plane and the reaction force vector and $\beta_{\text{surf},k}$ is a ratio between the normal moment and normal force which characterizes the rotational friction behavior. This last index provides an intuitive metric of rotational friction characteristics since it normalizes normal moments with respect to the normal force load. We determine the boundaries of the above indexes by using simple models involving static friction cones and a heuristic model for rotational friction, *i.e.*,

$$\alpha_{\text{surf},k} \ \varepsilon \ \text{fricCone(k)}, \tag{2.28}$$

$$\beta_{\text{surf},k} \ \varepsilon \ \text{rotIndex(k)}. \tag{2.29}$$

On the other hand, the boundaries of center of pressure locations are determined by the surfaces in contact, *i.e.*,

$$P_k \ \varepsilon \ \text{surfArea(k)}, \tag{2.30}$$

where P_k are contact CoPs and surfArea(k) is the surface area below the kth contact body. Overall, Equations 2.28–2.30 determine the boundaries of frictional and rotational stability that comply with the contact model defined in Equation 2.5.

As we mentioned earlier, we control contact CoPs to be located close to contact geometric centers. The choice of CoPs defines the anchors of the virtual linkage model and therefore determines the model of the grasp/contact matrix. Once G is defined, we use the boundaries defined in Equations 2.28–2.30 to obtain feasible values of CoM and internal forces by means of Equation 2.25. In this context the grasp/contact matrix G emerges as a powerful tool for the planning of advanced maneuverings in arbitrary 3D multi-contact environments.

2.3 Prioritized Whole-body Torque Control

In this section, we use the models previously described to provide a control framework that can (1) create whole-body compliant behaviors, (2) unify the control of force, motion, and multi-contact behaviors, (3) fulfill geometric and contact constraints, and (4) plan body displacements that are feasible in the terrain. A key aspect we focus on is in the fulfillment of geometrical and contact constraints. This problem has been traditionally addressed as a motion planning process. However, for fast response, we address it here as a reactive control problem. We seek control strategies that can adapt very quickly to geometric changes in the operational environment as well as to fast contact transitions.

To accomplish these goals, we describe here our collaborative work on model-based control for the creation and execution of whole-body compliant skills that comply with geometric and contact constraints. The highlights of this framework are (1) prioritized organization between task objectives that ensures that critical tasks are accomplished first, (2) fulfillment of geometric and contact constraints, (3) dynamically correct control of tasks and postures using all actuation redundancy, (4) multi-contact compliant control.

Historically, inverse kinematic control strategies have been widely used for manipulator control, because they enable the direct computation of trajectories according to geometric plans [47, 50]. Extensions for multi-objective control and task prioritization were later added to enable the control of redundant manipulators [16, 55]. However, with increasing demand on contact interactions, torque control strategies became highly relevant. In particular, the operational space control formulation [26] enabled the unified control of force and motion behaviors in mobile manipulators.

The control of humanoid robots has gone through a different path, since a differentiating issue in humanoids is the presence of a free moving base and contact forces on the supporting limbs. For many years, the main focus of humanoids had been on balance and on the generation of dynamic locomotion trajectories [18]. Full body motions of humanoids under geometric and balance constraints were later addressed as a motion planning problem [32]. A highly popular method to generate whole-body behaviors involves the generation of linear and angular momenta including task and balance constraints [22]. To enable whole-body compliant behaviors and exploit the redundancy of humanoids under geometric and contact constraints, prioritized torque control was developed [53]. Recently, a torque level control framework based on minimum norm distribution of contact forces has been introduced [20]. Accurate control of internal forces and contact centers of pressure in the context of prioritized whole-body control has been recently developed [54] and is summarized in this section. Important contributions in the field of constrained whole-body control have also been recently made by the AIST-CNRS Joint Robotics Laboratory [45, 56].

The topics discussed in this section include, representations of whole-body skills, prioritized torque control structures, real-time handling of dynamic constraints, task feasibility, and control of contact centers of pressure and internal tension and moments. We will also discuss several simulated examples.

2.3.1 Representation of Whole-body Skills

Although the goal of humanoid robots is to execute advanced dexterous and maneuvering tasks, much low-level coordination needs to take place to use efficiently their mechanical advantage. To create advanced skills we must first understand the many coordinated processes that need to take place to manipulate, maneuver, and respond to the environment.

From a pure physical standpoint, a robot is a parallel underactuated structure with motion and force interactions taking place with respect to the environment and between the limbs in contact. Displacements can be modeled as center of mass and end effector trajectories while manipulation tasks can be modeled in terms of force and motion interactions. It is logical to select the body's center of mass and the limb end effectors as operational points, each of them controlled as a separate, but cooperating, process. Another important objective is the coordination of internal forces and moments, which are characterized from the models of the previous section.

Not surprisingly, the list of coordinated processes that need to be simultaneously optimized for efficient humanoid maneuvers and interactions is long. Postural behavior to enhance workspace reachability and effort minimization are other important criteria to be controlled. A process that monitors and prevents joint limits from being violated is required for safety as is a process that monitors self-collisions between body parts and enforces a minimum separation.

Table 2.1 illustrates a list of control objectives that we use to define a manipulation skill while standing up. Task kinematics under supporting constraints were analyzed in the previous section. When a primitive defining a complex skill, such as

Table 2.1 Decomposition of a prototypical manipulation task

Task primitive	Coordinates	Control policy
Joint limits	joint positions	locking attractor
Self-collisions	distances	repulsion field
Balance	CoM	dynamic trajectory
Right hand	Cartesian	force and position
Gaze	head orientation	position
Standing posture	marker coordinates	captured sequences
Internal forces	tensions / moments	optimal contact

the one in Table 2.1, is considered, each descriptor can be kinematically characterized through unique coordinates and constrained Jacobians, *i.e.*,

$$x_k = T[x_b, q], \tag{2.31}$$

$$\dot{x}_k = J_k^* \, \dot{q}, \tag{2.32}$$

where $T[.]$ is a transformation defining the objective to be controlled, x_k is the coordinate representing the kth objective, J_k^* is the associated prioritized Jacobian, and x_b and q are base and joint coordinates.

2.3.2 Prioritized Torque Control

Torque control of humanoid robots represents an alternative to other popular control approaches, such as inverse kinematics and resolved momentum control [22]. In contrast with these two other approaches, prioritized torque control is based on two key feature that include, (1) using prioritized force models to unify whole-body motion and force interactions while ensuring the fulfillment of geometric and contact constraints, and (2) developing whole-body compliant multi-contact capabilities for effective interactions with the environment. Over the years, we have developed and matured prioritized torque control methods that can create complex interactive skills for operations of humanoids in highly constrained environments. The motivation behind prioritized structures is to develop mechanisms to temporarily override non-critical task processes in order to fulfill critical constraints as well as to exploit efficiently the actuation redundancy for postural behavior. we summarize here this work. Note that for proofs of the mathematical demonstrations, the reader should refer to the author's previously published work.

In general, for n_t arbitrary task objectives defining a skill we use the following recursive prioritized whole-body torque structure for control

$$\Gamma = \Gamma_1 + \Gamma_{2|\text{prec}(2)} + \cdots + \Gamma_{n_t|\text{prec}(n_t)} = \sum_{k=1}^{n_t} \Gamma_{k|\text{prec}(k)}, \tag{2.33}$$

where the subscript $k|\text{prec}(k)$ is used to indicate that the kth task operates in the null space of all higher priority tasks.

Definition 2.3 (Prioritized torques). The following expression determines the projection of lower priority tasks into the null space of higher priority tasks

$$\Gamma_{k|\text{prec}(k)} \triangleq N^T_{\text{prec}(k)} \Gamma_k, \tag{2.34}$$

where $N_{\text{prec}(k)}$ is the combined null space of all higher priority tasks (*i.e.*, all preceding tasks) to the kth level.

The hierarchical torque control structure Equation 2.33, unifies forces and motion by defining the recursive structure

$$\Gamma = \left(J_1^{*T} F_1\right) + \left(J_{2|1}^{*T} F_{2|1}\right) + \cdots + \left(J_{n_t|\mathrm{prec}(n_t)}^{*T} F_{n_t|\mathrm{prec}(n_t)}\right) + \left(N_t^{*T} \Gamma_{\mathrm{posture}}\right)$$
$$= \sum_{k=1}^{N} \left(J_{k|\mathrm{prec}(k)}^{*T} F_k\right) + N_t^{*T} \Gamma_{\mathrm{posture}}, \quad (2.35)$$

where the matrices $J^*_{k|\mathrm{prec}(k)}$ correspond to prioritized task Jacobians as defined in Equation 2.36, the vectors $F_{k|\mathrm{prec}(k)}$ correspond to operational forces to control the kth priority tasks, N_t^* corresponds to the product of all null spaces which defines the residual movement redundancy for posture control, and $\Gamma_{\mathrm{posture}}$ corresponds to the control policy for posture control which is described in detail in [52].

Definition 2.4 (Prioritized Jacobian). The following prioritized Jacobian is associated with the kth priority task:

$$J^*_{k|\mathrm{prec}(k)} \triangleq J^*_k N^*_{\mathrm{prec}(k)}, \quad (2.36)$$

where J^*_k is the constrained Jacobian associated with the kth task and $N^*_{\mathrm{prec}(k)}$ is the prioritizing null space of all preceding tasks.

Theorem 2.1 (Prioritized operational space control). *The following control vector yields linear control of forces for the kth prioritized task*

$$\Gamma_{k|\mathrm{prec}(k)} = J^{*T}_{k|\mathrm{prec}(k)} F_{k|\mathrm{prec}(k)}, \quad (2.37)$$

where $\Gamma_{k|\mathrm{prec}(k)}$ is the kth component of the prioritized torque control structure shown in (2.33), $J^*_{k|\mathrm{prec}(k)}$ is the prioritized Jacobian for the kth task point discussed in (2.36), and $F_{k|\mathrm{prec}(k)}$ is the vector of command operational forces.

Corollary 2.1 (Prioritized motion control). *In the absence of contact forces at the operational point, the following control command yields linear control of task accelerations:*

$$F_{k|\mathrm{prec}(k)} = \Lambda^*_{k|\mathrm{prec}(k)} a_k^{ref} + \mu^*_{k|\mathrm{prec}(k)} + p^*_{k|\mathrm{prec}(k)} -$$
$$\Lambda^*_{k|\mathrm{prec}(k)} J_k A^{-1} (S N_s)^T \sum_{i=1}^{k-1} \Gamma_{i|\mathrm{prec}(i)}. \quad (2.38)$$

Here $F_{k|\mathrm{prec}(k)}$ is the control force shown in Equation 2.37, a_k^{ref} is a feedback acceleration-level control policy for the kth priority task, and the remaining dynamic quantities for the above equation have the following expressions:

$$\mu^*_{k|\text{prec}(k)} \triangleq \Lambda^*_{k|\text{prec}(k)} J_k A^{-1} N_s^T b - \Lambda^*_{k|\text{prec}(k)} \dot{J}_k \begin{pmatrix} \vartheta_b \\ \dot{q} \end{pmatrix} + \Lambda^*_{k|\text{prec}(k)} J_k A^{-1} J_s^T \Lambda_s \dot{J}_s \begin{pmatrix} \vartheta_b \\ \dot{q} \end{pmatrix}, \tag{2.39}$$

$$p^*_{k|\text{prec}(k)} \triangleq \Lambda^*_{k|\text{prec}(k)} J_k A^{-1} N_s^T g. \tag{2.40}$$

When using the above structure for control we achieve linear control of task accelerations, i.e.,

$$\ddot{x}_k = a_k^{ref}. \tag{2.41}$$

Using a_k^{ref} we define a feedback control policy for motion control.

Corollary 2.2 (Prioritized force control). *During contact interactions, the following control vector yields feedback control of contact forces:*

$$F_{k|\text{prec}(k)} = -K_p(F_k - F_{\text{des}}) - K_v \dot{F}_k - K_i \int_0^t (F_k - F_{\text{des}}) dt, \tag{2.42}$$

where F_k is the vector of forces and moments measured at the operational point, F_{des} is the vector of desired forces and moments, and K_p, K_v, and K_i, are proportional, derivative, and integral gain matrices. When using the above control structure we directly control task forces at the operational point.

Corollary 2.3 (Dynamically consistent prioritized null space matrix). *The following matrix fulfills prioritization constraints:*

$$N^*_{\text{prec}(k)} = I - \sum_{i=1}^{k-1} \overline{J}^*_{i|\text{prec}(i)} J^*_{i|\text{prec}(i)}, \tag{2.43}$$

where $\overline{J}^*_{i|\text{prec}(i)}$ is the dynamically consistent generalized inverse of the prioritized Jacobian. Moreover, we use the conventions $N^*_{\text{prec}(1)} \triangleq I$ and $N^*_{1|\text{prec}(1)} \triangleq N^*_1 = I - \overline{J}^*_1 J^*_1$.

2.3.3 Real-time Handling of Dynamic Constraints

The goal of this section is to describe the implementation of prioritized torque control strategies to handle reactively dynamically changing geometric constraints. The use of prioritization provides an alternative to using high dimensional path planning strategies which scale poorly to highly redundant systems. Prioritization enables the response in real-time to dynamic constraints while optimizing the desired actions. It also prevents conflicting task objectives being accomplished if they violate geometric constraints acting on the robot. The constraint models described here are not

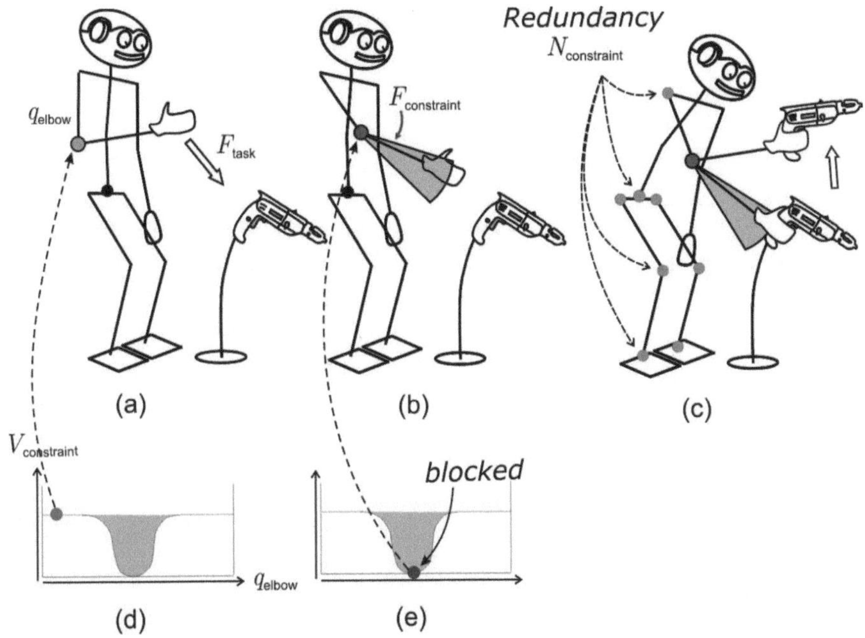

Figure 2.8 Handling joint limit constraints. (a) The robot's right hand is commanded to move towards the drill. (b) A constraint handling task is activated to prevent the right elbow reaching its physical limit. In turn, the reaching task is projected into the null space of the constraint. (c) The robot reaches the drill while complying with the elbow constraint. When a new hand command is issued, the elbow joint is unlocked

meant to replace the need for path planning, but to complement planners in behaviors that do not require global optimization.

Therefore, with the methods described here, whole-body control is reduced to the planning of the behavior of a few operational points regarding maneuvering and manipulation goals while other objectives such as fast response to geometric and contact constraints, and postural behavior are handled as reactive processes.

Internal constraints such as self-collision and joint limit avoidance are especially relevant when generating goal-based behaviors. Our approach to handle self-collisions will rely on implementing repulsion fields on proximity points of nearby links, creating dual repulsion forces on link pairs. self-collision constraints have been previously studied in the context of motion verification [23, 33]. However, our control approach goes further by providing support to modify the robot's pose in response to self-collisions. To handle joint limits, our approach consists of locking joints before they reach their physical limits. Strategies to handle joint limit constraints date back to [36]. With the implementation of visual servoing techniques, joint limit prevention has recently regained importance [12, 40, 39]. Our methods extend these approaches to operate in full humanoid systems, exploiting the overall mechanical redundancy to accomplish the desired actions given the internal and external constraints. In contrast with previous methods, our approach relies on enforcing constraints as priority tasks while other operational tasks operate in the residual

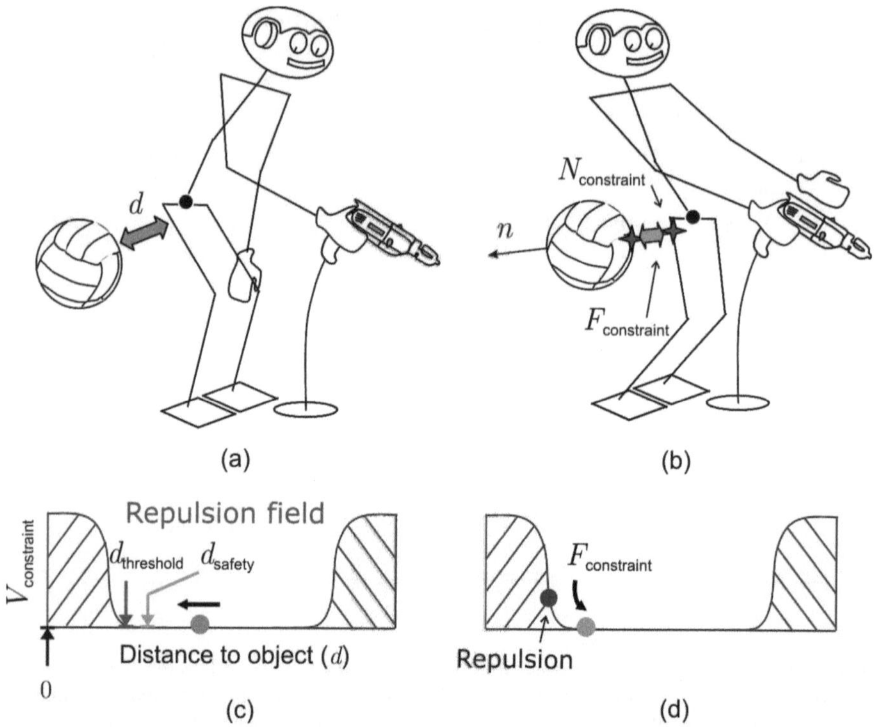

Figure 2.9 Obstacle avoidance illustration. When an incoming obstacle approaches the robot's body a repulsion field is applied to the closest point on the robot's body. As a result, a safety distance is enforced to avoid the obstacle

redundancy. This technique prevents operational tasks from violating constraints and provides a metric to determine task feasibility under the acting constraints.

Obstacle avoidance constraints are handled reactively via repulsion fields against incoming obstacles. Obstacle avoidance techniques have been popular in the context of path relaxation [31, 6, 3], reactive control [38, 25, 4], and collision free paths [43, 8, 37, 34, 35, 32]. Our techniques enhance and complement previous reactive and non-reactive techniques.

Real-time response to motion constraints has been extensively addressed as a secondary process. In contrast, our approach handles motion constraints as priority tasks and executes operational tasks in the null space that complies with the constraints. To illustrate our approach, we consider the whole-body behavior illustrated in Figure 2.8. The task decomposition to enforce the constraints while executing the behavior outlined in Table 2.1 is as follows:

Proposition 2.1 (Example of constraint-consistent whole-body control). *The following control structure creates whole-body behaviors that fulfill constraints while preventing lower priority tasks from violating the constraints,*

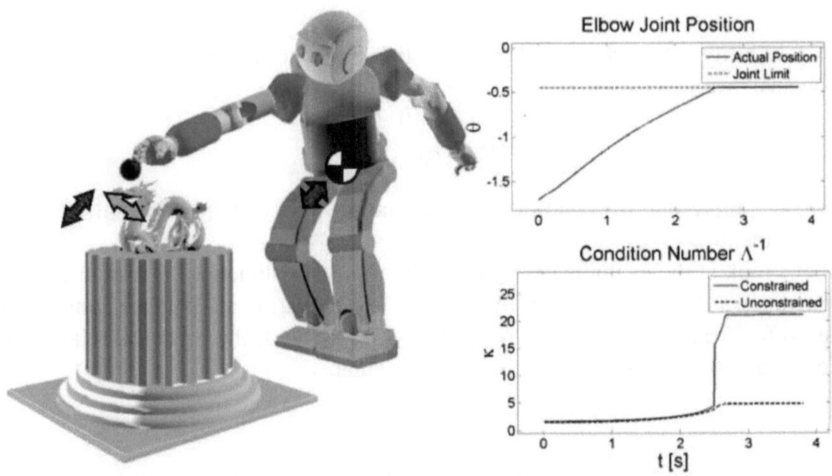

Figure 2.10 Task feasibility during a reaching task. The robot attempts to reach the object until no more dynamic redundancy is available under the motion constraints. At this point the Jacobian of the reaching task, determined through singular value decomposition, becomes singular. The singular directions determine the uncontrollable space

$$\Gamma = \Gamma_{\text{jlimits}} + \Gamma_{\text{scollision}|\text{p}(2)} + \Gamma_{\text{balance}|\text{p}(3)} + \Gamma_{\text{reaching}|\text{p}(4)} + \Gamma_{\text{gaze}|\text{p}(5)} + \Gamma_{\text{posture}|\text{p}(6)} + \Gamma_{\text{internal}|\text{p}(7)}. \quad (2.44)$$

Here the subscripts $\{\text{taskname}|\text{p(priority)}\}$ *indicate the priority order.*

As shown in Figure 2.8, implementing this ordering enables the right arm to stretch to reach the object, even when the right elbow joint is locked at the joint limit. The task becomes unfeasible when the current configuration of the robot uses all the dynamic redundancy to fulfill the constraints. Projecting operational tasks into the null space of the constraints is not only used to prevent constraint violations but also provides a metric to measure task feasibility under the constraints. Monitoring task feasibility is used as a mechanism to change behavior at runtime in response to dynamic constraints.

Similar control structures to the one shown in the above proposition can be used to handle obstacle constraints as priority tasks. To handle obstacles we apply repulsion fields in the direction of the approaching objects as shown in Figure 2.9.

2.3.4 Task Feasibility

Task feasibility can be measured by evaluating the condition number of prioritized Jacobians, *i.e.*,

$$\kappa\big(J^*_{k|\mathrm{prec}(k)}\big) \triangleq \frac{\sigma_1\big(J^*_{k|\mathrm{prec}(k)}\big)}{\sigma_{\dim(k)}\big(J^*_{k|\mathrm{prec}(k)}\big)}, \tag{2.45}$$

where $\kappa(.)$ represents the condition number and $\sigma_i(.)$ corresponds to the ith singular value of the prioritized Jacobian.

Let's look at the example of Figure 2.10. The robot stretches the arm and body to move toward the object. Joint limits at the hip and arm, and balance constraints take up the available movement redundancy. As a result, the prioritized Jacobian associated with the reaching task becomes singular.

2.3.5 Control of Contact Centers of Pressure and Internal Tensions/Moments

This section summarizes our recent collaborative work on internal force control [54]. In this work, a controller that governs the positions of contact centers of pressure and regulates internal tensions and normal moments to contact surfaces is described. This controller is integrated with the framework for whole-body prioritized control described earlier, unifying the control of center of mass maneuvers, operational tasks, and internal forces between contact closed loops.

The space of internal forces consists of torques that have no effect on the robot's motion, and therefore can be derived by analyzing the RHS of Equation 2.7. This condition leads to the following constraint on the torques:

$$(UN_s)^T \Gamma = 0, \tag{2.46}$$

which when plugged into (2.7) results in zero generalized accelerations. Therefore, the torque manifold that fulfills the above constraint belongs to the null space of (UN_s), defined by the projection

$$L^* \triangleq \left(I - UN_s\,\overline{UN_s}\right) \varepsilon\, \mathscr{R}^{n\times n}, \tag{2.47}$$

where we use the symbol L^* to denote contact closed loops, and the superscript * to indicate that the projection operates in contact space. The torques associated with internal forces are those that do not contribute to net movement, *i.e.*,

$$\Gamma = L^{*T}\Gamma_{\mathrm{int}}, \tag{2.48}$$

where Γ_{int} denotes the torque command to control internal forces and moments. Notice that plugging the above torques in the RHS of Equation 2.7 cancels out Γ_{int}, fulfilling the cancellation of acceleration effects.

We integrate the above structure with our prioritized controller discussed in Equation 2.35, leading to the unified torque structure

$$\Gamma = \sum_{k=1}^{N} \left(J_{k|\text{prec}(k)}^{*T} F_k \right) + N_t^{*T} \Gamma_{\text{posture}} + L^{*T} \Gamma_{\text{int}}. \tag{2.49}$$

In [54], we formulated the relationship between internal forces and moments with respect to contact forces as

$$F_{\text{int}} \triangleq \begin{pmatrix} f_t \\ \hline m_{\text{cop}} \\ \hline m_n \end{pmatrix} = W_{\text{int}} F_r \quad \varepsilon\, \mathscr{R}^{6(n_s-1)}, \tag{2.50}$$

where F_r is the cumulative vector of contact forces and moments described in Equation 2.19, W_{int} is the operator in the grasp matrix shown in Equation 2.20, that maps contact forces into internal force/moment variables, f_t is the $3(n_s-2)$-dimensional vector of tension forces, m_{cop} is the $2n_s$-dimensional vector of tangential moments computed at desired center of pressure points, and m_n is the n_s-dimensional vector of normal moments to contact surfaces. The above definition of internal forces implies defining a virtual linkage model anchored at contact CoP locations as described earlier in the chapter. We discussed the need to select contact CoP locations the closest possible to the geometric center of the contact surfaces to avoid unwanted rotations along edges and corners of the supporting links. To ensure that these locations become the actual CoPs we neutralize CoP moments at these points, *i.e.*, $m_{\text{cop}} = 0$.

We focus our attention on the problem of choosing internal forces that fulfill contact constraints. As we mentioned earlier, we control contact CoPs to be located close to contact geometric centers. The choice of CoPs defines the anchors of the virtual linkage model and therefore determines the form of the grasp/contact matrix. Once G is defined, we use the boundaries defined in Equations 2.28–2.30 to obtain feasible values of CoM and internal forces by means of Equation 2.25. In turn, the feasible solutions correspond to the range of values of F_{com} and F_{int} that fulfill contact constraints. The problem of choosing internal and CoM force control policies in optimal ways is a subject we are currently researching and therefore is not discussed here in detail.

The next step consists in implementing a controller that regulates internal force behavior to desired values, *i.e.*,

$$F_{\text{int}} \longrightarrow F_{\text{int,ref}} = \begin{pmatrix} f_{t,\text{ref}} \\ \hline [0]_{2n_s} \\ \hline m_{n,\text{ref}} \end{pmatrix}. \tag{2.51}$$

where $f_{t,\text{ref}}$ and $m_{n,\text{ref}}$ are desired internal force values obtained either through optimization processes or manually chosen.

To do that, in [54] we designed a new torque controller that can directly manipulate internal force behavior. Plugging Equation 2.49 into Equation 2.6 and using Equation 2.50 we obtain the equality

$$F_{\text{int}} = \overline{J}_{i|l}^{*T} \Gamma_{\text{int}} + F_{\text{int},\{t,p\}} - \mu_i - p_i, \tag{2.52}$$

where

$$\overline{J}_{i|l}^{*} \triangleq \left(L^* U \overline{J}_s W_{\text{int}}^T\right) \tag{2.53}$$

is a transformation matrix from torques to forces,

$$F_{\text{int},\{t,p\}} \triangleq W_{\text{int}} \overline{J}_s^T U^T \left[\sum_{k=1}^{N} \left(J_{k|\text{prec}(k)}^{*T} F_k \right) + N_t^{*T} \Gamma_{\text{posture}} \right] \tag{2.54}$$

are forces induced by task and postural behavior with torques shown in Equation 2.49, and μ_i and p_i are Coriolis/centrifugal and gravity terms defined as

$$\mu_i \triangleq W_{\text{int}} \mu_r, \tag{2.55}$$

$$p_i \triangleq W_{\text{int}} p_r. \tag{2.56}$$

Inverting Equation 2.52 we obtain the following internal force torque controller

$$\Gamma_{\text{int}} = J_{i|l}^{*T} \left(F_{\text{int,ref}} - F_{\text{int},\{t,p\}} + \mu_i + p_i \right), \tag{2.57}$$

where $J_{i|l}^{*}$ is a Moore–Penrose left pseudo-inverse of Equation 2.53 and the subscript $\{i|l\}$ denotes internal quantities operating in the space of contact closed loops. Plugging the above expression into Equation 2.52 and provided that $J_{i|l}^{*}$ is full row rank, we obtain the linear equality

$$F_{\text{int}} = F_{\text{int,ref}}. \tag{2.58}$$

To ensure that $J_{i|l}^{*}$ is full row-rank, L^* needs to span all internal force and moment quantities. This applies if there are at least six independent mechanical joints separating each contact extremity from the base link. A second required condition is to ensure that W_{int} defines independent internal quantities. This is already ensured in our derivation of the virtual linkage model. The term $J_{i|l}^{*}$ might be interpreted as a kinematic forward quantity mapping infinitesimal displacements of joint positions to infinitesimal displacements of virtual linkage displacements, *i.e.*,

$$\delta x_{\text{int}} = J^*_{i|l}\, \delta q. \tag{2.59}$$

Equation 2.57 will work in open loop, but to achieve accurate tracking of internal forces under actuator friction and mechanical modeling errors a feedback force control law involving proportional-integral-derivative feedback (PID) is needed. Given appropriate choice of the control law, the above linear relationship will ensure convergence to the desired internal force behaviors.

The above control structure provides a dynamically correct internal force controller that has no coupling effects on operational task, center of mass maneuvers, and postural behavior, hence enabling the efficient control of whole-body multi-contact interactions. It provides the support to simultaneously control the position of multiple contact centers of pressure and the internal tensions and normal moments acting between contact closed loops. A block diagram of the overall whole-body torque controller with internal force commands is shown in Figure 2.11.

2.4 Simulation Results

We study various examples on a simulated model of the child size humanoid robot Asimo. The objective of Asimo is to operate in offices and homes assisting humans in a variety of service and care giving chores. Recently, our colleagues at Stanford have developed a research version of Asimo that uses torque control commands [30, 66]. Torque controlled robots have the ability to execute whole-body compliant behaviors which is needed to operate efficiently and safely in human environments where contact interactions occur often. A dynamic simulation environment [7] and a contact solver based on propagation of forces and impacts [51] are used to simulate the execution of the methods described in this chapter. The whole-body controller described in Equation 2.49 is implemented on a task-based software environment that enables the online optimization of whole-body behaviors. Using this environment we create various behaviors involving biped and multi-contact stances as well as operational task interactions and center of mass maneuvers.

2.4.1 Multi-contact Behavior

In the example shown in Figure 2.12, we study a compliant behavior that emerges from controlling internal forces and moments. The robot is commanded to establish a four contact stance leaning against a pivoting table. We implement a multi-contact behavior that involves the following phases: (1) moving the robot's hands and its center of mass towards a table; (2) establishing four contacts; (3) perturbing the tabletop position by applying random external interactions; and (4) controlling the internal forces and moments to respond compliantly to the variations of the table. This sequence of actions is accomplished using a state machine where each state

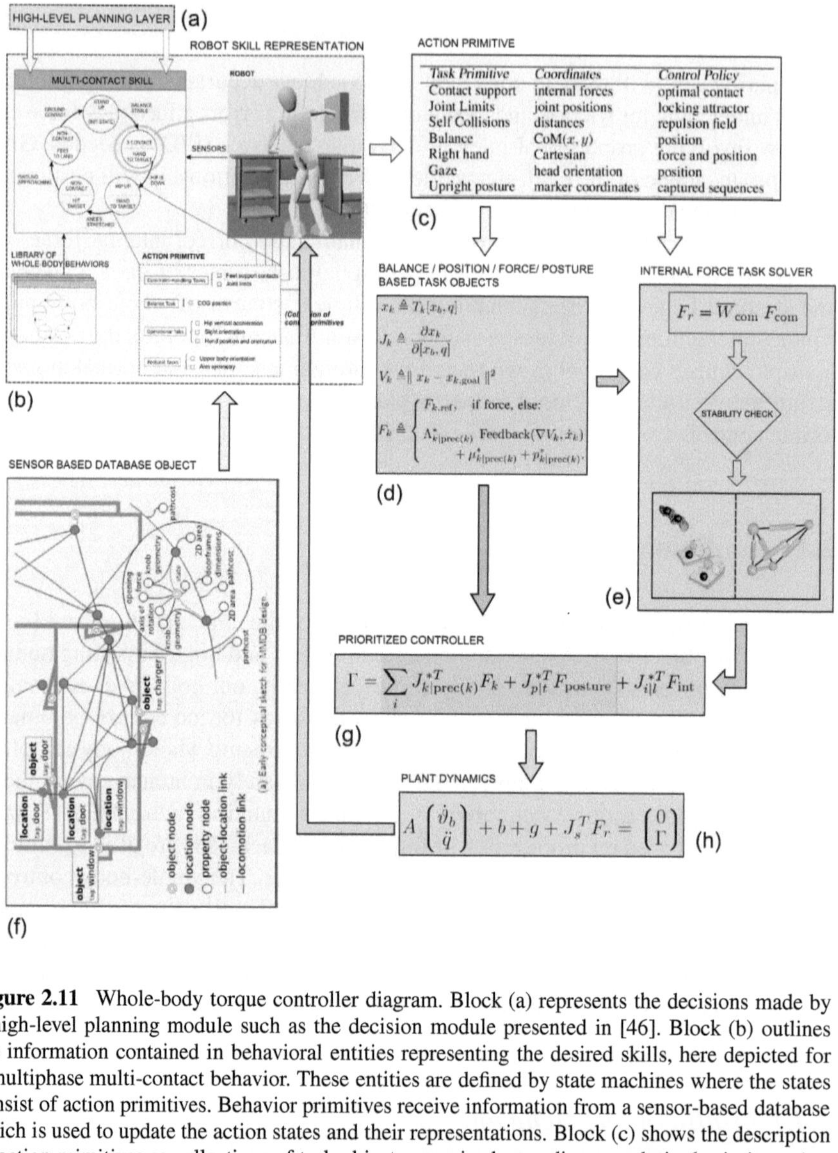

Figure 2.11 Whole-body torque controller diagram. Block (a) represents the decisions made by a high-level planning module such as the decision module presented in [46]. Block (b) outlines the information contained in behavioral entities representing the desired skills, here depicted for a multiphase multi-contact behavior. These entities are defined by state machines where the states consist of action primitives. Behavior primitives receive information from a sensor-based database which is used to update the action states and their representations. Block (c) shows the description of action primitives as collections of task objects organized according to a desired priority ordering. The action primitive shown above corresponds to one of the states of the desired multi-contact behavior. Block (d) describes task objects as entities containing coordinate representations, differential kinematic representations, goal-based potential functions, and force-based control policies. Block (e) represents a system currently under development that is used to automatically solve internal force behavior given a desired action representation. Block (g) represents our prioritized torque-based controller, which uses the previous behavior representations and control policies to create whole-body control torques. Block (h) represents the estimated dynamic behavior of the underactuated robot under multi-contact constraints in response to the previous torque commands

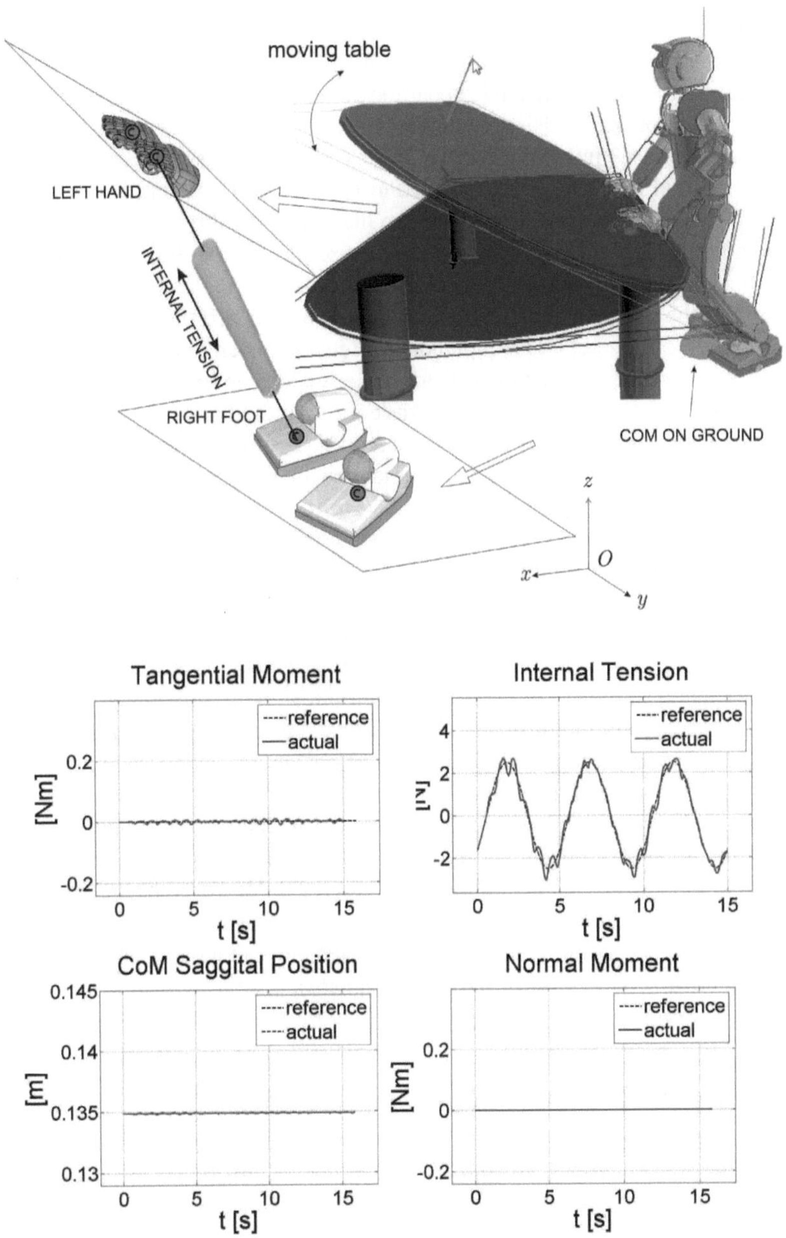

Figure 2.12 Compliant multi-contact behavior. A four contact stance is shown here. Contact centers of pressure are controlled to stay at the center of the bodies in contact. The center of mass is controlled to remain at a fixed location. The table is pivoted rapidly by an external user to challenge the pose. To add additional difficulty, the internal tension between the left hand and the right foot is commanded to track a sinusoidal trajectory. All other tensions and normal moments are commanded to become null. The robot was able to track effectively all tension behaviors with errors around 0.3 N. CoM positions were tracked with error values below 0.1 mm

involves optimizing multiple low-level task objectives. To achieve the desired global behavior we simultaneously control center of mass maneuvers, operational tasks, and postural stance as indicated in Proposition 2.1 as well as the internal force policy defined in Equation 2.57. Using Equation 2.18, we construct two virtual linkage models for the phases involving two and four contacts. For simplicity, all tension forces and normal moments are controlled to become null, except for one internal tension. To demonstrate force tracking at the internal level, the tension between the left hand and the right foot is commanded to track a sinusoidal trajectory. All other internal forces and tensions are commanded to be zero. At the same time, a user interacts with the pivoting table moving it up and down in arbitrary fast motions.

The position of contact CoPs on hands and feet is chosen to be at the geometric center of the extremities in contact. During the bipedal phase, the center of mass is commanded to stay between the feet while moving the hands towards the table, while CoM and CoP positions are simultaneously controlled. In the phases with four contacts, the center of mass is commanded to maneuver towards the table by tracking a trajectory that fulfills contact stability constraints as defined by the boundaries of Equations 2.28–2.30. A feedback controller to track this trajectory is implemented using force commands in CoM space. Postural behavior is controlled by optimizing a criterion that minimizes the distance with respect to human pre-recorded postures using a method similar to the one described in [11].

Because contact CoPs are commanded to stay at the center of the extremities in contact, the hands respond compliantly to table movement, remaining flat against the surface to maintain the desired CoP positions. The accompanying data graphs show tangential and normal moments, the tension between the left hand and the right foot, and the sagittal position of the CoM. The tracking error for the internal tension is small with a maximum value around 0.3 N. This error is mainly caused due to the unmodeled movement of the table. As we recall, our framework assumes that the table is static, which is implied in Equation 2.5. However, because the table undergoes fast accelerations the model is inaccurate. Despite this inaccuracy, the tracking behavior is very good. In contrast, if the tabletop remains at a fixed location, the tracking error converges to zero (not shown here).

2.4.2 Real-time Response to Dynamic Constraints

Let us briefly review another example demonstrating the response to rapidly moving obstacles as shown in Figure 2.13. More details on this example can be found in [52]. This behavior is accomplished using the controllers described in the section on geometric constraints. The graphs depict the response when the obstacle approaches the hand which is implementing an operational position task to remain fixed in place. Because the obstacle operates with highest priority, the desired safety distance is maintained. The prioritized Jacobian of the hand task becomes singular and a planner uses this condition to remove the hand task from the controller. When

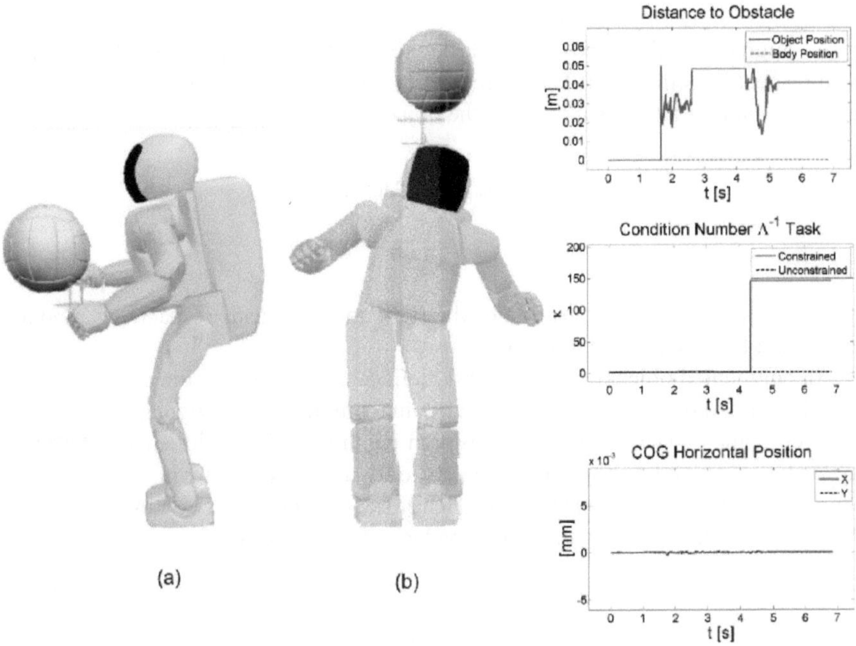

Figure 2.13 Response to a rapidly moving obstacles. An interactive behavior that responds to dynamically moving obstacles is simulated. Obstacles are handled as priority tasks with repulsion fields. Center of mass maneuvers, hand behavior, and postural behavior are controlled as lower priority tasks in the order listed. When the obstacle approaches the body (b), the posture changes since it operates with lower priority. In (a), the obstacle approaches the hand, which is controlled to remain at a fixed location. A conflict between objectives takes place, and as a result the hand task becomes unfeasible

the obstacle approaches the head, it is avoided using the whole-body redundancy. Stable balance is maintained at all times.

2.4.3 Dual Arm Manipulation

A third example shown in Figure 2.14 involving dual arm manipulation is briefly reviewed. The Asimo version we have at Stanford has only four degrees of freedom per arm. Therefore, the hands have been removed and replaced by cylinders to engage in point contact manipulation. The action primitive we use for this behavior is depicted below in order of priorities:

Task primitive	*Coordinates*	*Control policy*
Augmented object position	Cartesian	Proportional-derivative
Augmented object orientation	Cartesian	Proportional-derivative
Dual hand tension	Tension	Integral
Whole-body posture	Generalized coordinates	Proportional-derivative

An augmented object position task defines the Cartesian position of the object with respect to both hands. Therefore the coordinate representation and the associated Jacobian for a task that controls the object position are

$$x_{\text{object\_pos}} \triangleq \frac{1}{2}\left(p_{\text{rh}} + p_{\text{lh}}\right), \tag{2.60}$$

$$J_{\text{object\_pos}} \triangleq \frac{1}{2}\left(J_{\text{rh}} + J_{\text{lh}}\right), \tag{2.61}$$

where p_{rh} and p_{lh} are the position of the right and left hands respectively, and J_{rh} and J_{lh} are the associated Jacobians.

An augmented object orientation task defines the 2D orientation that can be realized with two point contacts. A virtual frame with the y axis aligned with the line connecting the two hands and the z axis pointing upwards perpendicular to the gravity vector is defined. Using point contacts, only the orientation with respect to the frame's x and z axis are controllable. Therefore, the coordinate representation and the associated Jacobian for an orientation task of that type are

$$x_{\text{object\_ori}} \triangleq \lambda_{\text{virtual\_frame}}, \tag{2.62}$$

$$J_{\text{object\_ori}} \triangleq {}^{0}R_{v}\, S_{xz}\, {}^{v}R_{0}\, Q_{\text{rl}}\, J_{\text{rl}}. \tag{2.63}$$

Here $\lambda_{\text{virtual\_frame}}$ is the quaternion associated with the orientation of the virtual frame connecting the two hands. Since the orientation of the frame is roughly defined by the position of the hands, the Jacobian of the orientation task involves only the cumulative position rows of the Jacobians of the right and left hands, J_{rl}, *i.e.*, ignoring the rotation components. In the above equation, the operator Q_{rl} defines the cross product operator of the right and left hands, *i.e.*,

$$Q_{rl} \triangleq \frac{1}{2}\left(\widehat{p}_{\text{rh}} \,|\, \widehat{p}_{\text{lh}}\right), \tag{2.64}$$

where the operator $(\widehat{.})$ indicates cross product. Because the y orientation of the task frame is not controllable, we rotate the quantity $(Q_{\text{rl}}J_{\text{rl}})$ in the Jacobian of Equation 2.63 to align with the virtual orientation frame using the rotation matrix from global to virtual frames ${}^{v}R_{0}$, and we eliminate by selection the y direction using the matrix S_{xz}. Finally, we rotate the Jacobian back to the global frame using ${}^{0}R_{v}$ to unify the task representation with respect to global frame.

Although the orientation task is composed of three coordinates, removing the y coordinate on the Jacobian will cause our whole-body controller to optimize only the two controllable orientation directions while discarding the non-controllable direction. To control both the position and orientation of the object we use simple PD controllers. In these controllers, the angular velocity of the object is needed and can be determined using the expressions,

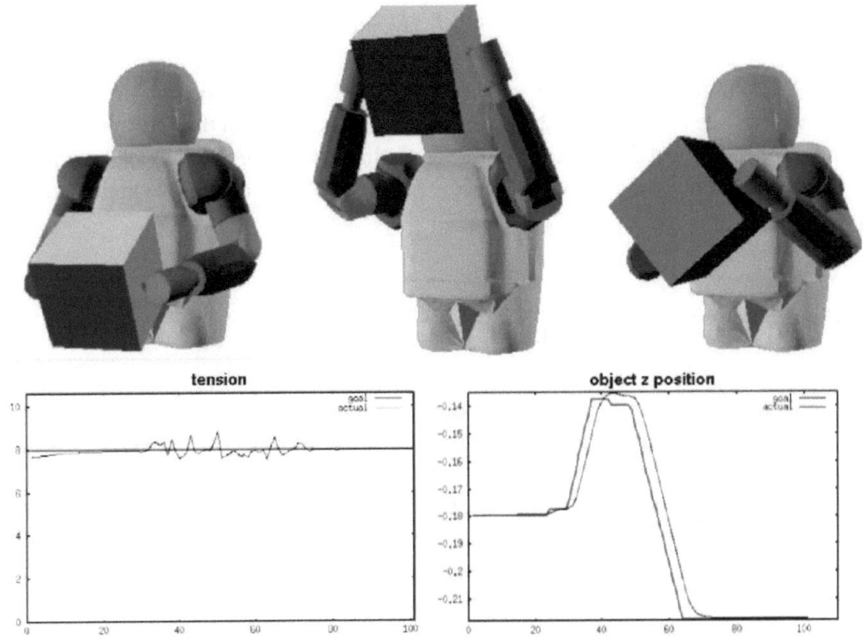

Figure 2.14 Dual arm manipulation. This sequence is taken from a simulation of a dual arm point manipulation behavior. The position and 2D orientation of the object are defined and controlled using PD control tracking. A tension task defined between the points of contact is controlled to grasp the object efficiently. A posture control law optimizes the configuration of the residual redundancy.

$$\omega \triangleq \frac{1}{2}\left(\omega_{\text{rh}} + \omega_{\text{lh}}\right), \tag{2.65}$$

$$\omega_{\text{rh}} = \frac{p_{\text{rh}} \times v_{\text{rh}}}{||p_{\text{rh}}||}, \quad \omega_{\text{lh}} = \frac{p_{\text{lh}} \times v_{\text{lh}}}{||p_{\text{lh}}||}, \tag{2.66}$$

where the ω_is correspond to angular velocities and the v_is correspond to linear velocities. Moreover, the operator $\times$ indicates cross product.

To grasp the object using forces, we simultaneously implement a dual hand tension task, which we define through the differential forces acting along the line that connects the points of contact. The Jacobian that represents this task is

$$J_{\text{tension}} \triangleq S_y \, {}^{v}R_0 \, \Delta_{\text{rl}} \, J_{\text{rl}}, \tag{2.67}$$

with

$$\Delta_{\text{rl}} \triangleq \left(I_{3\times 3} \,\middle|\, -I_{3\times 3} \right) \tag{2.68}$$

representing a differential operator of the coordinates of both hands. The matrix S_y selects only the y dimensions along the line connecting the robot hands. The control command that we use for the tension task is purely integral, *i.e.*,

$$F_{\text{tension}} = -K_i \int_0^t (F_k - F_{\text{des}})dt, \tag{2.69}$$

where F_k is the force read by the sensor and F_{des} is the desired force.

As we can see in the Figure 2.14, we tracked efficiently both the position and orientation of the object trajectories while regulating the tension to the desired value. In this experiment, object trajectories were provided by a user using keyboard commands.

2.5 Conclusion and Discussion

This chapter has summarized some of our contributions and collaborative work in the area of whole-body compliant control of humanoid robots. Developing whole-body torque control methods addresses several key aspects for advanced humanoid robot interactions, including compliant behaviors, handling of geometric and contact constraints, unified motion and force control, prioritized multi-objective optimization, and skill representation, among others.

Prioritization is a mechanism that substitutes the need for search-based or global optimization methods in the generation of several types of behaviors: (1) handling of geometric and contact constraints; (2) exploiting the available redundancy to execute operational tasks; and (3) exploiting the available redundancy to optimize postural performance. Using prioritized compliant control enables fast adaptation to contact interactions and fast response to dynamically changing geometric constraints.

By identifying the set of internal forces and moments between contact closed loops we enable the characterization and control of the complex multi-contact interactions of the robot with the environment. Enabling the precise control of contact centers of pressure, we create compliant contact behaviors and by placing contact CoPs near the center of contact bodies we prevent unwanted rotations along contact edges. Characterizing the behavior of internal tensions and moments as well as the behavior of the robot's center of mass with respect to contact reaction forces we provide tools to plan whole-body interactions and maneuvers that satisfy frictional constraints. Other methods solely based on ZMP modeling disregard the local interactions between contact bodies, hindering the ability to comply with contact constraints and to create compliant contact behaviors. Our methods are dynamically correct, enabling the simultaneous control of operational tasks, center of mass maneuvers, postural behaviors, and internal force behaviors with high performance. We have demonstrated these capabilities through whole-body multi-contact examples involving upper and lower extremities.

Using dynamic modeling enables safe open loop compliant interactions. It also enables the simultaneous control of multiple task points with high accuracy and without introducing coupling effects between task objectives. Moreover, dynamic modeling of the interactions between contact closed loops allows us to derive internal force controllers that operate simultaneously with motion objectives. The controllers developed here exhibit strong robustness with respect to model errors including errors in link masses and their associated center of mass positions.

In recent years, we have co-developed with colleagues at Stanford, a real-time embedded software architecture that implements the methodology described in this chapter. This software has been adapted to control a research version of the humanoid robot Asimo and several robotic arms including the PUMA and the Barrett WAM arms. To handle the computational complexity of the models, the architecture is organized in multiple processes running at different rates. A servo process implements the feedback force and position controllers and runs at fast rates. The model process computes the complex dynamic and prioritized operators and runs at slower speeds than the servo process. A layer to create complex skills as aggregations of task objectives is provided in this software in the form of behavioral entities. These behavioral abstractions are designed to be coordinated from high-level planning processes. Recently, we presented a position paper outlining our view for bridging the gap between semantic planning and continuous control of humanoid and mobile robots [46].

Future work includes the development of planners to dynamically maneuver using multi-contact interactions in unstructured terrains. The models of multi-contact and center of mass behavior we have developed in this chapter provide a powerful environment to sample potential contact transitions and feasible center of mass maneuvering trajectories. Another important area of research is compliant behaviors for multi-finger manipulation tasks. Here, the models of multi-contact interactions that have been presented can be extended for planning dexterous hand manipulation tasks.

Acknowledgments Many thanks to Dr Kensuke Harada, Dr Eiichi Yoshida, and Dr Kazuhito Yokoi for their effort to put this book together. Many thanks to Professor Oussama Khatib, Dr Jae-Heung Park, Dr Roland Philippsen, Taizo Yoshikawa, and Francois Conti for their contributions and advice. Many thanks to Dr Abderrhamane Kheddar for reviewing the manuscript. I am very grateful to Honda Motor Co. for supporting my research at Stanford during recent years.

References

[1] C.O. Alford and S.M. Belyeu. Coordinated control of two robot arms. In: proceedings of the IEEE international conference on robotics and automation, pp 468–473, March 1984.
[2] T. Bretl and S. Lall. Testing static equilibrium for legged robots. IEEE Trans Robotics, 24(4):794–807, August 2008.
[3] O. Brock and O. Khatib. Elastic strips: A framework for motion generation in human environments. Int J Robotics Res, 21(12):1031–1052, 2002.
[4] R. Brooks. A robust layered control system for a mobile robot. Int J Robotics Automat, 2(1):14–23, March 1986.
[5] J. Buchli, F. Iida, and A.J. Ijspeert. Finding resonance: Adaptive frequency oscillators for dynamic legged locomotion. In: proceedings of the IEEE/RSJ international conference on intelligent robots and systems, 2006.

[6] C.E. Buckley. The application of continuum methods to path planning. PhD thesis, Stanford University, Stanford, USA, 1986.
[7] K.C. Chang and O. Khatib. Operational space dynamics: Efficient algorithms for modeling and control of branching mechanisms. In: proceedings of the IEEE international conference on robotics and automation, April 2000.
[8] R. Chatila. Systeme de Navigation pour un Robot Mobile Autonome: Modelisation et Processus Decisionnels. PhD thesis, Universite Paul Sabatier, Toulouse, France, 1981.
[9] C. Collette, A. Micaelli, C. Andriot, and P. Lemerle. Robust balance optimization control of humanoid robots with multiple non coplanar grasps and frictional contacts. In: proceedings of the IEEE international conference on robotics and automation, Pasadena, USA, May 2008.
[10] S. Collins, A. Ruina, R. Tedrake, and M. Wisse. Efficient bipedal robots based on passive-dynamic walkers. Science Magazine, 307(5712):1082–1085, 2005.
[11] E. Demircan, L. Sentis, V. DeSapio, and O. Khatib. Human motion reconstruction by direct control of marker trajectories. In: Advances in robot kinematics (ARK), 11th international symposium, Batz-sur-Mer, France, June 2008. Springer.
[12] B. Espiau, F. Chaumette, and P. Rives. A new approach to visual servoing in robotics. IEEE Trans Robotics Automat, 8(3):313–326, 1992.
[13] R. Featherstone. Robot Dynamics Algorithms. Kluwer Academic Publishers, Norwell, USA, 1987.
[14] E.F. Fichter and E.D. McDowell. A nover design for a robot arm. In: Advancements of Computer Technology, pp 250–256. ASME, 1980.
[15] A.A. Frank. Control Systems for Legged Locmotion Machines. PhD thesis, University of Southern California, Los Angeles, USA, 1968.
[16] H. Hanafusa, T. Yoshikawa, and Y. Nakamura. Analysis and control of articulated robot with redundancy. In: proceedings of IFAC symposium on robot control, volume 4, pp 1927–1932, 1981.
[17] K. Harada, S. Kajita, K. Kaneko, and H. Hirukawa. Zmp analysis for arm/leg coordination. In: proceedings of the IEEE/RSJ international conference on intelligent robots and systems, pp 75–81, Las Vegas, USA, October 2003.
[18] K. Hirai, M. Hirose, Y. Haikawa, and T. Takenaka. The development of Honda humanoid robot. In: proceedings of the IEEE international conference on robotics and automation, volume 2, pp 1321–1326, Leuven, Belgium, 1998.
[19] R. Holmberg and O. Khatib. Development and control of a holonomic mobile robot for mobile manipulation tasks. Int J Robotics Res, 19(11):1066–1074, 2000.
[20] S-H. Hyon, J. Hale, and G. Cheng. Full-body compliant humanhumanoid interaction: Balancing in the presence of unknown external forces. IEEE Trans Robotics, 23(5):884–898, October 2007.
[21] A.J. Ijspeert, J. Buchli, A. Selverston, M. Rabinovich, M. Hasler, W. Gerstner, A. Billard, H. Markram, and D. Floreano, editors. Dynamical principles for neuroscience and intelligent biomimetic devices. Abstracts of the EPFL-LATSIS symposium, Laussane, Switzerland, 2006.
[22] S. Kajita, F. Kanehiro, K. Kaneko, K. Fujiwara, K. Harada, K. Yokoi, and H. Hirukawa. Resolved momentum control: Humanoid motion planning based on the linear and angular momentum. In: proceedings of the IEEE/RSJ international conference on intelligent robots and systems, pp 1644–1650, Las Vegas, USA, October 2003.
[23] F. Kanehiro and H. Hirukawa. Online self-collision checking for humanoids. In: proceedings of the 19th annual conference of the robotics Society of Japan, Tokyo, Japan, September 2001.
[24] O. Khatib. Commande Dynamique dans l'Espace Opérationnel des Robots Manipulateurs en Présence d'Obstacles. PhD thesis, l'École Nationale Supérieure de l'Aéronautique et de l'Espace, Toulouse, France, 1980.
[25] O. Khatib. Real-time obstacle avoidance for manipulators and mobile robots. Int J Robotics Res, 5(1):90–98, 1986.

[26] O. Khatib. A unified approach for motion and force control of robot manipulators: The operational space formulation. Int J Robotics Res, 3(1):43–53, 1987.
[27] O. Khatib. Object manipulation in a multi-effector robot system. In R. Bolles and B. Roth, editors, Robotics Research 4, pp 137–144. MIT Press, 1988.
[28] O. Khatib, O. Brock, K.C. Chang, F. Conti, D. Ruspini, and L. Sentis. Robotics and interactive simulation. Communications of the ACM, 45(3):46–51, March 2002.
[29] O. Khatib, L. Sentis, J.H. Park, and J. Warren. Whole body dynamic behavior and control of human-like robots. Int J Humanoid Robotics, 1(1):29–43, March 2004.
[30] O. Khatib, P. Thaulaud, and J. Park. Torque-position transformer for task control of position controlled robots. Patent, November 2006. Patent Number: 20060250101.
[31] B.H. Krogh. A generalized potential field approach to obstacle avoidance control. In: Proceeding of the Internatioal robotics research conference, Betlehem, USA, August 1984.
[32] J. Kuffner, K. Nishiwaki, S. Kagami, M. Inaba, and H. Inoue. Motion planning for humanoid robots. In: proceedings of the international symposium of robotics Research, Siena, Italy, October 2003.
[33] J. Kuffner, K. Nishiwaki, S. Kagami, Y. Kuniyoshil, M. Inabal, and H. Inouel. Self-collision detection and prevention for humanoid robots. In: proceedings of the IEEE international conference on robotics and automation, pp 2265–2270, Washington, USA, May 2002.
[34] J.C. Latombe. Robot Motion Planning. Kluwer Academic Publishers, Boston, USA, 1991.
[35] J.P. Laumond and P.E. Jacobs. A motion planner for nonholonomic mobile robots. IEEE Trans Robotics Automat, 10(5):577–593, 1994.
[36] A. Liegeois. Automatic supervisory control of the configuration and behavior of multibody mechanisms. IEEE Trans Syst Man Cybern, 7:868–871, 1977.
[37] T. Lozano-Perez. Spatial planning: a configuration space approach. IEEE Trans Comput, 32(2):108–120, 1983.
[38] A.A. Maciejewski and C.A. Klein. Obstacle avoidance for kinematically redundant manipulators in dynamically varying environments. Int J Robotics Res, 4(3):109–117, 1985.
[39] N. Mansard and O. Khatib. Continous control law from unilateral constraints. In: proceedings of the IEEE international conference on robotics and automation, Pasadena, USA, May 2008.
[40] E. Marchand and G. Hager. Dynamic sensor planning in visual servoing. In: proceedings of the IEEE/RSJ international conference on intelligent robots and systems, volume 3, pp 1988–1993, Leuven, Belgium, May 1998.
[41] T. McGeer. Passive dynamic walking. The Int J Robotics Res, 9(2):62–68, 1990.
[42] J.-P Merlet. Redundant parallel manipulators,. Laboratory of Robotics and Automation, 8(1):17–24, February 1996.
[43] H. Moravec. Obstacle Avoidance and Navigation in the Real World by a Seeing Robot Rover. PhD thesis, Stanford University, Stanford, USA, 1980.
[44] Y. Nakamura, H. Hanafusa, and T. Yoshikawa. Mechanics of coordinative manipulation by multiple robotic mechanisms. In: proceedings of the IEEE international conference on robotics and automation, pp 991–998, April 1987.
[45] K. Abderrahmane and A. Escande. Challenges in Contact-Support Planning for Acyclic Motion of Humanoids and Androids. In: international symposium on robotics, pp 740–745, 2008.
[46] R. Philippsen, N. Negati, and L. Sentis. Bridging the gap between semantic planning and continuous control for mobile manipulation using a graph-based world representation. In: proceedings of the workshop on hybrid control of autonomous systems, Pasadena, July 2009.
[47] D.L. Pieper. The Kinematics of Manipulators Under Computer Control. PhD thesis, Stanford University, Stanford, USA, 1968.
[48] M.H. Raibert, M. Chepponis, and H.B. Brown. Running on four legs as though they were one. IEEE J Robotics Automat, 2(2):70–82, June 1986.
[49] M.H. Raibert and J.J. Craig. Hybrid position force control of manipulators. ASME J Dyn Sys Measurement Contr, 103(2):126–133, 1981.

[50] B. Roth, J. Rastegar, and V. Sheinmann. On the design of computer controlled manipulators. In: First CISM IFToMM symposium, pp 93–113, 1973.

[51] D. Ruspini and O. Khatib. Collision/contact models for dynamic simulation and haptic interaction. In: The 9th international symposium of robotics research (ISRR'99), pp 185–195, Snowbird, USA, October 1999.

[52] L. Sentis. Synthesis and Control of Whole-body Behaviors in Humanoid Systems. PhD thesis, Stanford University, Stanford, USA, 2007.

[53] L. Sentis and O. Khatib. Synthesis of whole-body behaviors through hierarchical control of behavioral primitives. Int J Humanoid Robotics, 2(4):505–518, December 2005.

[54] L. Sentis, J. Park, and O. Khatib. Modeling and control of multi-contact centers of pressure and internal forces in humanoid robots. In: proceedings of the IEEE/RSJ international conference on intelligent robots and systems, St. Louis, USA, October 2009.

[55] B. Siciliano and J. Slotine. A general framework for managing multiple tasks in highly redundant robotic systems. In: proceedings of the IEEE international conference on advanced robotics, pp 1211–1216, Pisa, Italy, June 1991.

[56] O. Stasse, A. Escande, N. Mansard, S. Misossec, P. Evrard, and A. Kheddar. Real-time (self)-collision avoidance task on a hrp-2 humanoid robot. In: proceedings of the IEEE international conference on robotics and automation, pp 3200–3205, Pasadena, May 2008.

[57] A. Takanishi, M. Ishida, Y. Yamazaki, and I. Kato. The realization of dynamic walking robot wl-10rd. In: proceedings of the international conference of advanced robotics, 1985.

[58] A. Takanishi, H.O. Lim, M. Tsuda, and I. Kato. Realization of dynamic biped walking stabilized by trunk motion on sagitally uneven surface. In: proceedings of the IEEE/RSJ international conference on intelligent robots and systems, 1990.

[59] R. Tedrake, T.W. Zhang, M. Fong, and H.S. Seung. Actuating a simple 3d passive dynamic walker. In: proceedings of the IEEE international conference on robotics and automation, 2004.

[60] M. Vukobratović and B. Borovac. Zero-moment point thirty tive years of its life. Int J Humanoid Robotics, 1(1):157–173, 2004.

[61] J. T. Wen and L. Wilfinger. Kinematic manipulability of general constrained rigid multibody systems. In: proceedings of the IEEE international conference on robotics and automation, volume 15, pp 558–567, Detroit, USA, May 1999.

[62] E.R. Westervelt, J.W. Grizzle, C. Chevallereau, J.H. Choi, and B. Morris. Feebdack control of dynamic bipedal robt locomotion. CRC Oress, 2007.

[63] D. Williams and O. Khatib. The virtual linkage: A model for internal forces in multi-grasp manipulation. In: proceedings of the IEEE international conference on robotics and automation, pp 1025–1030, Atlanta, USA, October 1993.

[64] K. Yamane and Y. Nakamura. Dynamics filterconcept and implementation of online motion generator for human figures. IEEE Trans Robotics Automat, 19(3):421–432, 2003.

[65] Y. Yi, J. McInroy, and Y. Chen. General over-constrained rigid multi-body systems: Differential kinematics, and fault tolerance. In: proceedings of the SPIE international symposium on smart structures and materials, volume 4701, pp 189–199, San Diego, USA, March 2002.

[66] T. Yoshikawa and O. Khatib. Compliant and safe motion control of a humanoid robot in contact with the environment and humans. In: proceedings of the IEEE/RSJ international conference on intelligent robots and systems, Nice, Frane, September 2008.

Chapter 3
Whole-body Motion Planning – Building Blocks for Intelligent Systems

Michael Gienger, Marc Toussaint and Christian Goerick

Abstract Humanoid robots have become increasingly sophisticated, both in terms of their movement as well as their sensorial capabilities. This allows one to target for more challenging problems, eventually leading to robotic systems that can perform useful tasks in everyday environments. In this paper, we review some elements we consider to be important for a movement control and planning architecture. We first explain the whole-body control concept, which is the underlying basis for the subsequent elements. We then present a method to determine optimal stance locations with respect to a given task. This is a key element in an action selection scheme that evaluates a set of controllers within a parallel prediction architecture. It allows the robot to quickly react to changing environments. We then review a more global movement planning approach which casts the overall robot movement into an integral optimization problem, and leads to smooth and collision-free movements within interaction time. This scheme is then extended to cover the problem of grasping simple objects.

3.1 Introduction

While in its beginning, humanoid robotics research focused on individual aspects like walking, current systems have become increasingly sophisticated. Many humanoid robots are already equipped with full-body control concepts and advanced sensorial capabilities like stereo vision, auditory and tactile sensor systems. This is the prerequisite to tackle complex problems, such as walking and grasping in dy-

Dr.-Ing. Michael Gienger, Dr.-Ing. Christian Goerick
Honda Research Institute Europe GmbH, Carl-Legien-Strasse 30, 63073 Offenbach, Germany, e-mail: michael.gienger@honda-ri.de

Dr.rer.nat. Marc Toussaint
Technical University of Berlin, Franklinstr. 28/29, 10587 Berlin, Germany e-mail: mtoussai@cs.tu-berlin.de

namically changing environments. Motion planning seems to be a promising way to deal with this class of problem. State of the art planning methods allow one to flexibly account for different criteria to be satisfied. Further, many computationally efficient methods have been proposed (see [39, 38, 40] for a comprehensive overview), so that fast planning and replanning can be achieved in real-world, real-time problems.

In general, two problem fields in humanoid robot motion planning have emerged. One recent research focus is centered around solving the gait [27, 9] and footstep planning problem in dynamic environments [12, 45]. This is complemented by efforts to plan collision-free arm movements for reaching and grasping [4, 41], and to incorporate the dynamics of the objects to be manipulated [63].

However, there seems to be no approach to address all problem domains within a consistent architecture. In this article, we will present some steps in this direction. We start out with the whole-body control concept applied to our humanoid robot ASIMO in Section 3.2. We will explain the underlying robot model and derive the kinematics for the task and null space control. Based on the proposed model, we present a method to efficiently estimate an optimal stance location in Section 3.3. Reactive prediction and action selection is added with an architecture described in Section 3.4. It compares a set of different controller instances and selects the most suitable one according to their prediction result. However, this scheme has a limited time horizon. To generate movements that satisfy criteria throughout the whole trajectory, we present a controller-based optimization scheme in Section 3.5 in which we determine the attractor points describing the trajectory. The elements share a common basis, the whole-body control concept. The contribution finishes with the concept of *task maps* in Section 3.6. The fundamental idea is that there exists a space of feasible solutions for grasp problems that can be represented in a map. We will show how to generate and seamlessly integrate such maps into movement optimization, addressing the coupled problem of reaching and grasping in an integrated framework.

3.2 Models for Movement Control and Planning

While many movement planning approaches deal with navigation problems, this work will focus mainly on the problem of reaching and grasping with humanoid robots. Comprehensive kinematic models are particularly suited to describe the robot's end effector movement in an efficient and flexible way. In this section we briefly review the chosen redundant control concept: the general definition of task spaces, inverse kinematics and attractor dynamics to generate whole-body motion for high-dimensional humanoid robots.

Findings from the field of biology impressively reveal how efficiently movement is represented in living beings. Besides the well-known principle of movement primitives, it is widely recognized that movement is represented in various frames of reference, such as in eye centered, reach and grasp centered or object centered ones

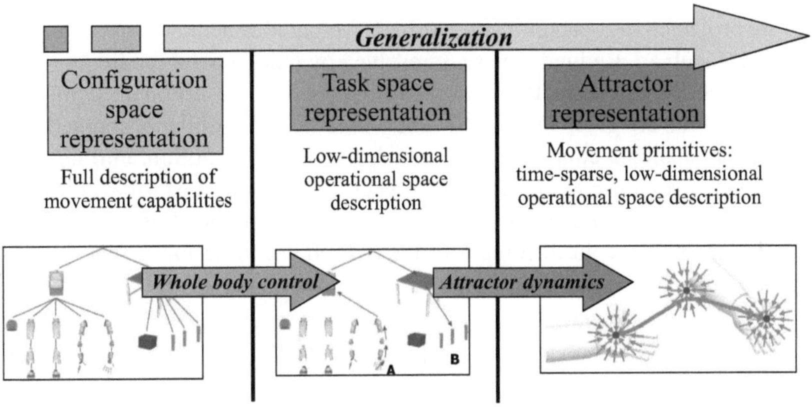

Figure 3.1 Task description

[15]. Egocentric frames describe movements with respect to the own body, and are a powerful representation when it comes to introducing invariance to a task. We borrow the above principle and use it as the underlying description of our control and planning models.

This work will focus on the large class of kinematically controlled robots. They differ from computed torque concepts such as [36] in that on the lowest level, the movement is represented in positions or velocities instead of torques. For this class of robots, the projection from a configuration space into task spaces is often done with redundant control schemes (e. g. *resolved motion rate control*, see [42, 49, 17]). Proper choice of the task description is a crucial element in humanoid robot control. Other than in other robotics domains, tasks may be carried out with two effectors.

Among the well-known trajectory generation methods, we chose a dynamical systems approach. This is closely related to the biological findings, and yields further advantages like robustness against perturbations and dynamical environments [48, 32]. The overall control architecture is summarized in Figure 3.1.

3.2.1 Control System

Kinematic control of redundant robots has been subject to extensive research. A popular method that allows one to split the control objective into a task and null space goes back to Liégeois [42] in the 1970s. Others have extended this approach towards introducing hierarchical task spaces [53, 3, 26, 16], to deal with collisions, singularities and ill-defined configurations [43, 44, 14, 61] and have formulated criteria to map into the null space of such systems [11]. References [49, 17] give a good overview on these approaches.

We employ a motion control system that is based on [42]. The task space trajectories are projected into the joint space using a weighted generalized pseudo-inverse of the task Jacobian. Redundancies are resolved by mapping the gradient of a joint limit avoidance criterion into the null space of the motion. Details on the whole-body control algorithm are given in [18, 19]. The whole-body controller is coupled with a walking and balancing controller [31], which stabilizes the motion. Further, a real-time collision avoidance algorithm [61] protects the robot against self-collisions.

Setting up the controller equations is done by flexibly augmenting a task Jacobian J_{task} holding row-wise the Jacobians of the desired task descriptors that we derive in Section 3.2.1.1 (see also [18]).

$$\dot{q} = J^{\#}\, \dot{x}_{task} - \alpha\, NW^{-1} \left(\frac{\partial H}{\partial q} \right)^{T}. \tag{3.1}$$

Matrix $J^{\#}$ is a weighted generalized pseudo-inverse of J with metric W and null space projector N:

$$J^{\#} = W^{-1} J^{T} (J W^{-1} J^{T})^{-1} \qquad N = E - J^{\#} J. \tag{3.2}$$

Matrix E is an identity matrix. We chose a diagonal matrix W with elements proportional to the range of the corresponding joint. Scalar H is an arbitrary optimization criterion. Its gradient is mapped into the null space with projection matrix N and scalar α defining the step width. Vector $\dot{x}_{task}$ comprises a feedback term to minimize the tracking error (closed loop inverse kinematics or "CLIK").

3.2.1.1 Task Kinematics

The robot's kinematics and dynamics are described in the form of a tree structure depicted in Figure 3.2. The individual links are connected by degrees of freedom (joints) or fixed transformations. Further, the tree may also comprise objects from the environment. This allows derivation of the inverse kinematics equations not only with respect to a heel or world reference frame, but also to formulate task descriptors accounting for robot–object relations. In the forward kinematics pass (left), the transformations of all bodies are computed according to the current configuration space vector. The connectivity of the tree is such that the computation can be carried out starting from a root node and descends the branches of the tree. In the inverse kinematics pass (right), the desired Jacobians are computed. Since they depend only on the degrees of freedom that connect the respective body to the root, the connectivity in this direction is the shortest path towards the root node. We employ an efficient algorithm that allows one to compute the Jacobians by re-using the results of the forward kinematics pass. We will skip the details for brevity and refer the interested reader to [10].

In the following, a task is defined as the movement of one body with respect to any other belonging to the tree. This allows, for instance, one to describe the position

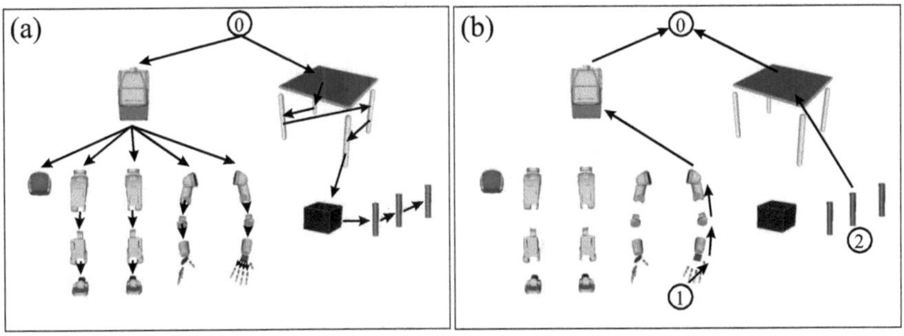

Figure 3.2 (a) Forward kinematics loop; and (b) loop for Jacobian computation

of one hand with respect to the other, the orientation of the head to the body, *etc.* It is also possible to describe robot link transformations with respect to objects in the environment, such as the position of the hand with respect to an object, or the direction of the gaze axis with respect to an object.

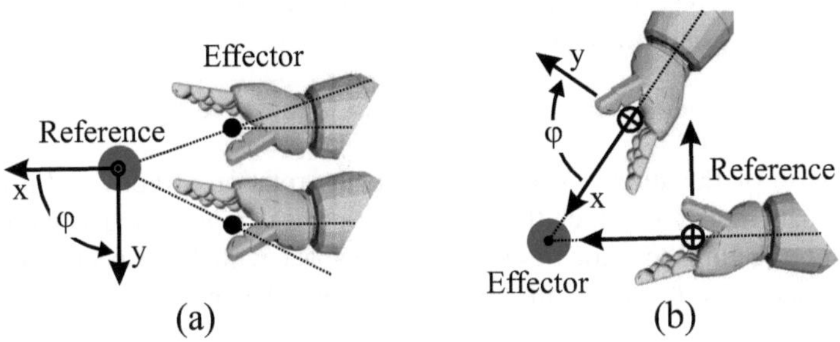

Figure 3.3 Relative hand–object task description: (a) with respect to the cylinder; and (b) with respect to the hand

The choice of the order of the relative coordinates yields some interesting aspects. This is illustrated in Figure 3.3 for a simple planar example. Representing the movement of the hand with respect to the cylinder results in Figure 3.3 (a). A coordinated hand–object movement has to consider three task variables ($x\ y\ \varphi$). Switching the frame of reference and representing the object movement with respect to the hand, such as depicted in Figure 3.3 (b), leads to a description of the movement in hand coordinates. In this example, this might be advantageous, since the object is symmetric and can be approached from any side. While in the first case the task variables are dependent, in the second case φ and y are invariant and can be set to zero. There are many other examples, such as representing a gazing controller as an object in head-centered coordinates which is "pointed" to by the focal axis, or a pointing controller in a similar way.

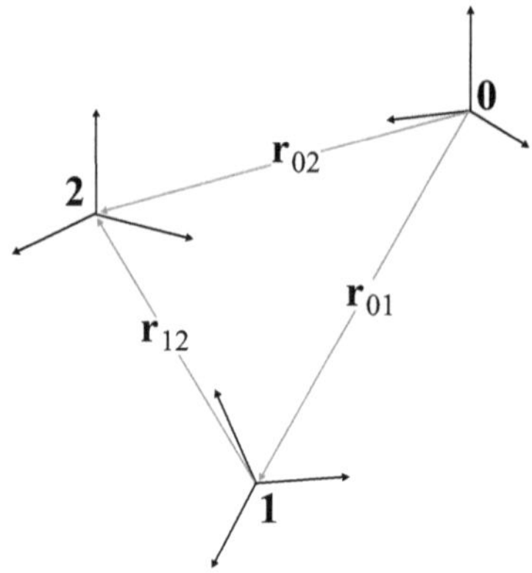

Figure 3.4 Relative effector kinematics

To mathematically formalize this concept, we look at the relative kinematics of an articulated chain, such as depicted in Figure 3.4. Coordinate frame 0 denotes its origin. Frame 1 is an arbitrary body which is connected to 0 through a set of joints. Body 2 shall be represented relative to body 1 with vector r_{12}. We now can write the (coordinate-free) kinematic equations as follows:

$$r_{12} = r_{02} - r_{01} \qquad \dot{r}_{12} = \dot{r}_{02} - \dot{r}_{01} + \omega_1 \times r_{12} \quad . \tag{3.3}$$

The last term of Equation 3.3 right is due to the rotation of body 1. Introducing the coordinate system in which the respective vector is represented as the left sub-index and projecting the velocities into the state space with the respective Jacobians $\dot{r}_i = J_{T,i}\,\dot{q}$ and $\omega_i = J_{R,i}\,\dot{q}$, the differential kinematics becomes

$$_1\dot{r}_{12} = A_{10}\left({}_0J_{T,2} - {}_0J_{T,1} + {}_0\tilde{r}_{12}^T\,{}_0J_{R,1}\right)\dot{q} = J_{T,rel}\,\dot{q}, \tag{3.4}$$

with J_T and J_R being the translational and rotational Jacobians, respectively, expression $\tilde{r} = r\times$ being a skew-symmetric matrix representing the outer product, and A_{10} being a rotation matrix from frame 0 to frame 1. If the reference ("1") body corresponds to a fixed frame, it has no velocity and the corresponding Jacobian is zero. In this case, we get the standard differential end effector kinematics $_1\dot{r}_{12} = A_{10}\,{}_0J_{T,2}\dot{q}$.

The task descriptors for attitude parameters are computed slightly differently. This is due to the fact that many parametric descriptions such as Euler angles have discontinuities and ill-defined configurations (gimbal lock). We therefore project the tracking error directly on the Jacobians for the angular velocities:

$$_1\omega_{12} = A_{10}\left({}_0J_{R,2} - {}_0J_{R,1}\right)\dot{q} = J_{R,rel}\,\dot{q}\ , \tag{3.5}$$

using the formulation of [13] for Euler angles. It is particularly elegant and avoids gimbal-locks. Similar feedback errors are formulated for 1D and 2D orientation task descriptors, for details see [18].

A well-investigated task descriptor is the overall linear momentum, which corresponds to the center of gravity velocity [60, 34]. It computes as

$$r_{cog} = \frac{1}{m}\sum_{i=1}^{bodies} m_i r_{cog,i} \qquad \dot{r}_{cog} = \frac{1}{m}\left\{\sum_{i=1}^{bodies} m_i J_{T,cog,i}\right\}\dot{q} = J_{cog}\,\dot{q}\,. \tag{3.6}$$

Similarly, we can derive the task descriptor for the overall angular momentum L with respect to a non-accelerated reference. Tensor I denotes the inertia tensor of the respective body link:

$$L = \sum_{i=1}^{bodies} m_i r_{cog,i} \times \dot{r}_{cog,i} + I\omega = \left\{\sum_{i=1}^{bodies} m_i\,\tilde{r}_{cog,i} J_{T,cog,i} + I_i J_{R,i}\right\}\dot{q} = J_{am}\,\dot{q}. \tag{3.7}$$

Another very simple but useful task descriptor is the movement of a single joint. The task variable is simply the joint angle, and the single joint Jacobian is a zero row vector with a single "one" entry at the column of the corresponding joint.

3.2.1.2 Null Space Control

In the following we present two of the null space criteria employed, namely a well-known joint limit avoidance criterion [42] and a collision avoidance criterion. The joint limit cost computes as

$$H_{jl}(q) = \frac{1}{2}\sum_{i=1}^{\text{dof}}\left(\frac{q_i - q_{0,i}}{q_{max,i} - q_{min,i}}\right)^2. \tag{3.8}$$

The contribution of each individual joint is normalized with respect to its joint range. To avoid collisions, we use the formulation in [62] and loop through all collision-relevant pairs of bodies, summing up their cost contributions. Each body is represented as a rigid primitive shape. Currently we use capped cylinders and sphere swept rectangles [61]. The cost associated with a pair of bodies is composed of two terms, one related to the distance between the closest points $d_p = |P_1 - P_2|$ and one related to the distance between their centers $d_c = |C_1 - C_2|$, see Figure 3.5 (a). To compute the closest point cost g_p, we set up three zones that are defined by the closest point distance d_p between two collision primitives. Figure 3.5 (b) shows the linear, the parabolic and the zero cost zones, respectively. In the region between contact ($d_p = 0$) and a given distance boundary d_B, the closest point cost g_p is determined as a parabolic function, being zero at $d_p = d_B$ and having the slope s for $d_p = 0$. It progresses linearly for $d_p < 0$, and for $d_p > d_B$, it is zero.

Similarly, the center point cost g_c shall only be active if the link distance has dropped below the distance d_B. The cost function will be scaled continuously with a factor zero at $d_p = d_B$ and one if $d_p = 0$. This cost adds an additional approximate avoidance direction, which is useful when the bodies are in deep penetration and the closest point direction is not well defined. Putting this together, the costs for one

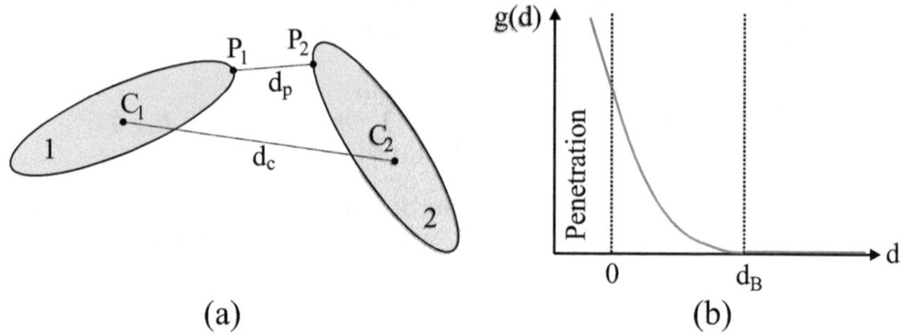

Figure 3.5 Zones for the collision cost function determination: (a) distance terms; and (b) cost zones

body pair become

$$g_p = \begin{cases} s d_B(d_B - 2d_p) & \text{for } d_p < 0 \\ s(d_p - d_B)^2 & \text{for } 0 \le d_p \le d_B \\ 0 & \text{for } d_p > d_B \end{cases} \quad g_c = \begin{cases} e^{-d_c} & \text{for } d_p < 0 \\ \left(1 - \frac{d_p}{d_B}\right) e^{-d_c} & \text{for } 0 \le d_p \le d_B \\ 0 & \text{for } d_p > d_B \end{cases} \tag{3.9}$$

with s defining the inclination of the gradient when penetrating. The overall collision cost is summed over all relevant body pairs as

$$H_{coll}(q) = \sum_i^{pairs} g_p(d_{p,i}) + g_c(d_{p,i}, d_{c,i}). \tag{3.10}$$

To derive the overall collision gradient, let us first derive the gradient of the distance $d_p = |p_1 - p_2|$ w.r.t. the joint configuration q. Differentiating with respect to the closest points p_1 and p_2 leads to

$$\frac{\partial d_p}{\partial p_1} = -\frac{1}{d_p}(p_2 - p_1)^T \qquad \frac{\partial d_p}{\partial p_2} = \frac{1}{d_p}(p_2 - p_1)^T. \tag{3.11}$$

If the collidable object is fixed to the environment, the partial derivative of the points with respect to the state is a $3 \times dof$ zero matrix. If it corresponds to a body part or is attached to the robot's body (*e.g.* held in the hand), we use the closest point Jacobians $\frac{\partial p_1}{\partial q} = J_{p_1}$ and $\frac{\partial p_2}{\partial q} = J_{p_2}$. With Equation 3.11 we get

$$\frac{\partial d_p}{\partial q} = \frac{1}{d_p}(p_2 - p_1)^T (J_{p_2} - J_{p_1}). \tag{3.12}$$

Analogously we can compute the gradient of $d_c = |C_1 - C_2|$. Differentiating Equation 3.9 with respect to the distance d_p, and inserting the distance gradient (Equation 3.12) leads to the closest point gradient

$$\left(\frac{\partial g_p}{\partial q}\right)^T = \begin{cases} -2s\frac{d_B}{d_p}(J_{p_2} - J_{p_1})^T(p_2 - p_1) & \text{for } d_p < 0, \\ 0 & \text{for } d_p > d_B, \\ 2s\frac{(d_p - d_B)}{d_p}(J_{p_2} - J_{p_1})^T(p_2 - p_1) & \text{else.} \end{cases} \quad (3.13)$$

The cost function g_c depends on the distance between the body centers d_c and on the closest point distance d_p, so we need to apply the chain rule to get the center point gradient:

$$\frac{\partial g_c}{\partial q} = \frac{\partial g_c}{\partial d_c}\frac{\partial d_c}{\partial q} + \frac{\partial g_c}{\partial d_p}\frac{\partial d_p}{\partial q} \quad (3.14)$$

where

$$\frac{\partial g_c}{\partial d_c} = -\frac{d_B - d_p}{d_B}e^{-d_c} \qquad \frac{\partial g_c}{\partial d_p} = -\frac{1}{d_B}e^{-d_c} \quad (3.15)$$

and the respective distance gradient is given in Equation 3.12. The overall collision gradient is

$$\frac{\partial H_{coll}}{\partial q} = \sum_{i}^{pairs} \frac{\partial g_d(i)}{\partial q} + \frac{\partial g_c(i)}{\partial q}. \quad (3.16)$$

3.2.2 Trajectory Generation

The quality of robot movements is very much related to the underlying trajectory generation. Popular trajectory generation methods use higher-order polynomials (splines) [51, 9], time-optimal profiles [29], or employ attractor dynamics [32]. Polynomials are particularly suited for movements that require precise timing, such as generating step patterns, *etc*. Dynamical systems represent the time implicitly and can form attractor systems or periodic solutions. They are closely related to the biological findings, and yield further advantages like robustness against perturbations and dynamic environments. We apply a simple attractor system [18, 62] to the task elements to be controlled. The same attractor dynamics are applied to other controllers that are not related to the inverse kinematics, such as "closing the fingers to a power grasp", *etc.*

Given two points x_k^* and x_{k+1}^* we shift the attractor point continuously from one to the other. This is captured by the linear interpolated trajectory $r_t \in \mathbb{R}^m$. In Figure 3.6 this is illustrated by the dashed line. Point r_t is taken as attractor point to a second order dynamics which generates the task trajectory $x_t \in \mathbb{R}^m$:

$$x_{t+1} = x_t + \pi(x_t, x_{t-1}, r_{t+1}) \quad (3.17)$$

$$\pi(x_t, x_{t-1}, r_{t+1}) = a(r_{t+1} - x_t) + b(x_t - x_{t-1}) \,. \quad (3.18)$$

The step response of the scheme is depicted as the solid line in Figure 3.6. We choose the coefficients a and b according to

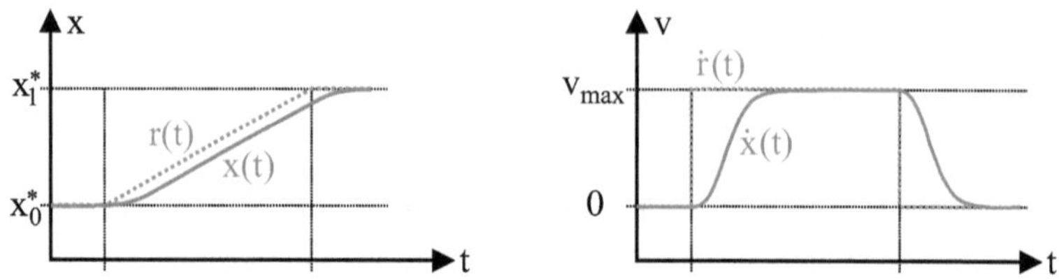

Figure 3.6 Step response of attractor system

$$a = \frac{\Delta t^2}{T_{\mathrm{mc}}^2 + 2T_{\mathrm{mc}}\Delta t\xi + \Delta t^2} \qquad b = \frac{T_{\mathrm{mc}}^2}{T_{\mathrm{mc}}^2 + 2T_{\mathrm{mc}}\Delta t\xi + \Delta t^2}, \tag{3.19}$$

with a relaxation time scale T_{mc}, the oscillation parameter ξ, and the sampling time Δt. We select $\xi = 1$, which leads to a smooth non-overshooting trajectory and an approximately bell-shaped velocity profile.

3.2.3 Task Relaxation: Displacement Intervals

In common classical motion control algorithms the tracking of the desired trajectories is very accurate. In many cases the tracking accuracy of a reference trajectory is not very critical, or there are at least some phases where the accuracy may be lower than in others. For example, "reaching" or "pointing" a humanoid robot to an object does not necessarily have to precisely follow the commanded trajectory. A certain impreciseness is permitted, and sometimes even desired, since machine-like motions look somewhat strange when performed by humanoid or animal robots.

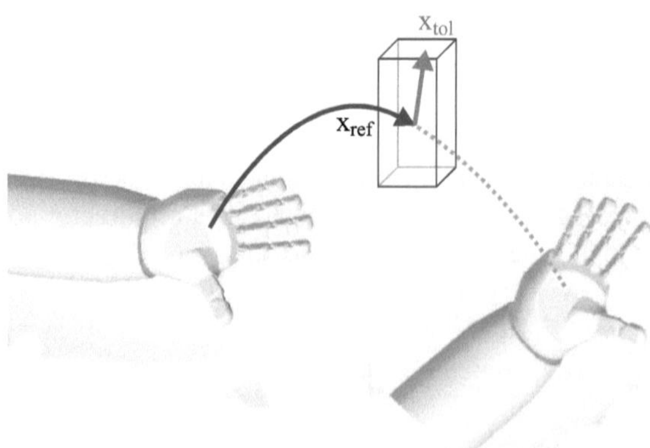

Figure 3.7 Displacement interval

In this section, we introduce the concept of displacement intervals [19] in task space. These intervals describe regions around a given task variable, in which the tracking has to be realized. Analogous to the null space motion, the displacement intervals are exploited to satisfy one or several cost functions. By choosing appropriate criteria, the motion can be influenced in almost arbitrary manners, *e.g.*, with respect to joint limit or collision avoidance, energy *etc*. In the following, the gradient of the joint limit avoidance cost function (Equation 3.8) is projected into the task interval. Its gradient with respect to the task space is

$$\frac{\partial H}{\partial x} = \frac{\partial H}{\partial q}\frac{\partial q}{\partial x} = \nabla H^T J^{\#}. \tag{3.20}$$

To describe the displacement interval in position coordinates, many solutions are thinkable: ellipsoids, cuboids or other 3D shapes. We choose a cuboid because the computational complexity is low and the interval can be described in a physically transparent way. The cuboid can be conceived as a virtual box around the reference point, in which the effector is allowed to move (see Figure 3.7). If one dimension of this box is set to zero, the effector may move on a plane. Similarly, setting two box-dimensions to zero, the effector may move on a straight line in the third, free direction. Setting all interval dimensions to zero leads to the standard motion control tracking the reference trajectory exactly. Therefore, the proposed approach can be seen as an extension to common trajectory generation methods. Figure 3.8 left illustrates the computation of the linear displacement in each iteration. It computes as

$$\delta x_{disp} = -\alpha_{pos}\left(\frac{\partial H}{\partial x}\right)^T. \tag{3.21}$$

Displacement x_{disp} is superposed with the reference trajectory, and it is checked if the updated effector command lies within the permitted boundary. If the boundary is exceeded, the displacement vector x_{disp} is clipped to stay within the permitted region. Figure 3.8 (a) illustrates this for a 2D example.

An interval formulation for the effector axis direction is depicted in Figure 3.8 (b). The commanded effector axis a_{cmd} is allowed to move within a cone with symmetry axis being the reference axis and opening angle φ being the displacement boundary. The cone edge is of unit length, so that the depicted circumference is the intersection of the cone and a unit sphere. The tangential displacement on the unit sphere results from the gradients $\frac{\partial H}{\partial \omega_x}$ and $\frac{\partial H}{\partial \omega_y}$:

$$\delta a = -\alpha_{att}\begin{pmatrix} \frac{\partial H}{\partial \omega_x} \\ \frac{\partial H}{\partial \omega_y} \\ 0 \end{pmatrix} \times a_{cmd}. \tag{3.22}$$

If the propagated command axis $a_{cmd} = a_{ref} + a_{disp}$ lies within the tolerance cone, no clipping has to be carried out. Otherwise, the command axis has to be clipped according to the lower parts of Figure 3.8.

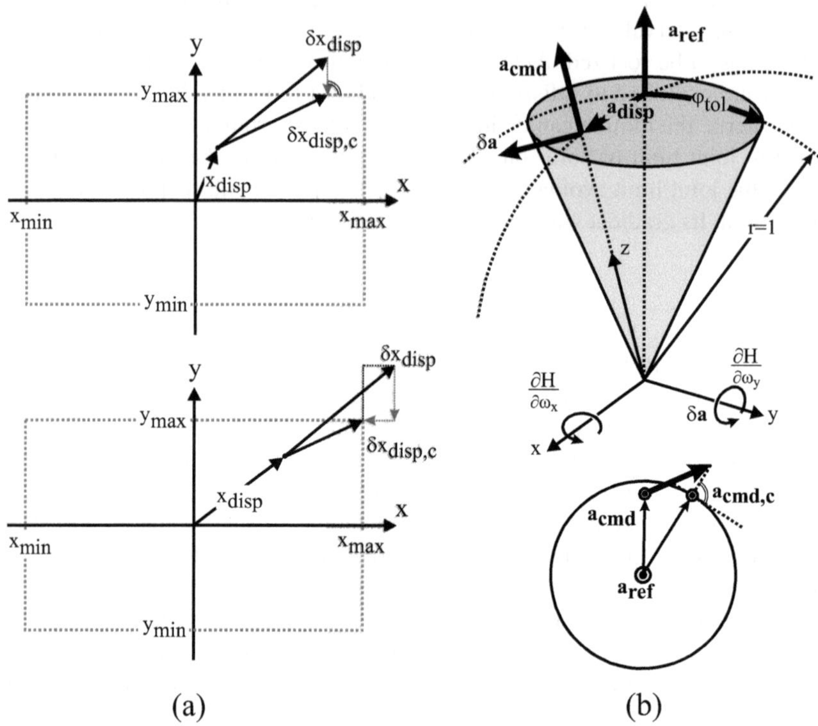

Figure 3.8 (a) Position and (b) attitude intervals

3.3 Stance Point Planning

When carrying out a task with a humanoid robot, it is crucial to determine a good stance position with respect to the object to grasp or manipulate. There exist some interesting approaches, which sample and evaluate a reachable space for feasible solutions [24, 25, 64]. In this section, we will explain a potential-field-based method to determine an optimal stance. The underlying kinematic model is depicted in Figure 3.9. We introduce a stance coordinate system that is described by the translation and rotation in the ground plane, corresponding to the stance poses the robot can reach. The upper body (body frame) is described with respect to this stance frame, the successive links (arms, legs, head) are attached to the upper body. Now we set up the controller equations according to Section 3.2. The important aspect is to unconstrain the three stance dofs, simply by assigning zeros to the corresponding column of the task Jacobian. This results in the stance dofs not being employed in the task space of the movement. However, they are still being utilized in the null space.

When assigning a target to the task vector, the controller equations will in each iteration make a step towards it, while shifting the stance coordinate system to a po-

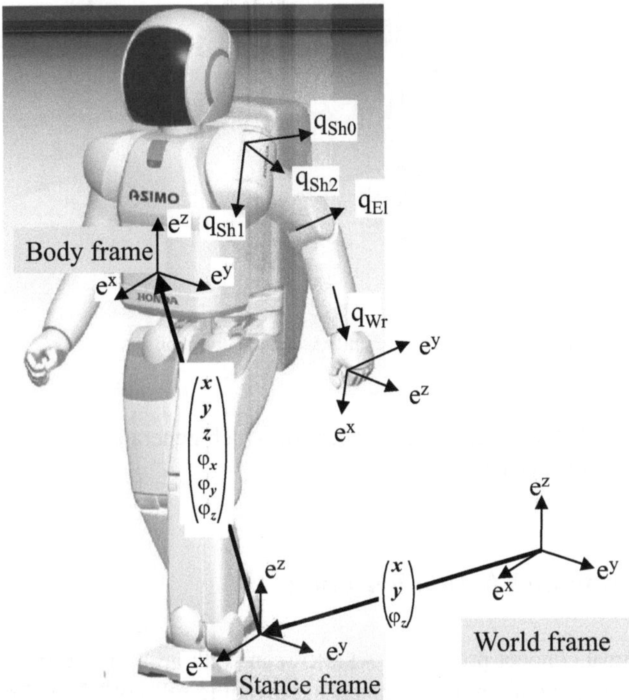

Figure 3.9 Kinematic model for stance pose optimization

sition and orientation that leads to a (local) minimum with respect to the employed null space criteria. A minimum can be found with regression techniques. Figure 3.10 illustrates this for the task of grasping a basket from a table. The task vector is composed of the following elements:

$$x_{task} = (x^T_{foot-l}\ \varphi^T_{Euler,foot-l}\ x^T_{foot-r}\ \varphi^T_{Euler,foot-r}\ x^T_{cog,xy}\ x^T_{hand-l}\ \varphi^T_{Polar,hand-l})^T . \tag{3.23}$$

The tasks for the feet are chosen to be in a normal stance pose. The horizontal components of the center of gravity lie in the center of the stance polygon. The left hand position and orientation are aligned with the handle of the basket. The null space of the movement is characterized by a term to avoid joint limits (Equation 3.8), and another term to avoid collisions between the robot links and the table (Equation 3.16). The weight of the latter is increased in Figure 3.10 left to right. It can be seen that the resulting stance pose has a larger body-to-table distance for a higher collision weight. This scheme is very general as it can be applied to arbitrarily composed task vectors. The resulting stance pose will always be a local optimum with respect to the null space criterion. Upon convergence, the resulting stance dofs

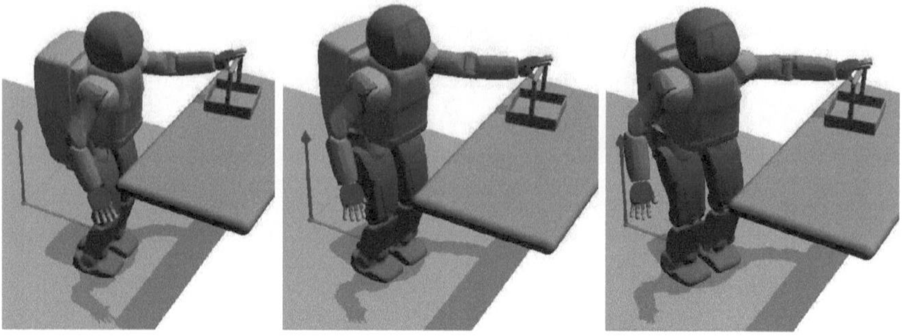

Figure 3.10 Optimal stance poses for differently weighted collision cost

can be commanded to a step pattern generator which generates a sequence of steps to reach the desired stance.

3.4 Prediction and Action Selection

With the control concept presented, ASIMO can walk around and safely reach towards objects while maintaining balance and avoiding self-collisions. However, the decision of *how to reach*, for instance, what hand to use or how to approach the object, is not tackled. In this section we will present an approach that solves this problem within a prediction and action selection architecture as depicted in Figure 3.11 (see [8, 22] for more details). The underlying idea is to connect the sensory (here visual) input to a set of predictors that correspond to simulated controllers of the robot. Each predictor solves the task associated with the sensory input in a different way. Within a strategy selection, these behavioral alternatives are continuously compared, and the command to the most suitable one is given to the physical robot.

First, we will explain the visual perception system employed which is based on a so called proto-object representation. Proto-objects are a concept originating from psychophysical modeling and can be thought of as coherent regions or groups of features in the field of view that are trackable and can be pointed or referred to without identification. Based on these stimuli, a prediction-based decision system selects the best movement strategy and executes it in real time. The internal prediction as well as the executed movements incorporate the control system presented.

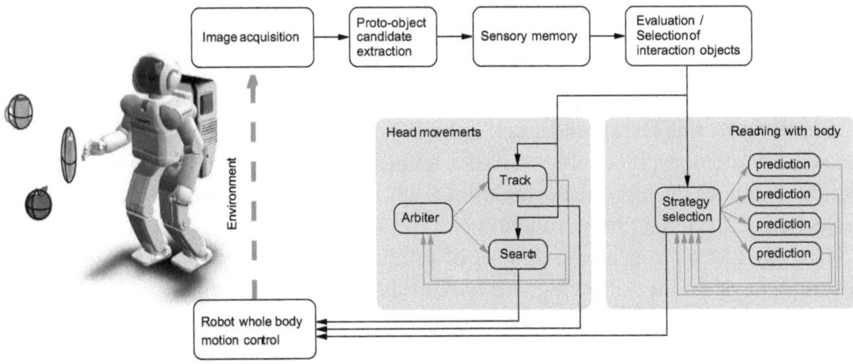

Figure 3.11 Overview of the system design

3.4.1 Visual Perception

To generate such proto-objects, we extract 3D ellipsoids from the visual input based on color segmentation and a disparity algorithm. The extracted blobs encode the position, metric size, and orientation of significant visual stimuli. They are stabilized and organized consistently as proto-objects in a *short term memory*. According to a set of extracted criteria, proto-objects are categorized into *found* if the object is seen in the visual scene, *memorized* if it has been found recently but is not seen currently, and *inaccurate* if it is only partially visible. Details of the chosen approach can be found in [8]. The 3D data and the above evaluation result are sent to the behaviors (search, track, reach). Each behavior can then extract the relevant information.

3.4.2 Behavior System

The output of the sensory memory is used to drive two different gazing behaviors: 1) searching for objects; and 2) tracking objects. Separate from these behaviors is a decision instance or *arbiter* [5] that decides which behavior should be active at any time. The decision of the arbiter is based on a scalar fitness value that describes how well a behavior can be executed. In this concrete case, tracking needs at least an inaccurate proto-object position to look at. Thus the tracking behavior will output a fitness of 1 if any proto-object is present and a 0 otherwise. The search behavior has no prerequisites at all and thus its fitness is fixed to 1.

The search behavior is realized by means of an inhibition of return map with a simple relaxation dynamics. If the search behavior is active and new vision data is available it will increase the value of the current gaze direction in the map and select the lowest value in the map as the new gaze target. The tracking behavior is realized as a multi-tracking of 3D points. The behavior takes all relevant proto-objects and object hypotheses into account and calculates the pan/tilt angles for centering them in the field of view. The two visual interaction behaviors together with the arbiter switching mechanism show very short reaction times and have proven to be efficient to quickly find and track objects.

Similarly to the search and track behaviors, the reaching behavior is driven by the sensory memory. It is composed of a set of internal predictors and a strategy selection instance. Each predictor includes a whole-body motion controller described in Section 3.2.1 and a fitness function.

The key idea is to evaluate this set of predictors, each solving the given task in different ways. In the following, we look at the task of reaching towards an object and aligning the robot's palm with the object's longitudinal axis. This corresponds to a pre-grasp movement, which brings the hand in a suitable position to grasp an object. In a first step, the visual target is split up into different motion commands, with which the task can be achieved. Four commands are chosen: reaching towards the target with the left and right hand, both while standing and walking. While the strategies that reach while standing assume the robot model to have a fixed stance position, we apply an incremental version of the stance point planning scheme introduced in Section 3.3 to the strategies that involve walking. This leads to a very interesting property: the control algorithm will automatically find the optimal stance position and orientation with respect to the given target and the chosen null space criteria. If a walking strategy is selected, the best stance pose is commanded to a step pattern generator, which generates appropriate steps to reach the desired stance position and orientation. In each time step, the strategies compute their motion and an associated fitness according to the specific command. The fitness is composed of the following measures:

- **Reachability:** penalty if the reaching target cannot be reached with the respective strategy.
- **Postural discomfort:** penalizes the proximity to the joint limits when reaching towards the target.
- **"Laziness":** penalizes the strategies that make steps. This way, the robot prefers standing over walking.
- **Time to target:** penalizes the approximate number of steps that are required to reach the target. This makes the robot dynamically change the reaching hand also during walking.

The costs are continuously evaluated, and the strategy with the highest fitness is identified. The corresponding command is given to the physical robot. The robot is controlled with the identical whole-body motion controller that is employed for the internal simulations. An interesting characteristic of the system is the temporal decoupling of real robot control and simulation. The strategies are sped up by a factor of 10 with respect to the real-time control, so that each strategy has converged to the target while the physical robot is still moving. From another point of view, the predictions could be seen as alternative results of a planning algorithm. A major difference is their incremental character. We use a set of predictors as continuously acting robots that each execute the task in a different way. The most appropriately acting virtual robot is mapped to the physical instance.

3.4.3 Experiments

The system as described above was tested many times with different people interacting with ASIMO with a variety of target objects. The scenario was always to have a human interaction partner who has an elongated object that was shown or hidden in various ways to ASIMO. The system is not restricted to only one object. If a number of objects are close to each other, the system will try to keep all objects in the field of view. If they are further apart, the objects leaving the field of view will be neglected after a short while and the system will track the remaining object(s).

Objects are quickly found and reliably tracked even when moved quickly. The robot will reach for any elongated object of appropriate size that is presented within a certain distance – from 20 cm to about 3 m. ASIMO switches between reaching with the right and left hand according to the relative object position with some hysteresis. It makes steps only when necessary. Figure 3.12 shows a series of snapshots taken from an experiment. From second 1–7, ASIMO is reaching for the bottle with its right hand. At second 8, the object becomes out of reach of the right hand, and the strategy selection mechanism selects the left hand reaching strategy, still while the robot is standing. At second 12, the object can not be reached with the left hand while standing. The strategy selection mechanism now selects to reach for the object with the left hand while walking towards it. The whole-body motion control generates smooth motions and is able to handle even extreme postures, which gives a very natural and human-like impression even to the casual observer. For more details of this system see [8].

3.5 Trajectory Optimization

The prediction architecture presented in the previous section allows the robot to dynamically act and react in a simple, but dynamically changing environment. However, it does not consider the overall movement throughout the trajectory, which is relevant when it comes to acting in a more difficult environment, with the potential danger to collide with objects, *etc.* In such cases more comprehensive planning and optimization schemes are required.

A lot of research in this field has focused on using spline-encoding as a more compact representation for optimization. This is particularly the case in the field of industrial robot trajectory optimization. Examples of such systems utilize cost functions that are formulated in terms of dynamics [30, 50], collision [56] or minimum jerk [1].

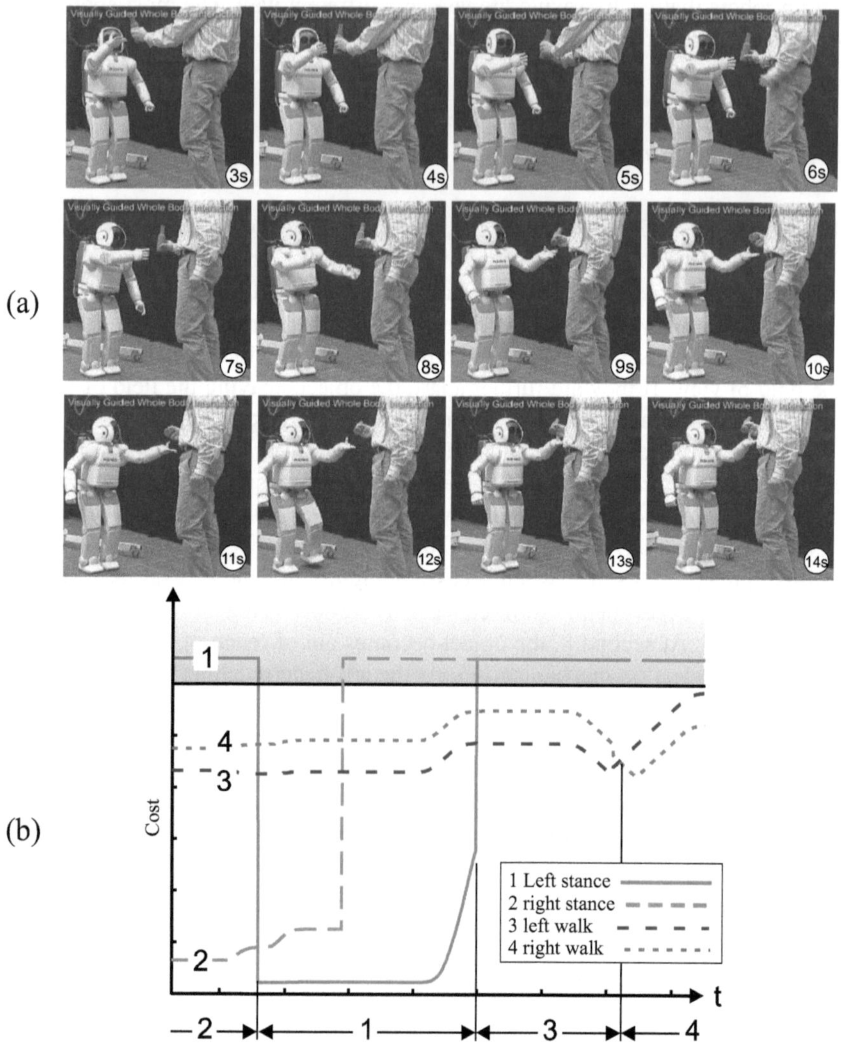

Figure 3.12 (a) Reaching towards a bottle; and (b) corresponding costs of predictor instances

General techniques like rapidly exploring random trees (RRTs) [37], or randomized road maps [35], have been shown to solve difficult planning problems like the alpha puzzle, generating a complex balanced reaching trajectory for a humanoid robot, or plan footstep trajectories. These techniques consider a direct representation of the trajectory and focus on finding a feasible solution rather than optimizing the trajectory w.r.t. additional criteria.

An alternative view on efficient movement representation is motivated from previous work on motor primitives in animals [48, 6]. Inspired by these biological findings several researchers have adopted the concept of motor primitives to the realm of robotic movement generation. For instance, Ijspeert et al. and Schaal et al. [32, 33, 54, 55] focus on non-linear attractors and learning the non-linearities, *e.g.*, in order to imitate observed movements. These approaches optimize the parameters of a single attractor system, *e.g.*, such that this single motor primitive imitates as best as possible a teacher's movement.

In this section, we will review an attractor-based optimization scheme [62]. It incorporates the robot's whole-body controller of Section 3.2.1 into the optimization process, and finds a sequence of task space attractors from Section 3.2.2 describing the optimal movement. The key idea is to optimize a scalar cost function by finding an optimal sequence of task space attractor vectors which determines the robot's motion. We consider an integral cost function over the movement in the general form of Equation 1 of Table 3.1. It is split into two terms. Function h subsumes costs for transitions in joint space and depends on the current and the previous time steps. It is suited to formulate criteria like the global length of the trajectory in joint space and the end effector velocity at the end of the trajectory:

1. costs $c_1 = \sum_{t=1}^{T} (q_t - q_{t-1})^T W (q_t - q_{t-1})$ for the global length of the trajectory in joint space;
2. costs $c_2 = |\tilde{\phi}(q_T) - \tilde{\phi}(q_{T-1})|^2$ for the end effector velocity at the end of the trajectory.

Function g subsumes cost criteria that depend on single time steps. It is suited to account for costs that depend on the posture of the robot. We formulate criteria to account for the offset of the final end effector state to a target, collisions and proximities between collidable objects throughout the trajectory, and joint limit proximities:

3. costs $c_3 = |\tilde{\phi}(q_T) - \hat{x}|^2$ for the offset of the final end effector state to a target $\hat{x}$;
4. costs $c_4 = \sum_{t=0}^{T} H_{coll}(q_t)$ for collisions and proximities between collidable objects throughout the trajectory, see Equation 3.10;
5. costs $c_5 = \sum_{t=0}^{T} H_{jl}(q_t)$ for joint limit proximities, see Equation 3.8.

The global cost function C is the linear combination of these terms, $C = \sum_{i=1}^{5} c_i$.

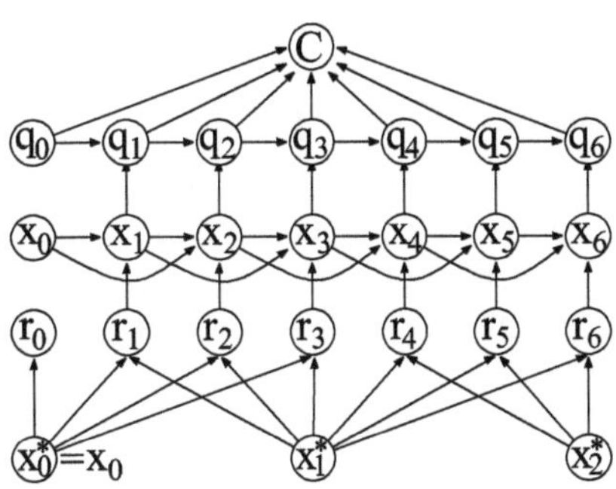

Figure 3.13 Functional network of the control architecture

The movement generation process can be summarized by Equations 5, 4, and 2 in Table 3.1. To derive analytic gradients, the whole movement generation process can be captured in the diagram in Figure 3.13 as a network of functional dependencies between variables. This is similar to a Bayesian network, but based on deterministic rather than probabilistic dependencies. The diagram tells us how to compute global gradients since the chain rule implies that for any global functional C the *total* derivative w.r.t. some arbitrary variable,

$$\frac{dC}{dx^*} = \sum_{\text{children } y_i \text{ of } x^*} \frac{\partial y_i}{\partial x^*} \frac{dC}{dy_i} . \tag{3.24}$$

The gradient computation is carried out in a forward and a backward computation step. In the *forward propagation step* we start with a given set of current attractor points $x^*_{1:K}$, then compute the task space trajectory $x_{0:T}$, then the $q_{0:T}$-trajectory, and finally the global cost C. In the *backward propagation step* we propagate the cost function gradients backward through the network using the chain rules. This involves first computing gradients dC/dq_t, then dC/dx_t, and finally $dC/dx^*_{1:K}$. Since all computations in the forward and backward propagation are local, the overall complexity is $O(T)$.

Figure 3.14 (a) shows a snapshot series of an experiment. The scenario has been chosen such that a purely reactive controller would fail. The robot holds a cylinder in the left hand and a box in the right hand. The target is to place the bottle into the box, which involves moving both, the bottle and the box, in a coordinated way without collision. The solution found by the robot is to move the bottle in an arc upwards and into the box while at the same time moving the box with the right hand downwards below the bottle. The task space in this experiment was defined 10D, comprising the positions of the left and right hand and the 2D polar orientation of the hand aligned axis for both hands. Figure 3.14 (b) displays the cost decay during optimization. A first collision-free solution is found after only 0.52 s, the final solution converged after 1.5 s. The method is particularly useful for human–robot interaction in complex environments, *e.g.*, when the robot has to reach around an obstacle that the human has just placed on the table. More experiments are given in [62].

3.6 Planning Reaching and Grasping

In this section, we will build on the movement optimization scheme presented and present an integrative approach to solve the coupled problem of reaching and grasping an object in a cluttered environment. While finding an optimal grasp is often treated independently from reaching to the object, in most situations it depends on how the robot can reach a pregrasp pose while avoiding obstacles. In essence, we are faced with the coupled problem of grasp choice and reaching motion planning.

Table 3.1 Costs and gradients underlying the optimization

cost function

$$C = \sum_{t=0}^{T} g(q_t) + \sum_{t=0}^{T-1} h(q_t, q_{t+1}) \ , \tag{1}$$

movement generation

$$q_{t+1} = q_t + J_t^{\#}(x_{t+1} - \phi(q_t)) - \alpha\ (I - J_t^{\#} J_t)\ W^{-1}\ (\partial_q H_t)^T \tag{2}$$

$$x_{t+1} = x_t + \pi(x_t, x_{t-1}, r_{t+1}) \tag{3}$$

$$\pi(x_t, x_{t-1}, r_{t+1}) = a(r_{t+1} - x_t) + b(x_t - x_{t-1}) \tag{4}$$

$$r_t = (1 - \tau) x_k^* + \tau x_{k+1}^* \ , \quad k = \lfloor tK/T \rfloor \ , \quad \tau = \frac{t - kT/K}{T/K} \tag{5}$$

chain rules following Equation 3.24

$$\frac{dC}{dq_t} = \frac{\partial C}{\partial q_t} + \frac{\partial q_{t+1}}{\partial q_t} \frac{dC}{dq_{t+1}} \tag{6}$$

$$\frac{dC}{dx_t} = \frac{\partial q_t}{\partial x_t} \frac{\partial C}{\partial q_t} + \frac{\partial x_{t+1}}{\partial x_t} \frac{dC}{dx_{t+1}} + \frac{\partial x_{t+2}}{\partial x_t} \frac{dC}{dx_{t+2}} \tag{7}$$

$$\frac{dC}{dr_t} = \frac{\partial x_t}{\partial r_t} \frac{dC}{dx_t} \tag{8}$$

$$\frac{dC}{dx_l^*} = \sum_t \frac{\partial r_t}{\partial x_l^*} \frac{dC}{dr_t} \tag{9}$$

partial derivatives

$$\frac{\partial C}{\partial q_t} = g'(q_t) + h'^1(q_t, q_{t+1}) + h'^2(q_{t-1}, q_t) \tag{10}$$

$$\begin{aligned} \frac{\partial q_{t+1}}{\partial q_t} &= I - J_t^{\#} J_t + (\partial_q J_t^{\#})(x_{t+1} - \phi(q_t)) \\ &\quad - \alpha\ (I - J_t^{\#} J_t)\ W^{-1}\ (\partial_q^2 H_t)^T + \alpha\ \partial_q (J_t^{\#} J_t)\ W^{-1}\ (\partial_q H_t)^T \end{aligned} \tag{11}$$

$$\frac{\partial q_t}{\partial x_t} = J_{t-1}^{\#} \tag{12}$$

$$\frac{\partial x_{t+1}}{\partial x_t} = 1 + \pi'^1(x_t, x_{t-1}, r_{t+1}) \tag{13}$$

$$\frac{\partial x_{t+2}}{\partial x_t} = \pi'^2(x_{t+1}, x_t, r_{t+2}) \tag{14}$$

$$\pi'^1(x_t, x_{t-1}, r_{t+1}) = -a + b \ , \quad \pi'^2(x_t, x_{t-1}, r_{t+1}) = -b \tag{15}$$

$$\frac{\partial x_t}{\partial r_t} = \pi'^3(x_{t-1}, x_{t-2}, r_t) \tag{16}$$

$$\frac{\partial r_t}{\partial x_l^*} = (1 - \tau)\delta_{l=k} + \tau \delta_{l=k+1} \ , \quad \tau \text{ and } k \text{ depend on } t \text{ as above} \tag{17}$$

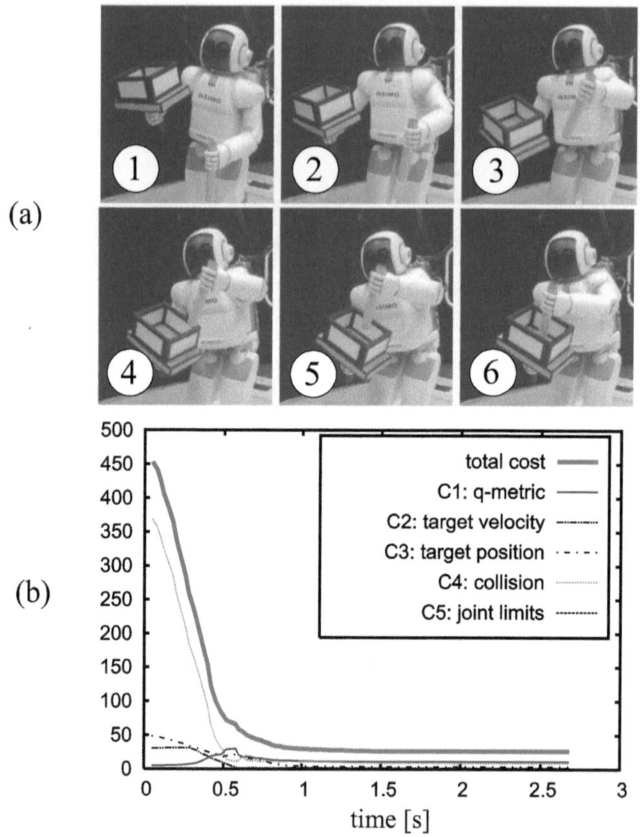

Figure 3.14 (a) Putting a cylinder into a box; and (b) cost decay

Most literature on grasp optimization focuses on the grasp itself, isolated from the reaching movement. For instance, [59] review the various literature on defining grasp quality measures, [57] learn which grasp positions are feasible for various objects, [28] efficiently compute good grasps depending on how the objects shall be manipulated, and [46] simplify the grasp computation based on abstracting objects into shape primitives. The coupling to the problem of reaching motion optimization is rarely addressed. In [52], reaching and grasping is realized by reactive control primitives. A recent approach [4] makes a step towards solving the coupled problem by including an "environment clearance score" in the grasp evaluation measure. In that way, grasps are preferred which are not prohibited by immediate obstacles directly opposing the grasp. Still, the full reaching motion is neglected in the grasp evaluation.

We approach this problem by proposing an object representation in terms of an object-specific task map which can be learnt from data and, during movement generation, efficiently coupled into a movement optimization process. Further, we generalise the optimization scheme presented in Section 3.5 to cope with such task maps [20, 21].

With the term *task map*, we refer to a map that comprises a set of sampled task coordinates, each associated with a scalar quality measure. In previous work [20], we proposed, for instance, to represent a set of hand–object pregrasp poses with respect to a failure/success criterion. These maps generally replace the concept of one explicit reaching target by the concept of a whole manifold of feasible targets in the task space. This relaxes the constraints imposed on the subsequent movement optimization process, which is particularly beneficial to improve other criteria governing the movement. If the chosen quality measure can be determined with the robot's sensors, it is further possible to build up or refine task maps in real experiments.

In the following, we will focus on simple "power grasps". However, the approach is not limited to a certain grasp. It is possible to represent different grasp types (*e.g.*, precision grasps, *etc.*) in several task maps. The concept even holds for bi-manual grasp strategies.

3.6.1 Acquisition of Task Maps for Grasping

Learning a task map requires exploring many different grasps on a specific object. The first question is how different grasp trials are sampled. The second, how each trial is evaluated. Previous approaches consider an exhaustive sampling over a grid [4]. To reduce the number of samples, we proposed to use RRTs [20]. While this technique is very fast, it is hard to generate a very dense set of samples. Further, when the set of feasible grasps separates into disjoint clusters, RRTs typically explore only one of the clusters. Here we will focus on an approach similar to the heuristics proposed in [4] and [58]. We assume that the robot can (visually) acquire a rough estimate of the object volume. Using the approximate shape information we can sample a random point from the volume and compute a pregrasp posture. For this, we initialize the hand inside the object volume. The hand is then retracted so as to get in palm contact with the object. Subsequently the finger joints are closed until they contact the objects surface. For this grasp, we collect the following data: (1) the hand position and orientation in coordinates relative to the object frame – this 6D point $g \in \mathbb{R}^6$ will become the vector of control parameters in the task space; (2) the contact points, normals and penetration between the finger segments and the object – this information is used to compute a quality measure based on force closure, which is a well-known measure determining the ability of a grasp to resist external forces [47].

While this learning phase can be executed on a real robot, we use realistic simulations to speed up the process. The force closure is computed from simulated tactile sensor positions and normals, excluding those that have zero pressure. In that way we collect a data set consisting of control parameters in $\mathbb{R}^6$ and the corresponding force closure scalars.

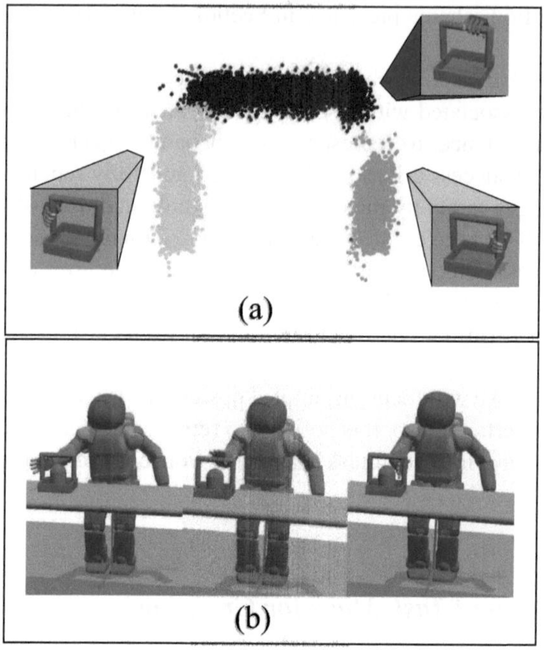

Figure 3.15 (a) Disjoint clusters in sampled task map. Solutions with the thumb and palm pointing upwards have been left out. (b) Initialization poses for different clusters.

For many realistic objects, the task map will consist of disjoint clusters which form qualitatively different solutions. Consider, for instance, a simple basket with handle that can be grasped on the top bar or on each of the side bars. The orientations and positions of feasible grasps change discontinuously from bar to bar, forming a set of clusters. We employ a Euclidean distance-based hierarchical clustering approach to extract a set of qualitatively different solutions. The chosen algorithm does not make any a priori assumption on the number of clusters.

Figure 3.15 (a) displays the extracted clusters for the given example. Clusters are formed by grasps applied to the left, right and top handle of the basket. In the figure, only the position elements of the 6D task map parameters are visualized. We only consider clusters of samples that have a significant size, eliminating outliers that comprise only a few samples.

3.6.2 Integration into Optimization Procedure

To find an appropriate initialization for the optimization problem, the target posture at the end of the movement is computed according to each cluster center, using inverse kinematics. The different target postures are then compared by an *end-state comfort* value, motivated from psychological studies. It is based on the joint limit and collision costs given in Equations 3.8 and 3.10. The best solution determines the target task vector to initialize the optimization problem. Figure 3.15 (b) illustrates the initializations for a given location of the basket on the table.

To incorporate the learnt task maps seamlessly into the optimization process, we formulate criteria to account for both the proximity to the nearest solution in the task map manifold and the quality of the resulting grasp: The *proximity criterion* enforces a smooth and collision-free movement into the manifold of feasible grasps represented in the map. It "pulls" the final grasp towards the manifold of valid preshape postures in the course of the optimization. The *quality criterion* evaluates the quality of the final grasp. It is based on the force-closure quality measure for each task map sample and guides the movement towards a preshape posture that leads to a high-quality grasp.

The proximity criterion will replace the distance of the target end-effector state and contribute to the cost function g in Equation 1 during motion optimization. Given the final task state at the last time step (*e.g.*, the hand position and orientation relative to the object), we compute the nearest element x_{map} in the task map. Now we define a cost:

$$g_p = (x_t^{\mathrm{rel}} - x_{\mathrm{map}})^T W_{\mathrm{map}} (x_t^{\mathrm{rel}} - x_{\mathrm{map}}) \ . \tag{3.18}$$

The metric W_{map} accounts for the difference in the linear and angular units. The nearest neighbor in the task map to the hand is computed with the approximate nearest neighbor algorithm described in [2]. For this, the hand position and orientation x_t^{rel} is represented in the reference frame of the object. The gradient is

$$\frac{\partial g_p}{\partial x_t^{\mathrm{rel}}} = 2(x_t^{\mathrm{rel}} - x_{\mathrm{map}})^T W_{\mathrm{map}}. \tag{3.19}$$

To account for the quality of the grasp that has been determined with the force closure measure, each sample of the task map is interpreted as a Gaussian:

$$f_i = |2\pi\Sigma_i^{-1}|^{-\frac{1}{2}} \exp(-\frac{1}{2} d_{\Sigma,i}), \tag{3.20}$$

with a mean vector μ and some small neighborhood $d_{\Sigma,i} = (x_i - \mu)^T \Sigma_i (x_i - \mu)$, determined by covariance matrix Σ^{-1}. The overall cost is computed as a weighted mixture of Gaussians considering the quality measure (force closure) w_i associated with each sample:

$$g_q = \frac{1}{\sum |f_i|} \sum w_i f_i. \tag{3.21}$$

We skip the gradient for brevity. This model has been chosen to account for the noise associated with the chosen force closure value. The noise is mainly due to the discontinuous number of contacts, the determination of the contact points and some other aspects. The mixture model smooths the criterion in a local region, assuring a smooth gradient.

While the proximity criterion tries to pull the final hand posture towards the task map manifold, the quality criterion becomes active only when the final hand posture is close to the manifold. It tries to level the target pose towards the best force closure. Figure 3.16 (b) shows the two cost terms over the time course of a trajectory. The proximity cost decreases as the final hand posture converges to the task manifold; the quality cost shows activity only when the hand posture is close to the manifold. To account for the formulated criteria during optimization, their gradient has to be derived with respect to the state vector according to term g' in Equation 10 of Table 3.1. For this, we can rewrite the differential kinematics as

$$\frac{\partial g_{\mathrm{map}}}{\partial q_t} = \frac{\partial (g_p + g_q)}{\partial x_t^{\mathrm{rel}}} \frac{\partial x_t^{\mathrm{rel}}}{\partial q_t} = \frac{\partial (g_p + g_q)}{\partial x_t^{\mathrm{rel}}} J_{\mathrm{rel}}, \tag{3.22}$$

with J_{rel} being the Jacobian of the task vector relating to the relative hand–object coordinates of the task map. This means that the task map can be represented in different coordinates than we chose to control the system. We can, for instance, represent the task map in relative hand–object coordinates and control the movement in global coordinates.

Both quality and proximity terms are only evaluated in the last time step of the trajectory, since this corresponds to the hand's final grasp pose. Their effect is back-propagated in time and on the attractor point locations with Equations 11 ff of Table 3.1. The nearest neighbor query needs to be carried only out once in each optimization iteration, which is advantageous in terms of computational efficiency.

3.6.3 Experiments

We have set up an experiment comprising a model of the humanoid robot ASIMO, a table and a basket with a U-shaped handle, see Figure 3.15 (b). To account for hand–object proximities, we consider collisions for the hand and finger segments, the lower arms, the body, the thighs and the environment objects table and basket. The overall collision model comprises 30 body pairs. The following task variables are subject to the optimization: right hand position and orientation (2D, polar angles) between hand and basket. Further, we constrained the horizontal components of the center of gravity and the foot transformations. Another constraint has been added to the gaze direction vector: it will continuously travel towards the center of the top basket handle. The employed task map has been sampled with 1000 valid grasps. The best initialization of the trajectory is determined according to Section 3.6.2.

The progression of the cost terms during the simulation run is depicted in Figure 3.16 (a). The costs massively decrease in the beginning, since the initial trajectory leads to many collisions, violates some joint limits and doesn't exactly reach the task map manifold. After a few iterations, it rapidly converges to a minimum. The enlarged subimage in the figure shows that (1) the target task manifold is exactly

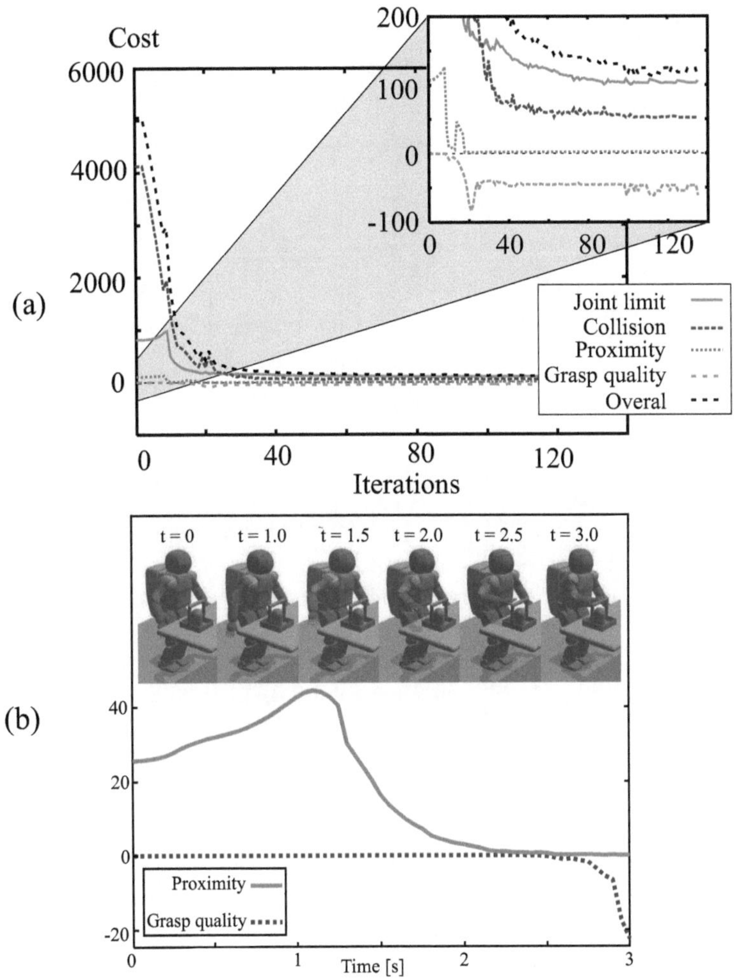

Figure 3.16 (a) Cost terms during optimization run; and (b) cost terms during movement

reached after approximately 15 iterations and (2) the grasp quality converges after approximately 25 iterations.

Figure 3.16 (b) shows the proximity and quality cost over the optimal trajectory. It can be seen that initially, the hand moves away from the table edge, and then moves upward in an arc around the table. The finally chosen grasp position is on the left side of the basket handle, which seems to lead to lower costs than grasping the handle at its center.

Figure 3.17 shows three representative movement sequences. In the upper row of the figure, the robot reaches and grasps for the inner handle of the basket. During reaching, it avoids to hit the other handles and finally shifts its hand to the lower

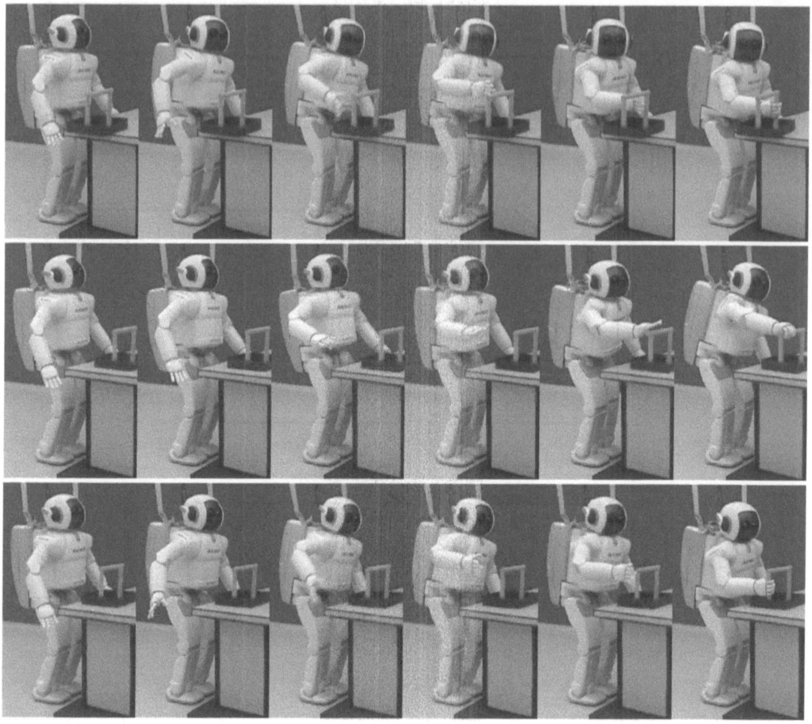

Figure 3.17 Movements for different basket locations

part of the handle before closing the fingers. In the middle row, the basket has been moved away in the frontal direction. The optimal pregrasp was found on the right end of the basket's top handle. In the lower row, the basket was moved to the left side. In this location, the movement optimization was initialized with the right handle. In this case, the hand did not have to move under the top handle, which resulted in a more "relaxed" movement.

3.7 Conclusion

In this work we presented elements towards a consistent control, prediction and movement planning architecture for humanoid robots. Movement control is achieved with a classical redundant control concept that has been extended with mechanisms to ensure balance stability and self-collision avoidance. This concept is the common basis of the subsequently presented methods that aim towards more movement intelligence. We presented a potential-field-based method to determine optimal stance poses for arbitrary end-effector tasks. It is an important element in the presented prediction and action selection architecture. This architecture predicts the outcome of a set of movement alternatives that solve a given task in different ways. The novel aspect is to continuously predict and evaluate the movement of a set of virtual controllers in parallel, these being able to quickly reorganize the movement behavior based on perceptual information. In order to cope with movements in more complex environments, a holistic trajectory optimization approach was presented. It operates on a somewhat slower time scale, but is still suited to be applied in interactive scenarios. Comparing this to the prediction architecture, it computes movements that satisfy criteria concerning the overall movement throughout a trajectory, such being able to reach towards objects while avoiding collisions and self-limits of the robot. This scheme has been extended to the coupled problem of reaching and grasping with the concept of object-specific *task maps*. These maps represent a functional characterization of objects in terms of their grasp affordances, *i.e.* a manifold of feasible pregrasp poses.

All elements have been verified in simulations and experiments with the humanoid robot ASIMO, and have been successfully integrated into large-scale systems such as [8, 23, 7].

Acknowledgments The authors thank A. Bendig, B. Bolder, M. Dunn, H. Janssen, N. Jetchev, J. Schmuedderich and H. Sugiura for their valuable contributions. Further, the authors thank the members of Honda's robot research teams in Japan for their support.

References

[1] Abdel-Malek K, Mi Z, Yang J, Nebel K (2006) Optimization-based trajectory planning of the human upper body. Robotica 24(6):683–696
[2] Arya S, Mount DM, Netanyahu NS, Silverman R, Wu AY (1998) An optimal algorithm for approximate nearest neighbor searching fixed dimensions. J ACM 45(6):891–923
[3] Baerlocher P, Boulic R (1998) Task-priority formulations for the kinematic control of highly redundant articulated structures. In Proceedings of the IEEE international conference on intelligent robots and systems (IROS), Canada
[4] Berenson D, Diankov R, Nishiwaki K, Kagami S, Kuffner J (2007) Grasp planning in complex scenes. In Proceedings of the IEEE-RAS/RSJ international conference on humanoid robots (humanoids), USA
[5] Bergener T, Bruckhoff C, Dahm P, Janssen H, Joublin F, Menzner R, Steinhage A, von Seelen W (1999) Complex behavior by means of dynamical systems for an anthropomorphic robot. Neural Netw
[6] Bizzi E, d'Avella A, P. Saltiel P, Tresch M (2002) Modular organization of spinal motors systems. Neuroscientist 8:437–442
[7] Bolder B, Brandl H, Heracles M, Janssen H, Mikhailova I, Schmdderich J, Goerick C (2008) Expectation-driven autonomous learning and interaction system. In Proceedings of the IEEE-RAS conference on humanoid robots (humanoids), Korea
[8] Bolder B, Dunn M, Gienger M, Janssen H, Sugiura H, Goerick C (2007) Visually guided whole body interaction. In Proceedings of the IEEE international conference on robotics and automation (ICRA), Italy

[9] Buschmann T, Lohmeier S, Ulbrich H, Pfeiffer F (2005) Optimization based gait pattern generation for a biped robot. In Proceedings of the IEEE-RAS conference on humanoid robots (humanoids), USA
[10] Buss SR. Introduction to inverse kinematics with jacobian transpose, pseudoinverse and damped least squares methods. Unpublished survey. http://math.ucsd.edu/ sbuss/ResearchWeb/ikmethods/index.html
[11] Chaumette F, Marchand E (2001) A redundancy-based iterative approach for avoiding joint limits: application to visual servoing. IEEE Trans Robotics Autom, 17(5):52–87
[12] Chestnutt J, Nishiwaki K, Kuffner J, Kagami S (2007) An adaptive action model for legged navigation planning. In Proceedings of the IEEE-RAS/RSJ international conference on humanoid robots (humanoids), USA
[13] Chiaverini S (1997) Singularity-robust task priority redundancy resolution for real-time kinematic control of robot manipulators. IEEE Trans Robotics Autom, 13(3)
[14] Choi SI, Kim BK (2000) Obstace avoidance control for redundant manipulators using collidability measure. Robotica 18:143–151
[15] Colby CL (1998) Action-oriented spatial reference frames in cortex. Neuron 20(1):15–24
[16] Dariush B, Gienger M, Arumbakkam A, Goerick C, Zhu Y, Fujimura K (2008) Online and markerless motion retargetting with kinematic constraints. In Proceedings of the IEEE international conference on intelligent robots and systems, France
[17] English JD, Maciejewski AA (2000) On the implementation of velocity control for kinematically redundant manipulators. IEEE Trans Sys Man Cybern:233–237
[18] Gienger M, Janssen H, Goerick C (2005) Task-oriented whole body motion for humanoid robots. In Proceedings of the IEEE-RAS/RSJ international conference on humanoid robots (humanoids), USA
[19] Gienger M, Janssen H, Goerick C (2006) Exploiting task intervals for whole body robot control. In Proceedings of the IEEE/RSJ international conference on intelligent robot and systems (IROS), China
[20] Gienger M, Toussaint M, Goerick C (2008) Task maps in humanoid robot manipulation. In Proceedings of the IEEE/RSJ international conference on intelligent robots and systems (IROS), France
[21] Gienger M, Toussaint M, Jetchev N, Bendig A, Goerick C (2008) Optimization of fluent approach and grasp motions. In Proceedings of the IEEE-RAS/RSJ international conference on humanoid robots (humanoids), Korea
[22] Gienger M, Bolder B, Dunn M, Sugiura H, Janssen H, Goerick C (2007) Predictive behavior generation - a sensor-based walking and reaching architecture for humanoid robots. Informatik Aktuell (AMS): 275–281, Germany, Springer (eds. Berns K, Luksch T)
[23] Goerick C, Bolder B, Janssen H, Gienger M, Sugiura H, Dunn M, Mikhailova I, Rodemann T, Wersing H, Kirstein S (2007) Towards incremental hierarchical behavior generation for humanoids. In Proceedings of the IEEE-RAS international conference on humanoid robots (humanoids), USA
[24] Guan Y, Yokoi K (2006) Reachable boundary of a humanoid robot with two feet fixed on the ground. In Proceedings of the international conference on robotics and automation (ICRA), USA, pp 1518 – 1523
[25] Guan Y, Yokoi K, Xianmin Zhang X (2008) Numerical methods for reachable space generation of humanoid robots. Int J Robotics Res 27(8):935–950
[26] Guilamo L, Kuffner J, Nishiwaki K, Kagami S (2005) Efficient prioritized inverse kinematic solutions for redundant manipulators. In Proceedings of the IEEE/RSJ international conference on intelligent robots and systems (IROS), Canada
[27] Harada K, Kajita S, Kaneko K, Hirukawa H (2004) An analytical method on real-time gait planning for a humanoid robot. In Proceedings of the IEEE-RAS international conference on humanoid robots (humanoids), USA
[28] Haschke R, Steil JJ, Steuwer I, Ritter H (2005) Task-oriented quality measures for dextrous grasping. In Proceedings of the IEEE international symposium on computational intelligence in robotics and automation (CIRA), Finland

[29] Haschke R, Weitnauer E, Ritter H (2008) On-line planning of time-optimal, jerk-limited trajectories. In Proceedings of the IEEE-RSJ international conference on intelligent robots and systems (IROS), France
[30] Heim A, van Stryk O (1999) Trajectory optimization of industrial robots with application to computer-aided robotics and robot controllers. Optimization 47:407–420
[31] Hirose M, Haikawa Y, Takenaka T, Hirai K (2001) Development of humanoid robot asimo. In Workshop of the IEEE/RSJ international conference on intelligent robots and systems (IROS), USA
[32] Ijspeert A, Nakanishi J, Schaal S (2002) Movement imitation with nonlinear dynamical systems in humanoid robots. In Proceedings of the IEEE international conference on robotics and automation (ICRA), USA
[33] Ijspeert A, Nakanishi J, Schaal S (2003) Learning attractor landscapes for learning motor primitives. Advances in Neural Information Processing Systems 15. MIT Press, Cambridge MA, USA, pp 1523–1530
[34] Kajita S, Kanehiro F, Kaneko K, Fujiwara K, Harada K, Yokoi K, Hirukawa H (2003) Resolved momentum control: humanoid motion planning based on the linear and angular momentum. In Proceedings of the IEEE/RSJ international conference on intelligent robots and systems (IROS), USA
[35] Kavraki L, Svestka P, Latombe J, Overmars M (1996) Probabilistic roadmaps for path planning in high-dimensional configuration spaces. IEEE Trans Robotics Automa (12):566–580
[36] Khatib O (1987) A unified approach for motion and force control of robot manipulators: the operational space formulation. IEEE Int J Robotics Autom, RA-3(1):43–53
[37] Kuffner J, LaValle S (2000) RRT-connect: An efficient approach to single-query path planning. In Proceedings of the IEEE international conference of robotics and automation (ICRA), USA
[38] Latombe JC (1991) Robot motion planning. Kluwer, Boston, MA, USA
[39] Laumond JP (1998) Robot motion planning and control. Springer-Verlag, Berlin, Germany. Available online at http://www.laas.fr/~jpl/book.html
[40] LaValle S (2006) Planning algorithms. Cambridge University Press, Cambridge, UK. Available at http://planning.cs.uiuc.edu
[41] Lebedev D, Klanke S, Haschke R, Steil J, Ritter H (2006) Dynamic path planning for a 7-dof robot arm. In Proceedings of the IEEE-RSJ international conference on intelligent robots and systems (IROS), China
[42] Liégeois A (1977) Automatic supervisory control of the configuration and behavior of multibody mechanisms. IEEE Trans Sys Man Cybern, SMC-7 (12)
[43] Maciejewski AA (1990) Dealing with the ill-conditioned equations of motion for articulated figures. IEEE Computat Graphics Applic, 10(3):63–71
[44] Marchand E, Chaumette F, Rizzo A (1996) Using the task function approach to avoid robot joint limits and kinematic singularities in visual servoing. In Proceedings of the IEEE/RSJ international conference on intelligent robots and systems (IROS), Japan
[45] Michel P, Chestnutt J, Kuffner J, Kanade T (2005) Vision-guided humanoid footstep planning for dynamic environments. In Proceedings of the IEEE-RAS conference on humanoid robots (humanoids), USA
[46] Miller AT, Knoop S, Christensen HI, Allen PK (2003) Automatic grasp planning using shape primitives. In Proceedings of the IEEE international conference of robotics and automation (ICRA), Taiwan
[47] Murray RM, Li Z, Sastry SS (1994) Robotic manipulation. CRC Press, FL, USA
[48] Mussa-Ivaldi FA, Giszter SF, Bizzi E (1994) Linear combinations of primitives in vertebrate motor control. Neurobiology 91:7534–7538
[49] Nakamura Y (1991) Advanced robotics: redundancy and optimization. Addison-Wesley
[50] Nakamura Y, Hanafusa H (1987) Optimal redundancy control of robot manipulators. Int J Robotics Res (6)
[51] Pfeiffer F (2008) Mechanical system dynamics. Springer Verlag

[52] Platt R, Fagg AH, Grupen R (2006) Improving grasp skills using schema structured learning. In Proceedings of the international conference on development and learning (ICDL), India
[53] Mas R, Boulic R, Thalmann D (1997) Interactive identification of the center of mass reachable space for an articulated manipulator. In Proceedings of the IEEE international conference of advanced robotics (ICAR), USA
[54] Schaal S (2003) Movement planning and imitation by shaping nonlinear attractors. In Proceedings of the 12th Yale workshop on adaptive and learning systems, Yale University, USA
[55] Schaal S, Peters J, Nakanishi J, Ijspeert AJ (2003) Control, planning, learning, and imitation with dynamic movement primitives. Workshop on bilateral paradigms on humans and humanoids, IEEE international conference on intelligent robots and systems (IROS), USA
[56] Schlemmer K, Gruebel G (1998) Real-time collision-free trajectory optimization of robot manipulators via semi-infinite parameter optimization. Int J Robotics Res, 17(9):1013–1021
[57] Schwammkrug D, Walter J, Ritter H (1999) Rapid learning of robot grasping positions. In Proceedings of the international symposion on intelligent robotics systems (SIRS), Portugal
[58] Steil JJ, Roethling F, Haschke R, Ritter H (2004) Situated robot learning for multi-modal instruction and imitation of grasping. Robotics Autonomous Sys 47 (Special issue on robot learning by demonstration):129–141
[59] Suárez R, Roa M, Cornellà J (2006) Grasp quality measures. Technical Report IOC-DT-P 2006-10, Universitat Politècnica de Catalunya, Institut d'Organització i Control de Sistemes Industrials
[60] Sugihara T, Nakamura Y (2002) Whole-body cooperative balancing of humanoid robot using COG jacobian. In Proceedings of the IEEE/RSJ international conference on intelligent robot and systems (IROS), Switzerland
[61] Sugiura H, Gienger M, Janssen H, Goerick C (2007) Real-time collision avoidance with whole body motion control for humanoid robots. In Proceedings of the IEEE-RSJ international conference on intelligent robots and systems (IROS), USA
[62] Toussaint M, Gienger M, Goerick C (2007) Optimization of sequential attractor-based movement for compact movement representation. In Proceedings of the IEEE-RAS/RSJ international conference on humanoid robots (humanoids), USA
[63] Yoshida E, Belousov I, Esteves C, Laumond JP (2005) Humanoid motion planning for dynamic tasks. In Proceedings of the IEEE-RAS/RSJ international conference on humanoid robots, USA
[64] Zacharias F, Borst C, Beetz M, Hirzinger G (2008) Positioning mobile manipulators to perform constrained linear trajectories. In Proceedings of the IEEE-RSJ international conference on intelligent robots and systems (IROS), France

Chapter 4
Planning Whole-body Humanoid Locomotion, Reaching, and Manipulation

Eiichi Yoshida, Claudia Esteves, Oussama Kanoun, Mathieu Poirier, Anthony Mallet, Jean-Paul Laumond and Kazuhito Yokoi

Abstract In this chapter we address the planning problem of whole-body motions by humanoid robots. The approach presented benefits from two cutting edge recent advancements in robotics: powerful probabilistic geometric and kinematic motion planning and advanced dynamic motion control for humanoids. First, we introduce a two-stage approach that combines these two techniques for collision-free simultaneous locomotion and upper-body task. Then a whole-body motion generation method is presented for reaching, including steps based on generalized inverse kinematics. The third example is planning of whole-body manipulation of a large object by "pivoting", by making use of the precedent results. Finally, an integrated experiment is shown in which the humanoid robot interacts with its environment through perception. The humanoid robot platform HRP-2 is used as the platform to validate the results.

4.1 Introduction

As progress in the hardware of humanoid robots has recently been accelerated, a number of applications are now expected. Their anthropomorphic shape is advan-

Eiichi Yoshida and Kazuhito Yokoi
CNRS-AIST JRL (Joint Robotics Laboratory), UMI 3218/CRT, National Institute of Advanced Industrial Science and Technology (AIST), Umezono 1-1-1, Tsukuba, Ibaraki, 305-8568, Japan, e-mail: \{e.yoshida, Kazuhito.Yokoi\}@aist.go.jp

Claudia Esteves
Facultad de Matematicas, Universidad de Guanajuato, Guanajuato, 36000 Gto., Mexico. e-mail: cesteves@cimat.mx

Oussama Kanoun, Mathieu Poirier, Anthony Mallet and Jean-Paul Laumond
LAAS-CNRS, 7 avenue du Colonel Roche, F-31077 Toulouse, and Universite de Toulouse ; UPS, INSA, INP, ISAE ; LAAS ; F-31077 Toulouse, France.
e-mail: \{okanoun, mpoirier, mallet, jpl\}@laas.fr

tageous in moving in environments designed for humans, using the machines or tools designed for humans, and also performing interactions and assistive tasks for humans. For the humanoid to execute effective tasks, whole-body motion coordination is indispensable. Humanoid motion is characterized by its redundancy and underactuation. The former is obvious, as humanoid robots have usually more than thirty degrees of freedom. The latter means the "base frame" of a humanoid robot, for example its waist, can only be controlled indirectly by articulated legs, unlike wheeled mobile robots. In this article, in order to tackle this planning problem of whole-body humanoid motions, we present the approaches that combine probabilistic geometric/kinematic motion planning and dynamic motion generation as follows. At the planning stage, the global humanoid motion is modeled by a kind of vehicle with bounding volume fully actuated to plan the basic path. It is then transformed into dynamically executable humanoid motion by the dynamic motion generator. The whole-body humanoid motions handled in this chapter include collision-free locomotion, reaching and manipulation.

In the rest of this section, we briefly introduce the basic planning tools and the software and hardware platform which is the common part throughout the chapter. In Section 4.2, a two-stage approach for collision-free locomotion is presented. Then we address dynamic whole-body motion generation by taking the example of reaching, including stepping, in Section 4.3. The whole-body manipulation is then dealt with in Section 4.4 by benefiting from both the collision-free path planning technique and whole-body motion generation. An experiment that integrates planning and perception is described in Section 4.5 using the humanoid robot HRP-2, before concluding the chapter.

4.1.1 Basic Motion Planning Methods

For complex robotic systems like humanoid robots, it is reasonable to employ efficient algorithms based on probabilistic planning methods such as diffusing methods like rapidly-exploring random trees (RRT) [11, 23] or sampling methods like probabilistic roadmap (PRM) [17] and all their variants (see [4, 22] for recent overviews). In those methods, the path search is usually made in configuration space such as the joint angles. The probabilistic methods compute a graph called a roadmap whose nodes are collision-free configurations chosen at random and whose edges model the existence of collision-free local paths between two nodes.

In probabilistic planning method, the graph is built incrementally by choosing configurations at random. Configurations and local paths between them are included in the graph as soon as they are collision-free. This construction of the graph is called the learning phase. Once the graph is built, then the query phase consists in first adding both starting and goal configurations of the given problem to the roadmap, and then searching the graph for a path.

On the other hand, in diffusing methods the graph is built gradually by expanding the tree from the start and/or goal configurations. After a configuration is randomly generated, the nearest configuration is identified. Then a new configuration is computed that advances by a given distance from the nearest node towards the randomly generated one. If this new configuration and the local path to it are collision-free, then they are added to the graph as a new node and a new edge. This diffusion is repeated until a path is found that connects the start and goal configurations.

In this way, the probabilistic motion planner eventually finds collision-free paths as connected components of the roadmap, which is proven as probabilistic completeness. There are several important components in implementing this probabilistic motion planning: collision checkers, steering methods and path optimizers. Collision checkers validate configurations and paths. Any available free or commercial libraries can be used for this function. A "steering method" is a method that computes an admissible path from an initial configuration to a final one in the absence of obstacle. In this article, dedicated steering methods are devised depending on the planning problems. Finally, since paths planned by probabilistic motion planners are often redundant, a "path optimizer" is necessary to remove the redundant parts to obtain a shorter path. For instance, we can employ the "adaptive shortcut" path optimization algorithm proposed by [11].

As described later in Section 4.2, we generally adopt a two-stage approach for whole-body motion planning. At the first stage, the basic geometric and kinematic motion planner described here is applied to collision-free path planning for a simplified model of the humanoid, like its bounding volume. This is because the computation cost would be too high if random sampling is directly applied to a complex dynamic system with many degrees of freedom.

4.1.2 Hardware and Software Platform

We implement motion planners presented in this article in a common software framework "Humanoid Path Planner" [36] on the basis of the motion planning software kit KineoWorks™ [21] as shown in Figure 4.1. As illustrated in the lower part of Figure 4.1, this software kit provides the basic methods of planning, like PRM or RRT, as "roadmap builders", as well as the aforementioned basic functions of collision checking mechanism, and a template for steering method and path optimizers.

The object-oriented architecture allows users to define the specific planner depending on the problem tackled as an inherited class of a general robot motion planner. The planner takes care of interaction with basic functions introduced in 4.1.1. The problem-specific components, especially steering methods for walking and manipulation are inherited from the template including basic functions. Since the basic interface of class definition of the general framework is common, it is relatively easy to implement those problem specific parts once the planning problem is defined.

The HPP framework also includes a walking pattern generator presented in Section 4.2 and a dynamic whole-body motion generator in Section 4.3.

As the hardware platform, we utilize the humanoid robot HRP-2 [16] shown in Figure 4.1 which is 1.58 m tall and weighs 58 kg, with 30 DOFs. It has two rota-

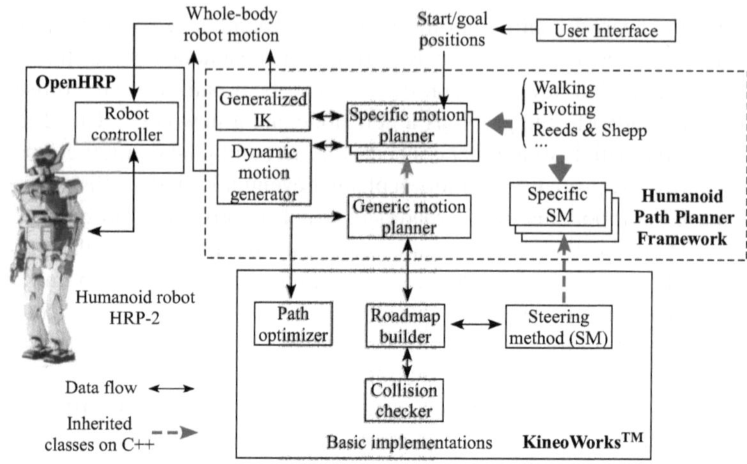

Figure 4.1 Architecture of Humanoid Path Planner framework that facilitates implementation of robotic motion planner according to specific problems

tional joints (pitch and yaw) at the chest that provides a wide upper-body workarea. Six-axis force sensors at the ankles, a rate gyroscope and an accelerometer for attitude estimation are equipped for stability control. Once the whole-body humanoid motion is generated, it is passed to the robot simulator and controller OpenHRP [15].

4.2 Collision-free Locomotion: Iterative Two-stage Approach

We have proposed a general and practical planning framework that integrates a geometric path planner and a dynamic motion generator [34]. It is based on an iterative two-stage motion planning method, by exploiting the efficiency of probabilistic planning and advanced dynamic motion generation. Within the classification proposed in [18], our two-stage approach fits alongside state-space and sensor-based approaches. In the first stage, a "path" is generated by geometric and kinematic planning, which is transformed into a dynamically executable "trajectory" through appropriate dynamic motion generators in the second stage. Due to dynamic effects, the path supporting the new trajectory may differ slightly from the initial path. Then the output trajectory is again verified with respect to collision avoidance by the first stage and reshaped if necessary. This process is iterated until a valid dynamic trajectory is obtained.

Although the necessity of reshaping for dynamic collision-free motion has been recognized [19], it has not been systematically addressed. Our method is inspired by a technique for key frame editing in the context of computer animation [8]. This approach places more emphasis on the gradual transition from the colliding trajectory by maintaining the motion timing in constrained environments. In this work we also emphasize the practical aspect of our approach through realistic simulations and experiments.

4.2.1 Two-stage Planning Framework

The proposed general framework of dynamic collision-free motion planning based on the two-stage planning method is illustrated in Figure 4.2. It outputs dynamic collision-free trajectories from the inputs of initial and goal configurations together with the geometry information of the environment. The resulting dynamic trajectories should be ready to be given to low-level robot controllers.

The path planner finds a geometric and kinematic collision-free path in 3D at the first stage (upper part of Figure 4.2). Any available planning method can be used for this part. We utilize PRM in our case for the first stage. The second stage is the dynamic motion generator to transform the given path into a dynamically executable robot trajectory (lower part in Figure 4.2). A dedicated dynamic controller can be used depending on the application.

The generated trajectory may deviate from the planned path due to robot dynamics, which may cause unpredicted collisions with obstacles. The reshaper is placed in the first stage as a mechanism that interacts with the dynamic motion generator

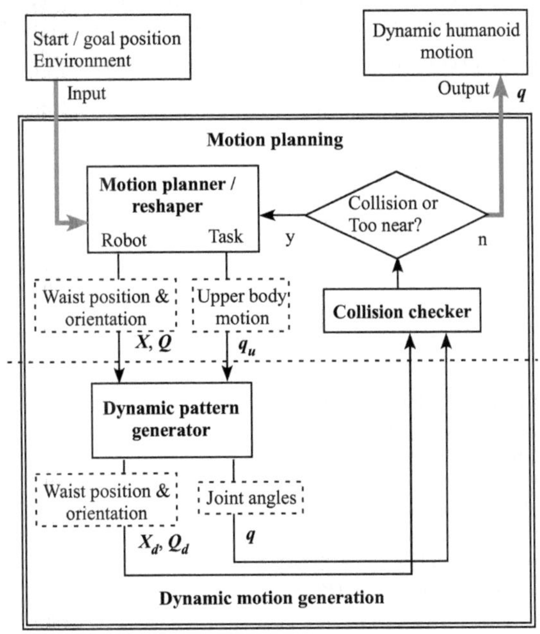

Figure 4.2 Two-stage motion planning framework. In the first stage the geometric and kinematic planner plan the collision-free path that is transformed into dynamic motion in the second stage. If collisions are detected the path is sent back to the first stage. This process is repeated until a collision-free dynamic trajectory is obtained

iteratively to remove these local collisions. Practically, if collisions are detected in the dynamic trajectory, the colliding portion is locally deformed by increasing the "tolerance" of the obstacle for the robot to move away from the obstacles.

If the dynamically colliding local paths become blocked by the "grown" obstacles, a replanning process is activated. In this process, the path planner searches for another path that avoids the blocked passage.

We utilize here a functional decomposition of the robot body that has already been applied to motion planning for virtual mannequins [6]. At the first stage, the robot is modeled as a geometric parallelepiped bounding box (Figure 4.3). Only that box and the object to be manipulated are considered with respect to collision avoidance. The robot motion is expressed by the planar position and orientation of its waist $r(x,y,\theta)$ (3 DOF) and the object motion by its position and orientation $R_o(x_o,\Theta_o)$ (6 DOF) with respect to a global coordinate system Σ_0. The configuration space to be searched is then 9-dimensioned.

In our case, the robot path is planned as Dubins curves composed of line segments and arcs of a circle [5]. Given the configuration of the robot waist and object, the joint angles (q_u) of the upper-body motion are derived by using inverse kinematics described later.

4.2.2 Second Stage: Smooth Path Reshaping

At the second stage, the planned motions r and q_u are given to the dynamic pattern generator [13] of humanoid robots to transform the input planar path into a dynamically executable motion. The walking pattern generator is based on preview control

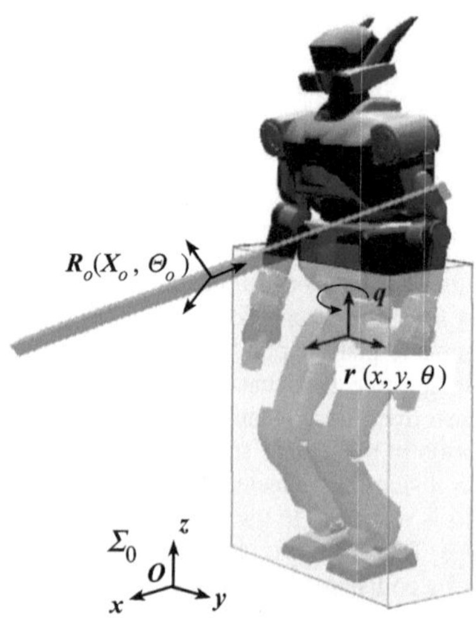

Figure 4.3 Humanoid modeled by rectangle box with a bar. In the first stage the geometric and kinematic path planner generates collision-free path for the 9 DOF system including robot waist (r, 3DOF) and object (R_o, 6DOF)

of zero moment point (ZMP) proposed by Kajita *et al.* [13]. The reference ZMP trajectory is derived from the foot placements obtained from the planned planer robot path. Based on preview control of ZMP position for an inverted pendulum model, this method is able to generate dynamically stable biped walking motion that always maintains the ZMP inside the support polygon formed by the foot (feet). Moreover, the pattern generator can combine upper-body motion q_u as auxiliary input to compute the mixed whole-body motion.

If collisions are found within the upper part of the body, the following reshaping procedure is applied. After identifying the endpoints of each colliding portion, a collision-free configuration is found in a free space within a nearby reachable area after increasing the "tolerance", which grows the obstacle. All the configurations within the portion are then replaced by this collision-free configuration. Next, anticipation and regaining motions are computed to smoothly connect the reshaped portion with the collision-free part of the original trajectory. Finally, inverse kinematics (IK) is applied to satisfy the constraints of the hands at each sample of the reshaped trajectory that synchronizes the upper body task with the lower body motion. As a result, this reshaping eliminates the collision locally as shown in Figure 4.4.

We have applied the proposed method to plan a motion to carry a bulky object in an environment with several obstacles as shown in Figure 4.5. The proposed method is implemented as an off-line planner on the assumption that the environment is completely known. In this case, what matters is not the weight of the object but its geometric complexity. Figure 4.6 shows the experimental results of the planned motion.

Since the distance between the two lamps is shorter than the bar length, the bar should pass through at an angle. At the beginning of the motion, the computed trajectory for the bar makes the robot move to the left, then walk forward with a certain angle to the path through the gap (Figure 4.6(a)(b). Here the motion of the

Figure 4.4 Transition of robot configurations during the reshaping. The colliding part of the object carried moves away from the obstacle by increasing tolerance

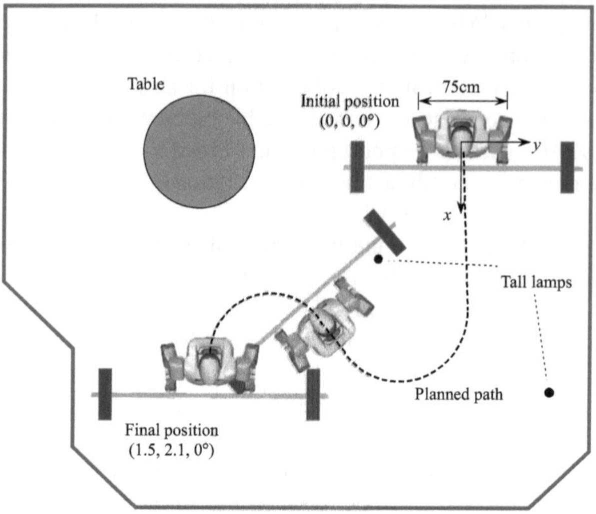

Figure 4.5 Top view of the simulation and experiment environment with two poles and a table. The initial and final configurations of the robot are also shown

upper part of the robot is computed using a generalized inverse kinematics and the chest is moved consequently to complete both tasks.

This example also shows that the complete 3D geometry of the object has been considered in the collision detection and path planning procedure and no bounding box has been used (see Figure 4.6(d)) where the concave part of the disk and the bar is close to the lamp. The complete trajectory execution time is around 28 s.

4.3 Reaching: Generalized Inverse Kinematic Approach

We here address the problem of how to re-position the humanoid body when performing reaching or grasping tasks for a target far away. The proposed method is based on reshaping the support polygon of the humanoid robot to increase its workarea by coupling generalized inverse kinematics and a dynamic walking pattern generator [35]. While using inverse kinematics, the global motion is guaranteed to be dynamically stable. Such a property is a direct consequence of ZMP control provided by the pattern generator we use.

The generation of whole-body dynamic motion is closely related to motion planning. Whereas motion planning takes charge of the global plan from initial to goal configurations, whole-body motion generation concerns how to make valid local motions by taking account of several constraints. So it is important in order to cre-

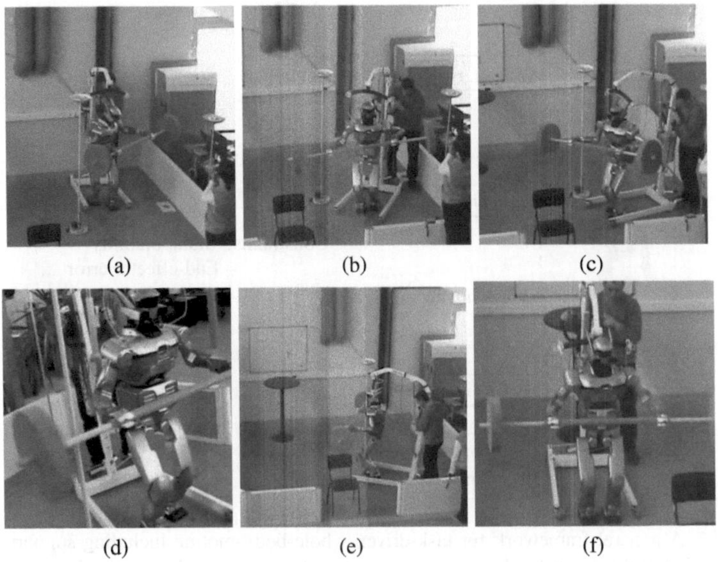

Figure 4.6 Experiment of 3D collision-free motion for bar-carrying task at JRL-France; the robot (a) starts walking turning the object; (b) moves the object between the lamps while walking; (c) passes completely through the gap; (d), (e) avoids collision against the table; and (f) arrives at the goal position

ate feasible dynamic trajectories from motions that have been provided by the global motion planner.

Khatib and his colleagues have been working on dynamic motion generation for humanoid robots by using a task specification in operational space approach [27]. In their work a hierarchical controller synthesizes whole-body motion based on prioritized behavioral primitives including postures and other tasks in a reactive manner. Kajita *et al.* proposed a "resolved momentum control" to achieve specified momentum by whole-body motion [14]. Mansard *et al.* [24] proposed a task sequencing scheme to achieve several tasks including walking and reaching at the same time.

Figure 4.7 illustrates the proposed motion generation framework with an example of a reaching task [35]. Priorities are given to the target task as well as to other tasks such as the position of the center of mass (CoM). We employ generalized inverse kinematics to generate a whole-body motion for those tasks based on the given priorities [25]. During the motion, several constraints are monitored which are expressed by such measures as manipulability for whole-body, end-effector errors from target, or joint limits.

If the task cannot be achieved because the monitored constraints are not satisfied, a reshaping planner for the support polygon is activated automatically to increase accessible space for the robot, keeping the inverse kinematics working to achieve the tasks. The reshaping is performed based on geometric planning to deform the support polygon in the direction required by the specified task. Thanks to the use

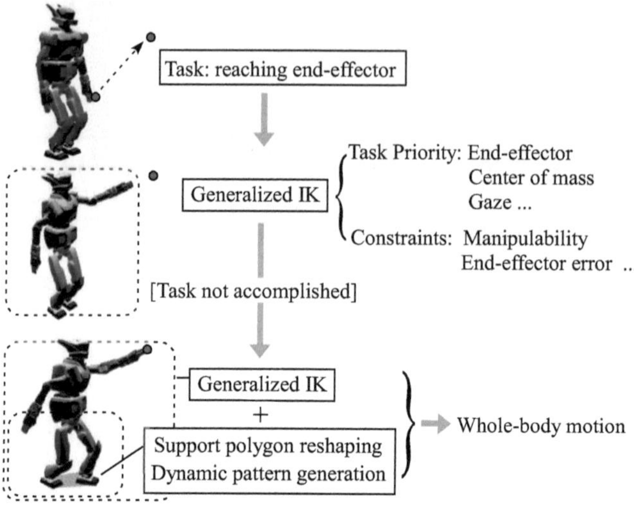

Figure 4.7 A general framework for task-driven whole-body motion including support polygon reshaping [35]. If the desired tasks cannot be achieved, the support polygon is reshaped to increase the workspace

of a free-floating base, the changes in support phase can easily be integrated in the computation. As a result, the stepping motion is generated using a biped walking pattern generator [13] and the blended whole-body motion, including the target task, is recalculated.

Our contribution is to consider the possibility of reshaping the support polygon by stepping to increase the accessible space of the end-effectors in 3D space. Our approach makes use of the whole-body 3D space as opposed to 2D in [39]. Moreover, in spite of our reasoning being based on inverse kinematics and simple geometric support polygon reshaping, our method guarantees that the motion is dynamically stable. This property is a consequence of the pattern generator [13] we use to generate the stepping behavior.

4.3.1 Method Overview

Support polygon reshaping integrates two important components, generalized inverse kinematics and the dynamic walking pattern generator. The former provides a general way to deal with whole-body motion generation to perform the prioritized tasks. The latter takes charge of the stepping motion to change the foot placements.

Figure 4.8 shows an overview of the method. The task is specified in the workspace as $\dot{x}_j$ with priority j from which the generalized IK solver computes the whole-body motion as joint angles $\dot{q}$ of the robot. Meanwhile, several criteria such

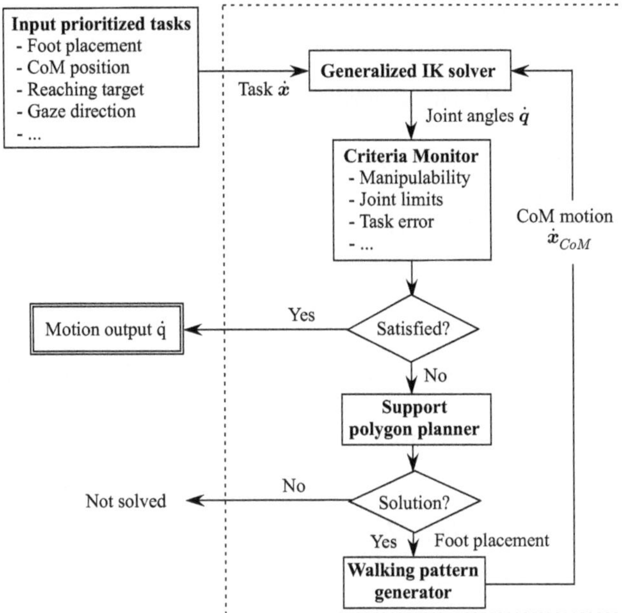

Figure 4.8 Method overview of the whole-body motion generation. Once the support polygon is reshaped, the dynamic stepping motion is computed as the CoM ($\dot{x}_{CoM}$) from the foot placement and again given to the generalized IK solver.

as manipulability or joint limits are monitored if they do not prevent the desired whole-body motion.

As long as the criteria are satisfied, the computation of whole-body motion continues until the task target is achieved. If the task cannot be achieved due to unsatisfied criteria, the support polygon planner is triggered in order to extend reachable space. A geometric module determines the direction and position of the deformation of the support polygon so that the incomplete task is fulfilled. The position of a foot is then derived to generate the CoM motion $\dot{x}_{CoM}$ using a dynamic walking pattern generator [13].

Using this CoM motion, the original task is then redefined as the whole-body motion including stepping that is recalculated using the same generalized IK solver.

The generalized IK solver benefits from the redundancy of the mechanism to choose the solution that best solves the task according to some constraints. Among this work, inverse kinematics algorithms that project tasks with lower priority into the null space of the Jacobian of the higher priority tasks have been widely studied (*e.g.*, [3, 25, 28, 32]).

4.3.2 Generalized Inverse Kinematics for Whole-body Motion

4.3.2.1 Inverse Kinematics for Prioritized Tasks

Let us consider a task $\dot{x}_j$ with priority j in the workspace and the relationship between the joint angle velocity $\dot{q}$ is described using Jacobian matrix, like $\dot{x}_j = J_j\dot{q}$. For the tasks with the first priority, using pseudoinverse $J_1^{\#}$, the joint angle velocity that achieves the task is given by

$$\dot{q}_1 = J_1^{\#}\dot{x}_1 + (I_n - J_1^{\#}J_1)y_1, \tag{4.1}$$

where y_1, n and I_n are an arbitrary vector, the number of joints and identity matrix of dimension n, respectively.

For the task with second priority $\dot{x}_2$, the joint velocities $\dot{q}_2$ is calculated as follows [25]:

$$\begin{aligned} \dot{q}_2 &= \dot{q}_1 + \hat{J}_2^{\#}(\dot{x}_2 - J_2\dot{q}_1) + (I_n - J_1^{\#}J_1)(I_n - \hat{J}_2^{\#}\hat{J}_2)y_2, \\ &\text{where} \quad \hat{J}_2 \equiv J_2(I_n - J_1^{\#}J_1) \end{aligned} \tag{4.2}$$

and y_2 is an arbitrary vector of dimension n. It can be extended to the task of jth $(j \geq 2)$ priority in the following formula [3, 28]:

$$\begin{aligned} \dot{q}_j &= \dot{q}_{j-1} + \hat{J}_j^{\#}(\dot{x}_j - J_j\dot{q}_{j-1}) + N_j y_j, \\ N_j &\equiv N_{j-1}(I_n - \hat{J}_j^{\#}\hat{J}_j),\ \hat{J}_j \equiv J_j(I_n - \hat{J}_{j-1}^{\#}\hat{J}_{j-1}). \end{aligned} \tag{4.3}$$

4.3.2.2 Monitoring Task Execution Criteria

While the motion is being computed by the generalized IK, several properties are monitored.

One of the important measures is the manipulability [40] defined as:

$$w \equiv \sqrt{\det\{JJ^T\}}. \tag{4.4}$$

This measure is continuously tracked during the motion generation as well as others such as joint angle limits or end-effector errors from the target. If it becomes below a certain value, it means that it is difficult to achieve the task.

Joint limit constraints can be taken into account by introducing a selection diagonal matrix $S = \text{diag}\{S_1, \ldots S_n\}$ ($S_i = 0$ or 1) to be multiplied to Jacobian to select the activated joints if the corresponding joint reaches a limit angle. The selection matrix is I_n if all the joints are used to achieve the task.

As shown in Figure 4.8, when one or more monitored measures go out of the admissible range to prevent the task from being achieved, the support polygon reshaping is launched to extend the accessible space as detailed next.

4.3.2.3 Support Polygon Reshaping

Figure 4.9 shows the proposed support polygon reshaping scheme. This simple algorithm allows the humanoid robot to make a step motion, keeping a large margin of accessible area for the task by facing the upper body to the target direction.

Then the CoM motion $\dot{x}_{CoM}$ is computed from the new foot position by the walking pattern generator based on the preview control of ZMP [13]. The basic idea is to calculate the CoM motion by anticipating the desired future ZMP positions derived from the footsteps.

Finally the original task is redefined as another problem of whole-body task using this newly generated CoM motion with an additional task of CoM, which is represented by the CoM Jacobian [29]. The same generalized IK solver framework is used to incorporate the motion required for the task and the stepping motion at the whole-body level.

The manipulability measure of the arm during the forward reaching task is provided in Figure 4.10. Without reshaping, the arm approaches a singular configuration where the manipulability becomes lower than the threshold at 2.3 s and the computation keeping the same support polygon is discarded. The reshaping starts at this moment to recalculate the overall whole-body motion including stepping motion. We can see that the manipulability regains higher value at the final position.

4.3.3 Results

We have conducted experiments on the generated motion using a humanoid platform HRP-2 for front and sideways reaching tasks that require stepping as shown in Fig 4.11. As can be seen, the robot successfully performs the desired reaching task through whole-body motion that unifies the reaching task and stepping motion by

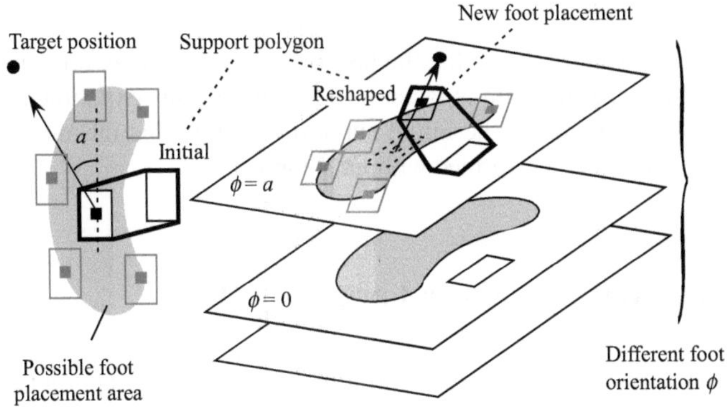

Figure 4.9 Support polygon reshaping method

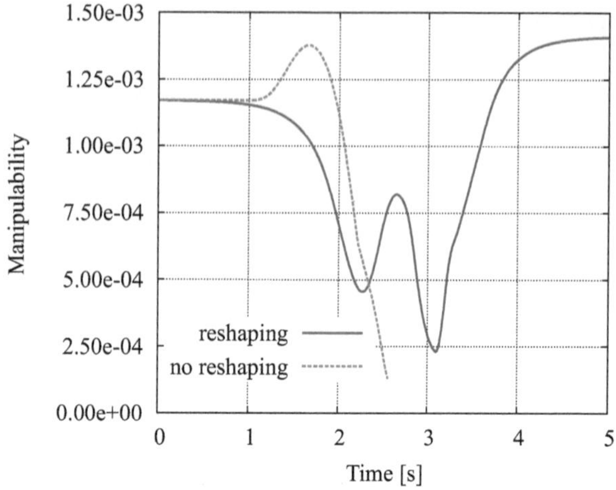

Figure 4.10 Manipulability for front reaching. Without support polygon reshaping, the manipulability measure decreases below the threshold. Although it also decreases with reshaping, the manipulability increases in the course of the stepping motion

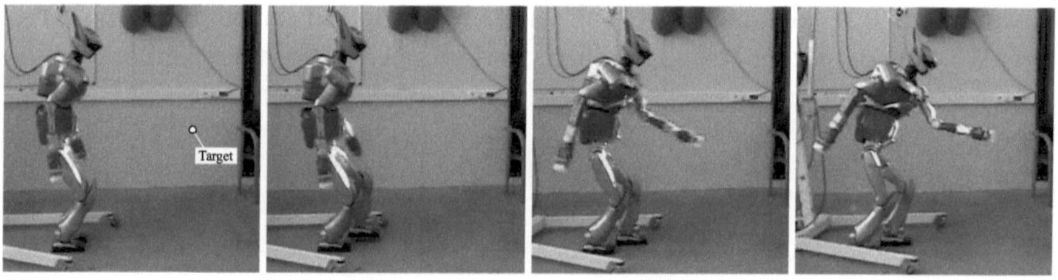

(a) (b) (c) (d)

Figure 4.11 Experimentation of front reaching task. The robot goes through a posture that is not statically stable (b) to finish stepping in (c). The final goal of the end effector is achieved at (d). The gaze direction is always maintained in the direction of the end-effector goal

keeping dynamic balance. Note that the task of keeping the gaze direction towards the end-effector target position is taken into account in this experiment. The final CoM apparently goes out of the initial support polygon: this means the reaching task could not have been performed without stepping.

4.4 Manipulation: Pivoting a Large Object

Manipulation requiring whole-body motion is one task that is appropriate for humanoid robots. In order to manipulate cumbersome object, humans often manipulate without lifting but by using contact with the ground (Figure 4.12). In this research, we apply such a manipulation method to humanoid robots.

There are several task-specific whole-body motions that have been intensively investigated: pushing [12, 9, 31], and lifting [10], and pivoting [33, 37]. Currently, many researchers are working to integrate these recent developments with a global motion planner.

Among them, the pivoting manipulation has several advantages such as precise positioning, stability and adaptability over other methods like pushing or lifting. For these reasons, pivoting-based manipulation has the potential to widen the capacity of manipulation of humanoid robots.

We introduce here a whole-body motion planner that allows a humanoid robot to autonomously plan a pivoting strategy that accounts for the various constraints: collision avoidance, legs–arms coordination and stability control.

The motion planning algorithm we propose considers a two-stage approach: a first collision-free path is computed, and then it is iteratively approximated by a sequence of pivoting motions.

4.4.1 Pivoting and Small-time Controllability

The robot starts inclining the box to realize a single contact point between the box and the floor. The contact point is a corner of the box. Then the robot performs a rotation of the box around the vertical axis at that corner. Then it sets the object horizontally along the box edge. Such an edge is said to be the supporting edge. We model the problem of 3D box pivoting as the problem of pivoting a 2D segment around its endpoints.

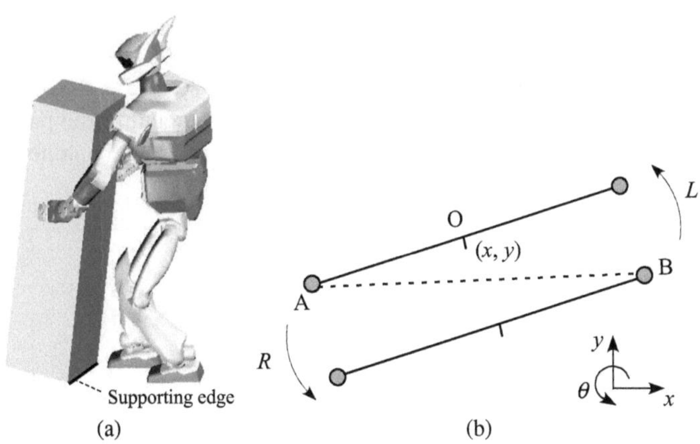

Figure 4.12 Supporting edge and pivoting problem modeling. (a) The pivoting sequence is planned using rotation of the endpoints of this edge. (b) The 3D pivoting problem is reduced to how to displace a line segment on vertices *A* or *B*

A system is said to be controllable if it may reach any arbitrary configuration q_{goal} from any other q_{init} [30]. It is said to be small-time controllable if the set of admissible configurations $Reach_q(T)$, which can be reached from configuration q before a given time $T(>0)$, contains a neighborhood of q. This property should hold at any configuration q for any T. It means that the system can move anywhere in the area η without leaving an imposed neighborhood V as shown in the left of Figure 4.13.

Small-time controllability is a critical property in path planning. The main consequence is depicted on the right side of Figure 4.13: any collision-free path can be approximated by a sequence of both collision-free and admissible motions as follows. Starting from the initial configuration q_{init}, take any collision-free neighborhood V_1. Then the system can advance to a configuration q_1 on the path within $Reach_{q_1}(T)$ without going out of V_1. The same procedure is repeated until the system reaches the goal configuration q_{goal} (Figure 4.13). This type of analysis plays an important role in nonholonomic motion planning [20].

We have proven that the considered pivoting system is small-time controllable by using Lie Algebra Rank Condition (LARC) [30]. First, the vector field of the motion in the space (x, y, θ) is considered for the rotation motions R and L turning around the corner A and B respectively. Then the Lie Bracket $[L, R]$ is computed to demonstrate that the three vector fields L, R and $[L, R]$ span a three-dimensional space. For details, the readers are referred to the reference [37].

4.4.2 Collision-free pivoting sequence planning

We here take into account the naturalness of the targeted solution: we want the robot to walk either forward or backward and to avoid sideways steps. When walking forward or backward the robot direction remains tangent to the path it follows as a wheeled mobile robot does. Such constraints are known to be nonholonomic. It

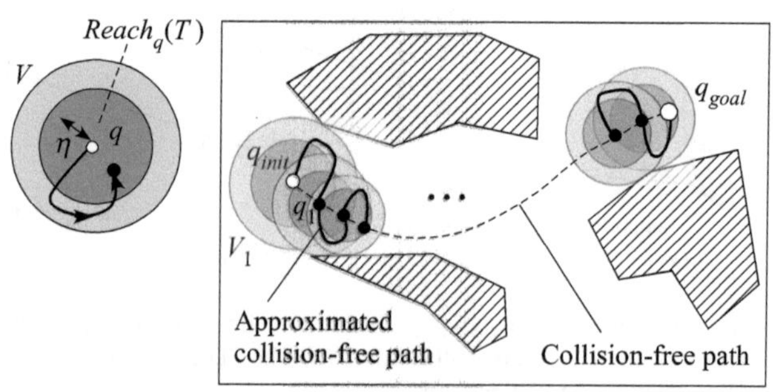

Figure 4.13 A system is small-time controllable from q if $Reach_q(T)$ contains a neighborhood of q for all neighborhoods V for any time $T > 0$.

has been demonstrated recently that they account for the natural human locomotion [2]. Our motion planning point of view then benefits from well experienced approaches in nonholonomic motion planning [4, 20, 22]. Among them we have chosen the probabilistic sampling approach with a steering method computing Reeds and Shepp curves [26], composed of arcs of a circle and straight line segments. Reeds and Shepp curves possess a geometric property accounting for small-time controllability, a critical property for planning method completeness.

By applying the general scheme composed of collision-free probabilistic motion planning, and path optimization in the first stage, we can obtain a path for the bounding volume as shown in Figure 4.14.

The manipulated object is placed near the wall and supposed to be displaced on the other side of an obstacle. As can be seen, the backward motion of the Reeds and Shepp curve is utilized appropriately to move the object away from the wall. Then the path switches to forward motion to reach the goal by avoiding the obstacle.

The collision-free path computed in the first stage should be converted into a sequence of collision-free pivoting sequences. The pivoting sequence generation is then based on two elementary operators: pivoting along a straight line segment and along an arc of a circle to follow Reeds and Shepp curves.

The computation of the pivoting sequence along a straight line segment is illustrated in Figure 4.4.2. Let D be the length of the straight line segment of the path to follow. As defined earlier, the length of the supporting edge is $2l$. Considering the constraint of the reachable area of robot arms, we introduce an angle β such that the robot is able to perform an elementary pivoting motion of total angle 2β.

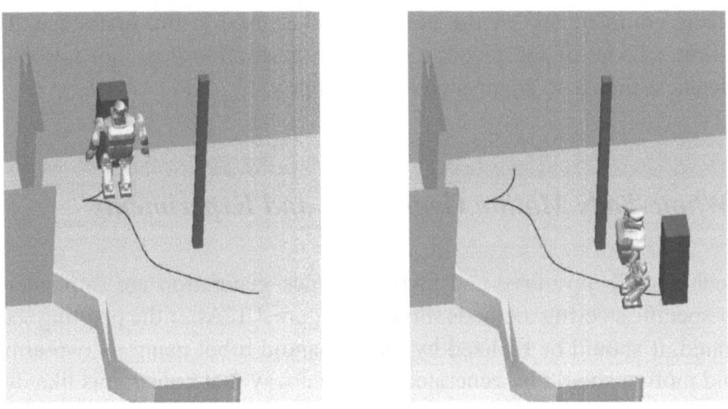

(a) (b)

Figure 4.14 Optimized collision-free path for a manipulated box object and the humanoid robot using Reeds and Shepp curves. The path allows the humanoid to move the object away from the wall starting from the initial state (a) by taking advantage of backward motion. Then the path switches to forward motion to avoid the obstacle and to move the object to the goal state (b)

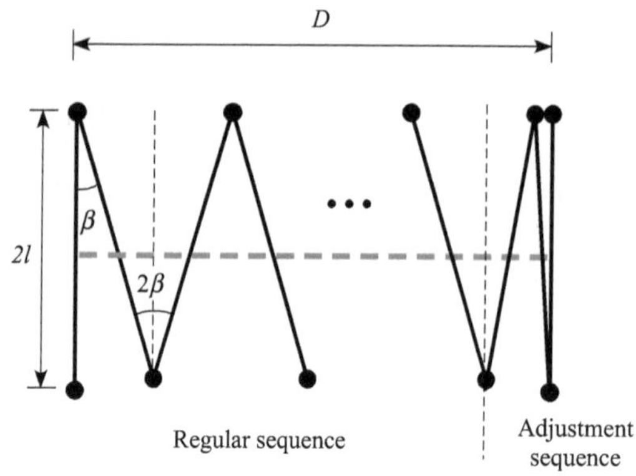

Figure 4.15 Transforming a straight line segment path into a pivoting sequence. The pivoting sequence is planned using rotation of the endpoints of the supporting edge. During the regular sequence, rotations of same angles are repeated before adjustment sequence that positions the line segment at the endpoint

After initializing the process by pivoting an angle β, we then apply N times the elementary pivoting motion of angle 2β, N being defined as the integer satisfying $D > 2Nl\sin\beta$. The same principle applies to the arcs of a circle.

We should notice that the rotation angle β may be tuned for obstacle avoidance. Indeed, the first stage of the algorithm provides a collision-free path that guarantees no collisions for the sliding supporting edge. As this rotation angle decreases, the final swept volume of the generated pivoting sequence converges to the one swept by the supporting edge when sliding along the Reeds and Shepp path. This property accounts for the small-time controllability of the pivoting system we have considered in Section 4.4.1. The 3D collision detection can be done by estimating the swept volume of the object attached to the supporting edge during the rotational motion.

As a consequence, the two-stage strategy we have developed follows from the probabilistic completeness of the motion planner used at the first stage. The approximation scheme of the pivoting sequence generation does not introduce any incompleteness thanks to small-time controllability.

4.4.3 Whole-body Motion Generation and Experiments

The Reeds and Shepp curve and pivot sequence generation are implemented as problem-specific steering methods shown in Figure 4.1. After the pivoting sequence is generated, it should be realized by the humanoid robot using its two arms. The humanoid motion should be generated in such a way that constraints like dynamic balancing and arm manipulation motion are satisfied at the same time. Moreover, stepping motion should be added in order to continue the manipulation when necessary.

For this purpose we adopt the dynamic whole-body motion generation of Section 4.3. Since all the joints are involved to make these complicated combined motions, we can expect better performance in the sense of reachable space than the functional decomposition utilized in [33].

The method is illustrated in Figure 4.16. The whole-body motion manager receives the desired hand trajectories and keeps track of the current robot configuration. Then it computes the trajectories or constraints, which are supplied to the generalized IK solver. The solver is also given trajectories of feet or CoM as well as their position and orientation constraints as prioritized tasks. With these inputs, the generalized IK solver computes the whole-body motion as joint angle trajectories.

When the pivoting rotation requires large horizontal displacement, a stepping motion is planned at the same time as the hand motion. The stepping foot is determined depending on the rotation direction. Then the new foot position is computed in such a way that the foot keeps its orientation parallel with the base Reeds and Shepp curves with an appropriate distance to the object. The same dynamic whole-body motion generator presented in Section 4.3 is applied to compute the coordinated arm and footstepping motions.

We have conducted experiments with the humanoid robot platform HRP-2 in the environment shown in Figure 4.14 for whole-body motion planning for pivoting of a box-shape object.

The execution time of the entire pivoting sequence is 281 s, which corresponds to 56200 command outputs at the rate of 5 ms for each of 30 joints. The computation time was 291.7 s on a PC with Intel Core2 Duo CPU at 2.13 GHz, which is comparable with the actual task execution time.

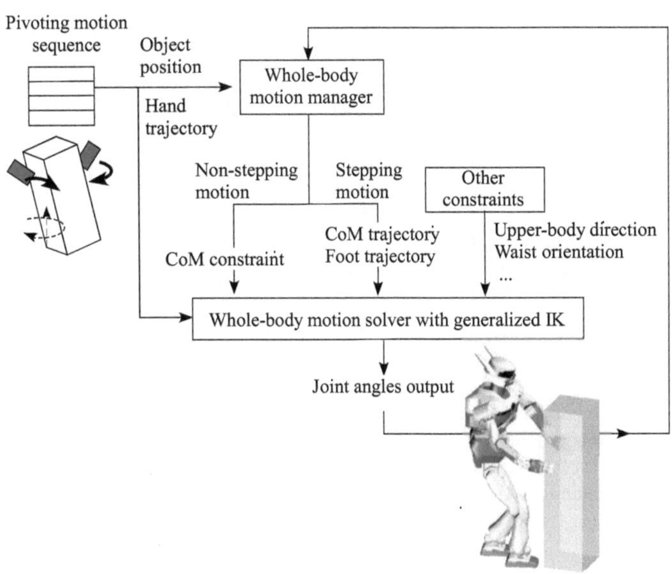

Figure 4.16 Use of generalized inverse kinematics for whole-body motion for pivoting-based manipulation

The experimental results are shown in Figure 4.17 to validate the proposed method. The motion has been planned offline with prior knowledge of the object and environment. The humanoid robot executes the complex pivoting manipulation with a coordinated whole-body motion including simultaneous manipulation and foot-stepping. As can be seen, the robot accomplished the long pivoting sequence.

The average error in the final position of the object carried was 0.46 m (0.31 m and 0.35 m short in x and y directions, respectively), and 5° in orientation θ. The error 0.46 m represents 9% of the length 5.1 m of the whole trajectory of the object carried. This confirms that the manipulation has been executed relatively accurately considering the lack of sensor feedback of object location during the manipulation.

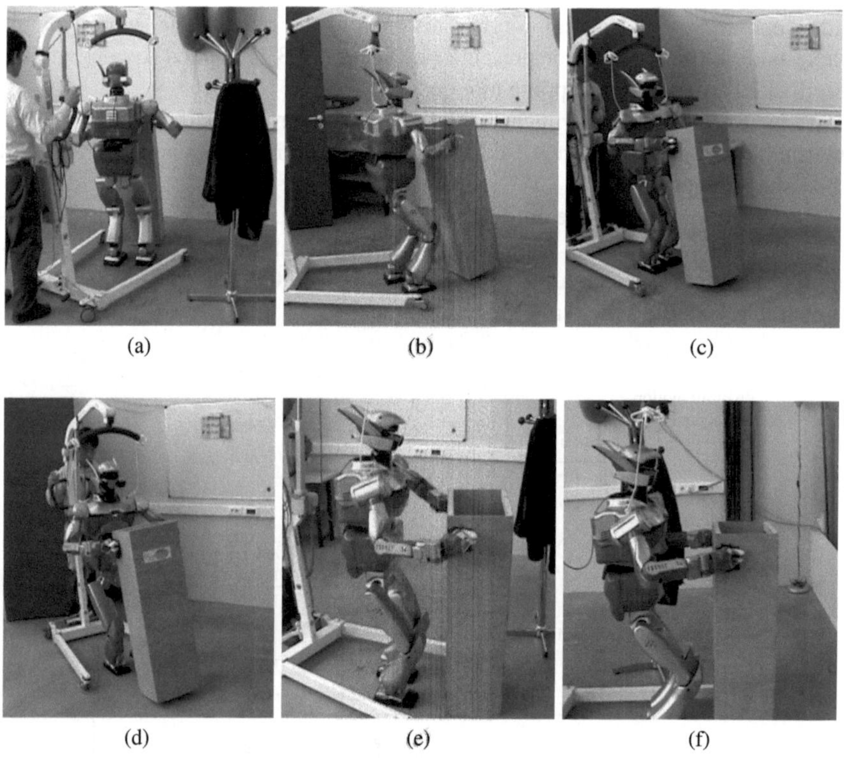

Figure 4.17 Experimental results. Starting from the initial position (a) with obstacle at right-hand side, the humanoid robot manipulates the object backwards away from the wall (b). After switching motion direction to forward (c), the robot continues to manipulate the object to the goal position by avoiding the obstacle (d)–(f)

4.4.4 Regrasp Planning

So far we have developed the whole-body manipulation method on the assumption that the robot can carry the object without changing the grasping points. However, when there are narrow spaces, the robot should sometimes release the object and hold it in another position according to the situation.

We provide a humanoid robot with more flexibility in whole-body pivoting manipulation by including regrasp planning. The robot releases the object when it cannot go further towards the goal position and grasps it again to continue manipulation.

The difficulty resides in finding narrow passages for the robot and object together and in combining the paths with different grasping positions to plan a manipulation motion to achieve the goal. We address the regrasp planning problem for pivoting manipulation through a roadmap-multiplexing approach [38].

Figure 4.18 shows on overview of the planning scheme. Several grasping positions are possible for a given object position. The roadmaps are built for the combined bounding volumes of both the robot and the object. We suppose that there are different grasping positions that allow the robot to hold the object. There are two types of roadmap: the first is the "manipulation roadmap" $\mathscr{G}^i_{manip}$ for the ith grasping position r^i_{grasp} expressed as the relative robot position with respect to the object. In this roadmap, the robot and the object move together. The second type of roadmap is the "regrasping roadmap" $\mathscr{G}_{reg}$ where the robot moves alone between different grasping positions.

As can be seen in the figure, manipulation roadmaps $\mathscr{G}^1_{manip}$ and $\mathscr{G}^2_{manip}$ for different grasping positions are interconnected via the regrasping roadmap $\mathscr{G}_{reg}$. For

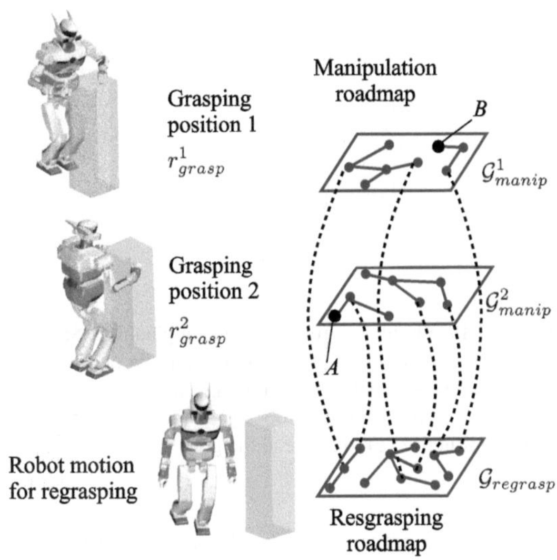

Figure 4.18 Roadmap multiplexing. Different manipulation roadmaps $\mathscr{G}^1_{manip}$ and $\mathscr{G}^1_{manip}$ are connected by way of the regrasping roadmap $\mathscr{G}_{reg}$

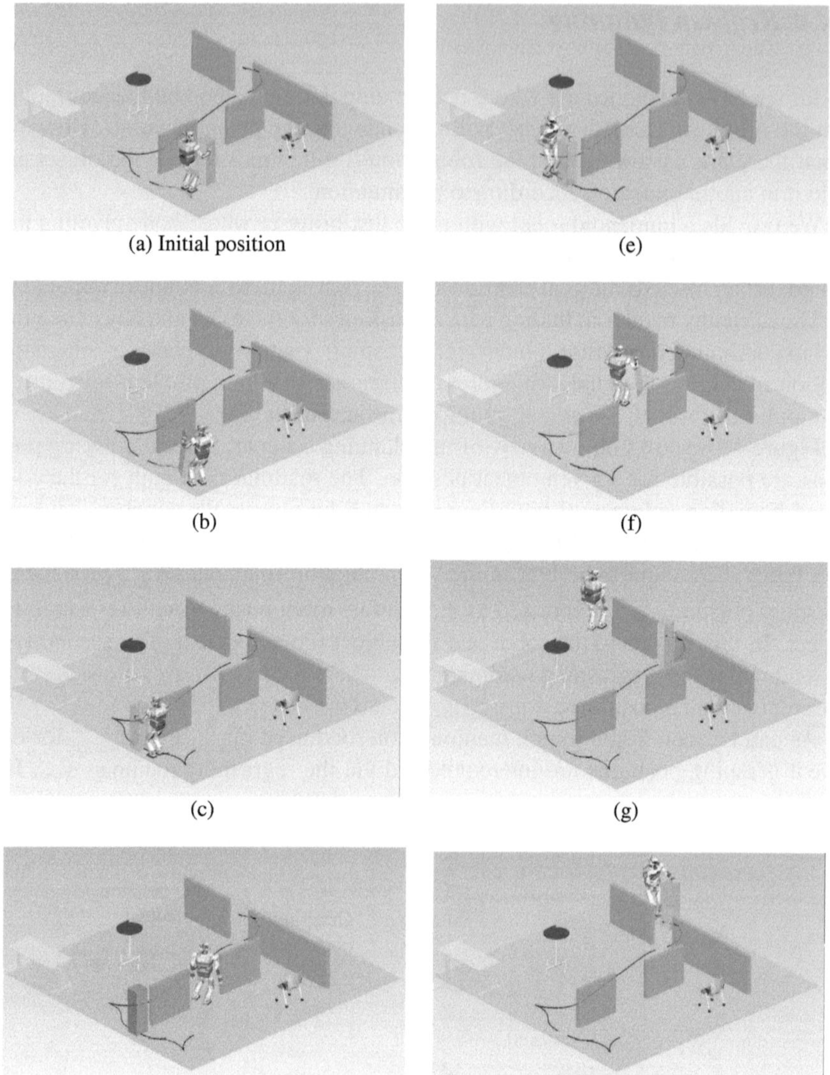
(a) Initial position (e)
(b) (f)
(c) (g)
(d) (h) Goal position

Figure 4.19 Simulation result of regrasp planning. Starting from initial position (a), the humanoid robot makes pivoting sequences (b) to put the object at the entry to the passage (c). It leaves the object and walks freely by combining forward and sideways walking (d) to regrasp the object on the other side (e). Then the robot goes towards another narrow passage (f) and performs another regrasp sequence (g) to arrive at the goal (h)

instance, the path from A to B is possible only by alternating the different grasping positions.

Figure 4.19 shows the result of regrasp planning. The humanoid robot HRP-2 is required to carry a box-share object from the initial position (Figure 4.19(a)) to its goal (Figure 4.19(h)). The humanoid robot places the object at the entry of a narrow passage (Figure 4.19(b),(c)). Then it releases the object and walk to the other side of the wall (Figure 4.19(d)). By combining backward and forward manipulation, the humanoid goes to another narrow passage (Figure 4.19(e),(f)). After another regrasping, the object is carried to the goal position (Figure 4.19(g),(h)).

4.5 Motion in Real World: Integrating with Perception

In this section the propsed motion planning methods are integrated with perception, principally vision, to make actions in the real world. This integration allows the robot to execute such commands as "go to the yellow table" and "take the orange ball."

4.5.1 Object Recognition and Localization

The HRP-2 robot is equipped with two pairs of firewire digital color cameras, configured as two independent stereo-vision camera pairs. We utilize standard state of the art components to implement a simple function of object recognition and localization.

For detection, models of the objects to be detected are previously learned using a 2D histogram in the $\{Hue, Saturation\}$ color space by taking a sample image with a color space.

Object detection is performed by *back projecting* the object histogram onto a video image. The back projection image is obtained by replacing each $\{H, S, V\}$ pixel value by the corresponding value in the histogram, leading to a probability image where each pixel value is the probability of that pixel belonging to the object model. The Continuously Adaptive Mean Shift *CamShift* algorithm then locates the object center and orientation in the back projection image.

A stereo-vision algorithm using pixel correlation is applied to the stereo image pairs, and produces a dense 3D image of the current scene. Even though pixel correlation is known to give poor results in indoor environments, the objects to localize are sufficiently textured so that precise enough 3D points can be obtained in the vicinity of the objects.

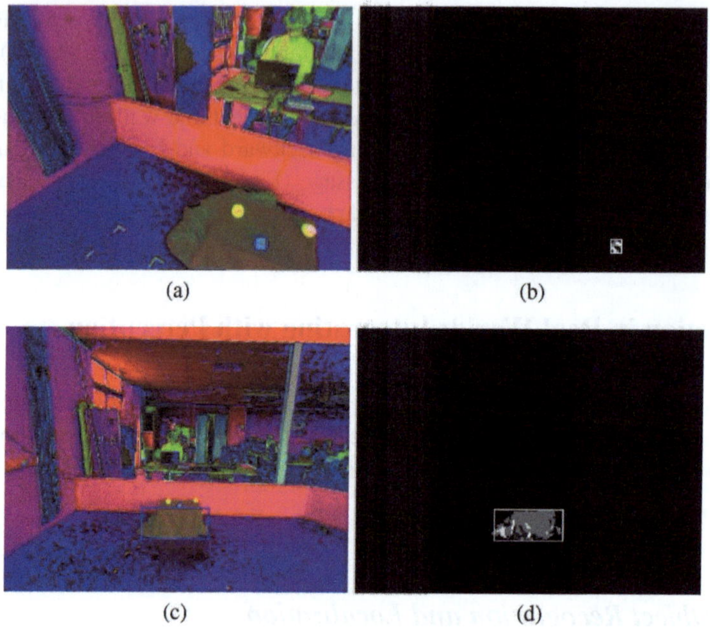

Figure 4.20 Object (a, b) and table (c, d) detection. Images show (a) and (c) the HSV image and images (b) and (d) are the back projection of the table color model in the source image. The rectangle is the result of execution of the CAMSHIFT algorithm on the back projection image.

4.5.2 Coupling the Motion Planner with Perception

The motion planners presented in previous sections are integrated with the vision system so that the robot can execute a task composed of navigation and object grasping.

For navigation, we apply the same type of two-stage motion planner for navigation planning presented in Section 4.2. At the first stage, a collision-free smooth locomotion path is calculated for the approximated bounding box. It is desirable for the robot to walk forward in order to look at the object and to take it. This preference can be modeled as a nonholonomic constraint in the same way as in Sections 4.2 and 4.4 and we can benefit from the well-developed planning method for a smooth path for a car-like robot [20]. Then the path is transformed into dynamic humanoid locomotion at the second stage by applying the dynamic walking pattern generator in the same way as in Section 4.2. This navigation planner allows the humanoid robot to go in front of the visually located colored table several meters away by avoiding known obstacles, as shown in Figure 4.21.

Finally, the whole-body motion generator presented in Section 4.3 is used for the grasping task. Given the object location from the vision system, the whole-body

Figure 4.21 A planned smooth walking trajectory to a target position

motion generator computes automatically a reaching motion, including stepping if necessary, depending on the detected object location.

All the software for perception, motion planning, dynamic motion generation, and controller is installed on the computers on board. In order to build the necessary software components, we used standard LAAS control architecture tools. In particular, we used the GenoM [7] tool that is able to generate robotics components. GenoM components can encapsulate C or C++ source code into an executable component that provides requests that can be invoked through simple scripts or through more complex supervision software. The components can also be dynamically interconnected together at run-time, providing a modular and programmable control architecture.

Figure 4.22 shows a subset of important components that have been defined for the experiment. All the components but the real-time control (OpenHRP [15]) runs on a Linux 1.8 GHz Pentium-M processor. The real-time part is operated by Art-Linux [1] on a similar hardware.

The vision processing chain is made up of three components: image acquisition (`camera`), stereo-vision by pixel correlation (`stereo`) and object detection and localization (`hueblob`). The two motion-related software components for navi-

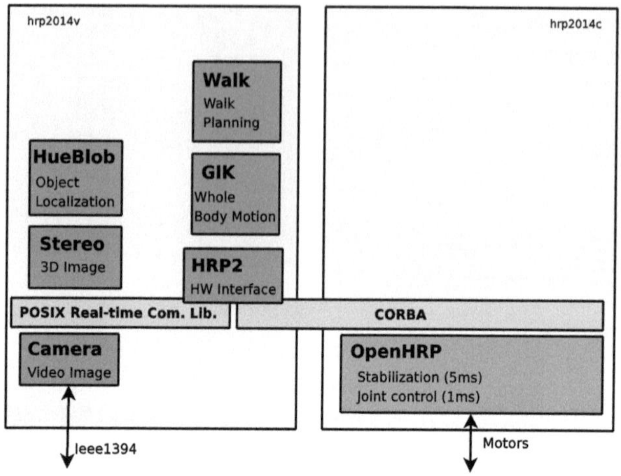

Figure 4.22 Software components running onboard the robot

gation and whole-body motion generation are implemented as components `walk` and `gik`, respectively. Finally, an interface component (`hrp2`) makes the connection with the OpenHRP software and bridges the CORBA communication bus of OpenHRP to the GenoM communication bus (Posix real-time communication library on Figure 4.22).

All these components define requests that can be invoked by a human operator, by supervision software or by a natural language processing system.

4.5.3 Experiments

We have conducted experiments to validate the integrated system. The humanoid robot is given a task to take a colored ball and put it at another place. The task is decomposed into several generic action commands, such as detection and localization of a learned object, locomotion to a location, and hand reaching to a position in 3D, with other simple tasks like turning on the spot and gripper opening and closing.

A simple supervision system that can invoke the actions with scripts is utilized to manage the robot behavior easily. Each action can report failures (*e.g.* failure in grasping an object). It is thus possible to implement error recovery strategies by analyzing the reports of the actions. In the following experiment, each action is associated with a vocal command to allow the user to give a sequence of commands to the robot in an interactive manner.

Figure 4.23 shows snapshots of experiments. Since the ball is too far away to be detected with the camera at the initial position, the humanoid robot first localizes the box on which the balls are placed (Fig 4.23(a)). The robot walks with a smooth trajectory to the front of the box (Fig 4.23(b)) and localizes precisely the colored

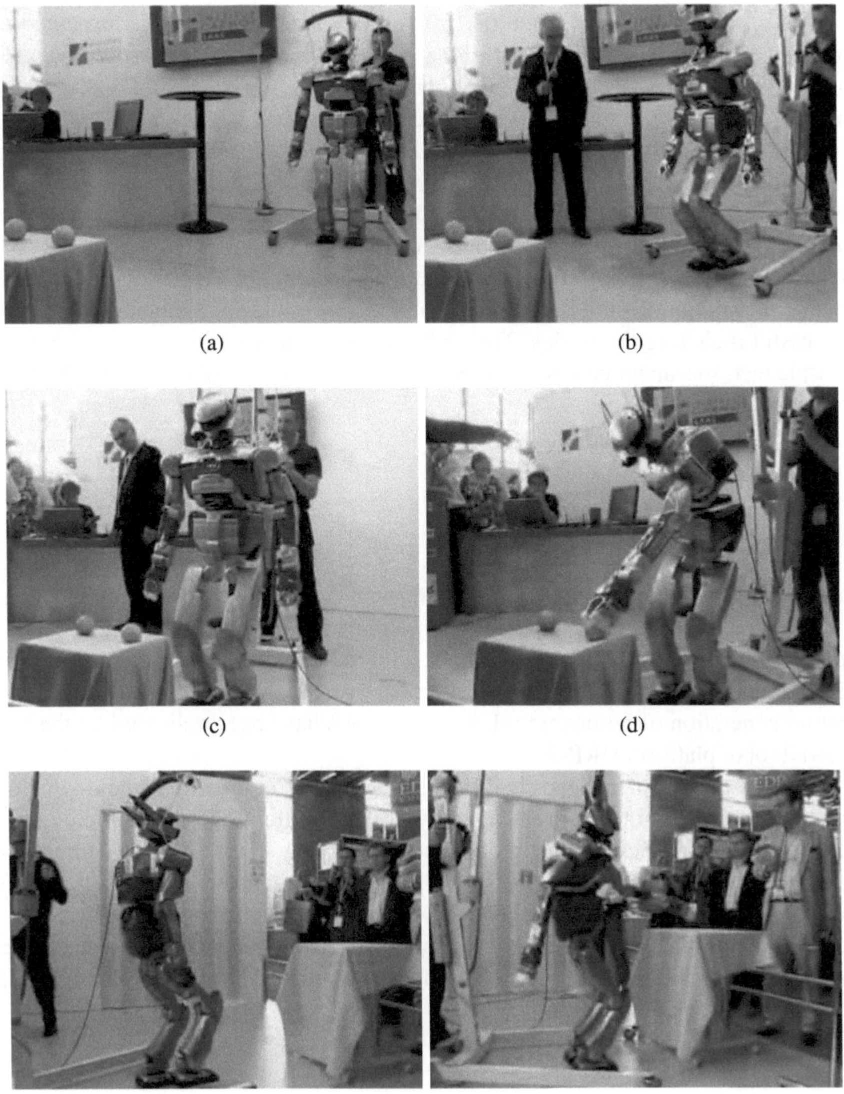

Figure 4.23 Ball-fetching task using visual perception and motion planner. (a) Localization of the box. (b) Robot walking to the box. (c) Detection and localization of the ball. (d) Whole-body reaching for grasping. (e) Locomotion to another location. (f) Robot putting the ball on the detected table

ball to grasp (Fig 4.23(c)). Then the whole-body reaching motion is executed to grasp the ball (Fig 4.23(d)). After turning, the robot is told to detect a colored table and walks towards it always with a smooth trajectory (Fig 4.23(e)). Finally it puts the ball on the table, again with whole-body motion (Fig 4.23(f)).

This experiment was conducted more than ten times in front of the public using vocal interaction by a human operator. Since the location of the robots and objects were different at every demonstration, it happened that the robot failed to grasp in the presence of unexpected disturbances or localization errors. However, the task was executed again successfully thanks to the generality of the action commands, by just repeating the same action command. As a result, all demonstrations were successful including the retries. This validates the reliability of the proposed motion planner, the integrated perception system and also the robustness of the task execution framework.

4.6 Conclusion

In this chapter, we have presented research results on whole-body motion planning from several aspects: collision-free locomotion with upper-body motion, whole-body reaching and manipulation. To cope with the redundancy and underactuation of humanoid robots, we have integrated probabilistic motion planning and dynamic motion generation of humanoid robots. The results have been validated by the humanoid robot platform HRP-2.

In future developments, we shall pursue our aim of improving the autonomy of humanoid robots by increasing the variety of their motions and thus enriching the possible behaviors. Besides this motion autonomy, we will also have to address reactivity in real environments. In the last part of this chapter we demonstrated an example of a closed perception–behavior loop. However, fast and precise environment recognition as well as a reactive motion planning scheme still needs to be investigated for the humanoid robot to adapt to more complex situations. We will address these problems in future work.

References

[1] ART linux. `http://sourceforge.net/projects/art-linux`

[2] Arechavaleta, G., Laumond, J.P., Hicheur, H., Berthoz, A.: An optimality principle governing human walking. IEEE Trans Robotics **24**(1), pp 5–14 (2008)

[3] Baerlocher, P., Boulic, R.: An inverse kinematics architecture enforcing and arbitrary number of strict priority levels. The Visual Computer **20**, pp 402–417 (2004)

[4] Choset, H., Lynch, K., Hutchinson, S., Kantor, G., Burgard, W., Kavraki, L., Thrun, S.: Principles of Robot Motion: Theory, Algorithms, and Implementation. MIT Press (2006)

[5] Dubins, L.E.: On curves of minimal length with a constraint on average curvature and prescribed initial and terminal positions and tangents. American J Mathematics **79**, pp 497–516 (1957)

[6] Esteves, C., Arechavaleta, G., Pettré, J., Laumond, J.P.: Animation planning for virtual characters cooperation. ACM Trans Graphics **25**(2), pp 319–339 (2006)
[7] Fleury, S., Herrb, M., Chatila, R.: Genom: a tool for the specification and the implementation of operating modules in a distributed robot architecture. In: proceedings of 1997 IEEE international conference on intelligent robots and systems, pp 842–849 (1997)
[8] Gleicher, M.: Comparing constraint-based motion editing method. Graphical Models **63**, 107–134 (2001)
[9] Harada, H., Kajita, S., Kanehiro, F., Fujiwara, K., Kaneko, K., Yokoi, K., Hirukawa, H.: Real-time planning of humanoid robot's gait for force controlled manipulation. In: proceedings of 2004 IEEE international conference on robotics and automation, pp 616–622 (2004)
[10] Harada, H., Kajita, S., Saito, H., Morisawa, M., Kanehiro, F., Fujiwara, K., Kaneko, K., Hirukawa, H.: A humanoid robot carrying a heavy object. In: proceedings of 2005 IEEE international conference on robotics and automation, pp 1712–1717 (2005)
[11] Hsu, D., Latombe, J.C., Sorkin, S.: Placing a robot manipulator amid obstacles for optimized execution. In: proceedings of 1999 international symposium on assembly and task planning, pp 280–285 (1999)
[12] Hwang, Y., Konno, A., Uchiyama, M.: Whole body cooperative tasks and static stability evaluations for a humanoid robot. In: proceedings of 2003 IEEE/RSJ international conference on intelligent robots and systems, pp 1901–1906 (2003)
[13] Kajita, S., Kanehiro, F., Kaneko, K., Fujiwara, K., Harada, K., Yokoi, K., Hirukawa, H.: Biped walking pattern generation by using preview control of zero-moment point. In: proceedings of 2003 IEEE international conference on robotics and automation, pp 1620–1626 (2003)
[14] Kajita, S., Kanehiro, F., Kaneko, K., Fujiwara, K., Harada, K., Yokoi, K., Hirukawa, H.: Resolved momentum control: Humanoid motion planning based on the linear and angular momentum. In: proceedings of 2993 IEEE/RSJ international conference on intelligent robots and systems, pp 1644–1650 (2003)
[15] Kanehiro, F., Hirukawa, H., Kajita, S.: OpenHRP: Open architecture humanoid robotics platform. Int J of Robotics Res **23**(2), pp 155–165 (2004)
[16] Kaneko, K., Kanehiro, F., Kajita, S., Hirukawa, H., Kawasaki, T., M. Hirata, K.A., Isozumi, T.: The humanoid robot HRP-2. In: proceedings of 2004 IEEE international conference on robotics and automation, pp 1083–1090 (2004)
[17] Kavraki, L., Svestka, P., Latombe, J.C., Overmars, M.: Probabilistic roadmaps for path planning in high-dimensional configuration spaces. IEEE Trans Robotics Automat **12**(4), pp 566–580 (1996)
[18] Kuffner, J.: Motion planning with dynamics. Physiqual (1998)
[19] Kuffner, J., Kagami, S., Nishiwaki, K., Inaba, M., , Inoue: Dynamically-stable motion planning for humanoid robots. Autonomous Robots **12**(1), pp 105–118 (2002)
[20] Laumond, J.P. (ed.): Robot Motion Planning and Control, *Lectures Notes in Control and Information Sciences*, vol. 229. Springer (1998)
[21] Laumond, J.P.: Kineo cam: a success story of motion planning algorithms. IEEE Robotics & Automation Magazine **13**(2), pp 90–93 (2006)
[22] LaValle, S.: Planning Algorithm. Cambridge University Press (2006)
[23] LaValle, S., Kuffner, J.: Rapidly-exploring random trees: Progress and prospects. In: K.M. Lynch, D. Rus (eds.) Algorithmic and Computational Robotics: New Directions, pp 293–308. A K Peters (2001)
[24] Mansard, N., Stasse, O., Chaumette, F., Yokoi, K.: Visually-guided grasping while walking on a humanoid robot. In: proceedings of 2007 IEEE international conference on robotics and automation, pp 3042–3047 (2007)
[25] Nakamura, Y.: Advanced Robotics: Redundancy and Optimization. Addison-Wesley Longman Publishing, Boston (1991)
[26] Reeds, J.A., Shepp, R.A.: Optimal paths for a car that goes both forwards and backwards. Pacific Journal of Mathematics **145**(2), pp 367–393 (1990)

[27] Sentis, L., Khatib, O.: Synthesis of whole-body behaviors through hierarchical control of behavioral primitives. Int J of Humanoid Robotics **2**(4), pp 505–518 (2005)

[28] Siciliano, B., Slotine, J.J.E.: A general framework for managing multiple tasks in highly redundant robotic systems. In: proceedings of IEEE international conference on advanced robotics, pp 1211–1216 (1991)

[29] Sugihara, T., Nakamura, Y., Inoue, H.: Realtime humanoid motion generation through zmp manipulation based on inverted pendulum control. In: proceedings of 2002 IEEE international conference on robotics and automation, pp 1404–1409 (2002)

[30] Sussmann, H.: Lie brackets, real analyticity and geometric control. In: R. Brockett, R. Millman, H. Sussmann (eds.) Differential Geometric Control Theory, *Progress in Mathematics*, vol. 27, pp 1–116. Michigan Technological University, Birkhauser (1982)

[31] Takubo, T., Inoue, K., Sakata, K., Mae, Y., Arai, T.: Mobile manipulation of humanoid robots – control method for com position with external force –. In: proceedings of 2004 IEEE/RSJ international conference on intelligent robots and systems, pp 1180–1185 (2004)

[32] Yamane, K., Nakamura, Y.: Natural motion animation through constraining and deconstraining at will. IEEE Trans Visualization and Computer Graphics **3**(9), 352–360 (2003)

[33] Yoshida, E., Blazevic, P., Hugel, V., Yokoi, K., Harada, K.: Pivoting a large object: whole-body manipulation by a humanoid robot. J of Applied Bionics and Biomechanics **3**(3), 227–235 (2006)

[34] Yoshida, E., Esteves, C., Belousov, I., Laumond, J.P., Sakaguchi, T., Yokoi, K.: Planning 3D collision-free dynamic robotic motion through iterative reshaping. IEEE Trans Robotics **24**(5), pp 1186–1198 (2008)

[35] Yoshida, E., Kanoun, O., Esteves, C., Laumond, J.P., Yokoi, K.: Task-driven support polygon reshaping for humanoids. In: proceedings of 2006 IEEE-RAS international conference on humanoid robots, pp 827–832 (2006)

[36] Yoshida, E., Mallet, A., Lamiraux, F., Kanoun, O., Stasse, O., Poirier, M., Dominey, P.F., Laumond, J.P., Yokoi, K.: "give me the purple ball" – he said to HRP-2 N.14. In: proceedings of 2006 IEEE-RAS international conference on humanoid robots (2007)

[37] Yoshida, E., Poirier, M., Laumond, J.P., Alami, R., Yokoi, K.: Pivoting based manipulation by humanoids: a controllability analysis. In: proceedings of 2007 IEEE/RSJ international conference on intelligent robots and systems, pp 1130–1135 (2007)

[38] Yoshida, E., Poirier, M., Laumond, J.P., Kanoun, O., Lamiraux, F., Alami, R., Yokoi, K.: Regrasp planning for pivoting manipulation by a humanoid robot. In: proceedings of 2009 IEEE international conference on robotics and automation, pp 2467–2472 (2009)

[39] Yoshida, H., Inoue, K., Arai, T., , Mae, Y.: Mobile manipulation of humanoid robots – a method of adjusting leg motion for improvement of arm's manipulability. In: proceedings of 2001 IEEE/ASME international conference on advanced intelligent mechatronics, pp 266–271 (2001)

[40] Yoshikawa, T.: Manipulability of robotic mechanisms. Int J Robotics Res **4**(2), 3–9 (1985)

Chapter 5
Efficient Motion and Grasp Planning for Humanoid Robots

Nikolaus Vahrenkamp, Tamim Asfour and Rüdiger Dillmann

Abstract The control system of a robot operating in human-centered environments should address the problem of motion planning to generate collision-free motions for grasping and manipulation tasks. To deal with the complexity of these tasks in such environments, different motion planning algorithms can be used. We present a motion planning framework for manipulation and grasping tasks consisting of different components for sampling-based motion planning, collision checking, integrated motion planning and inverse kinematics as well as for planning single arm grasping and dual-arm re-grasping tasks. We provide an evaluation of the presented methods on the humanoid robot ARMAR-III.

5.1 Introduction

Motion planning for humanoid robots with a high number of degrees of freedom (DoF) requires computationally efficient approaches to determine collision-free trajectories. The planning algorithms have to fulfill several requirements, such as low runtime, short path length or reasonable distance to obstacles. Since the motion planning problem is known to be PSPACE-hard [24], complete algorithms [9, 26] are time consuming and therefore less suitable for real-time tasks of highly redundant humanoid robots which operate in a human-centered environment. Over recent years efficient probabilistic, sampling-based approaches have been developed [19]. These approaches are probabilistically complete, which means that if the motion planning problem does not have a solution the algorithm will run forever. To overcome this problem, an implementation will usually stop the search after a specified time and will report that a solution does not exist. This drawback is compensated by the efficiency of probabilistic planners. Efficient and problem adapted imple-

Nikolaus Vahrenkamp, Tamim Asfour and Rüdiger Dillmann
Institute for Anthropomatics, Karlsruhe Institute of Technology (KIT), Adenauerring 2, 76131 Karlsruhe, Germany, e-mail: \{vahrenkamp, asfour, dillmann\}@kit.edu

mentations of probabilistic planning approaches can be applied to real-world robot systems and they are suitable for robots operating in a human-centered environment.

5.1.1 RRT-based Planning

Rapidly-exploring random trees (RRTs) belong to the category of sampling-based, randomized planning algorithms [16, 20]. They are widely used for single-query path planning because of their simplicity and efficiency as well as the possibility of involving differential constraints and many degrees of freedom. Several variations of the basic RRT algorithm for problem-adapted planners have arisen in the last few years due to the multifarious application range of RRTs [18]. The key advantages of the basic RRT construction, described in [16], are that the expansion of a RRT is biased toward unexplored state space and only few heuristics and parameters are needed. Since RRT-based algorithms are probabilistically complete, the coverage of C-space gets arbitrarily close to any point in the free C-space with increasing iterations.

In principle, the basic RRT can be used as a planner because of the fact that its vertices will eventually cover the whole collision-free configuration space C_{free}, coming arbitrarily close to any specified goal configuration c_{goal}. But such a basic planner would suffer from slow convergence and bad performance, so that problem-adapted improvements are reasonable.

5.1.2 The Motion Planning Framework

A motion planning framework should provide different planning algorithms for selected problems like navigation, approaching, grasping and manipulating. Planning with specialized algorithms allows use in real-world applications, e.g. for service or humanoid robots. These kinds of robots must be able to plan motions in reasonable time, so that human-robot interaction is not affected by long planning times. In the optimal case the user should never be aware of the internally running planning system.

The motion planning framework shown in Figure 5.1 is embedded in the robot control software and offers a pool of motion planning services which can be configured depending on the requested planning operation. The planning system selects and configures the specialized planning services, provided by the planning framework, in order to generate a collision-free trajectory that fulfills the needs of a higher level task planning. Therefore, the motion planning component communicates with inverse kinematics (IK) solvers, a collision checker and grasp planner that provides feasible grasps for manipulation tasks. On ARMAR-III, the motion planning component is based on Simox, a simulation and motion planning toolbox [29]. The

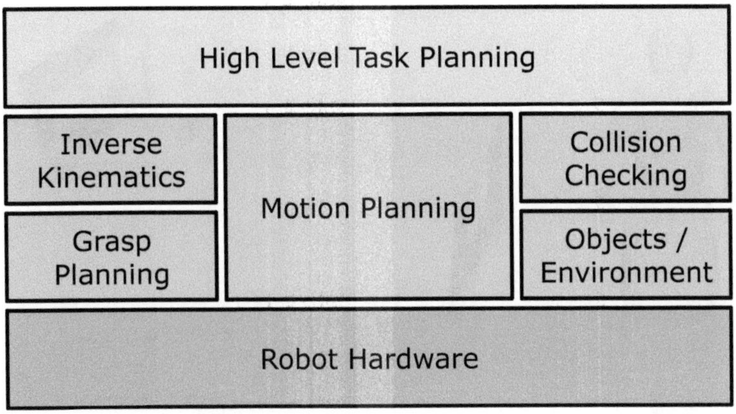

Figure 5.1 The motion planning framework is based on Simox, a simulation and motion planning toolbox

trajectories generated are delivered to the lower level control system, which is responsible for execution on the robot hardware.

5.2 Collision Checks and Distance Calculations

Motion planning approaches need collision and/or distance computation methods that operate on 3D models of the robot and the environment. Several libraries are available and can be used for collision detection [21, 14, 17]. In our framework [29] we are using the PQP library, which uses swept sphere volumes to test the collision status of 3D models [17], because of it's fast and robust implementation and the included distance computation routines.

Typically the collision checking of sampled configurations is the most time consuming task in RRT planners. A typical planning query requires thousands of collision checks. Thus a RRT-planner greatly benefits from speeding up the collision check routines [30]. To achieve a faster collision computation we are using simplified collision models of the robot and the environment in which it is operating (Figure 5.2). This reduces the number of triangles from 32,500 to 650 in the case of the robot model and from 24,000 to 4,200 in the case of the environment model. The reduction leads to an average collision query time of 0.20 ms and an average distance query time of 0.65 ms. Compared with 0.32 ms and 3.58 ms for a setup with full models of the robot and the environment we could achieve a speedup of 37.5% and 81.8%. These performance evaluations have been carried out for motion planning tasks of the humanoid robot ARMAR-III in a kitchen environment (see [3]) on a Linux-Pentium4 system with 3.2 GHz processor.

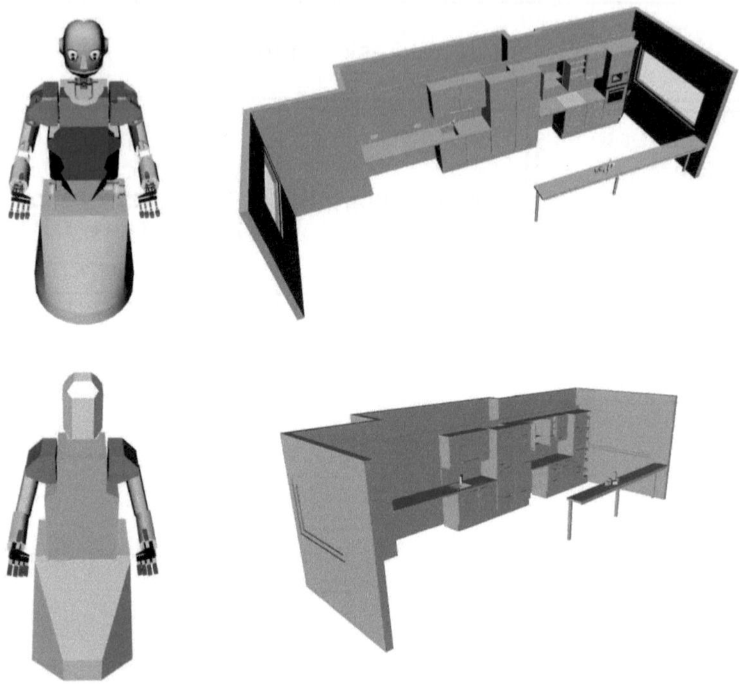

Figure 5.2 Full CAD models of the humanoid robot ARMAR-III and the kitchen environment (top row). The simplified models (bottom row) are used for collision checks and distance calculations

5.3 Weighted Sampling

The dimensions of the configuration space C differ in the effect of the robot system in workspace. Since each dimension of C describes a different effect in workspace, the dimensions have to be weighted in order to generate uniform sampling.

There are several ways of doing joint weighting. A simple technique is a manual and fixed weighting with $w_i > w_j$, if the center of the axis of joint i is further away from the tool center point of the robot manipulator than from the center of the axis of joint j. These distances, however, are different in each single configuration and depending on the actual robot state, they have more or less effect on the movement of the robot, so dynamically adapted joint weights would be a better choice than fixed weighting. But dynamic weight adaption increases the computation time since the distances between all joints have to be computed for every single configuration. Therefore, an upper bound for the workspace movements of each limb is used for an efficient and approximated uniform sampling.

A change ε_{trans} in a translational component of the C-space moves the robot in workspace by ε_{trans}. All other dimensions of the C-space have to be investigated explicitly to derive the upper bound of the robot's workspace movement. Table 5.1 gives an overview of the maximum displacement of a point on the robot's surface

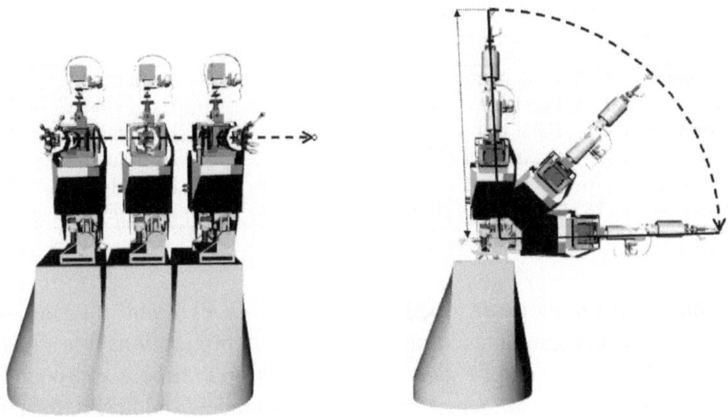

(a) (b)

Figure 5.3 (a) The humanoid robot ARMAR-III moving in a translational dimension. (b) The effect in workspace when changing the *C*-space value for the dimension associated with the torso pitch joint

when changing one unit in C. The effects of moving one unit in the different dimensions can be seen in Figure 5.3.

The different workspace effects are considered by using a weighting vector w whose elements are given by the values of the workspace movements from Table 5.1. In Equation 5.1 the maximum workspace movement $d_{WS}(c)$ of a C-space path $c = (c_0, ..., c_{n-1})$ is calculated:

$$d_{WS}(c) = \sum_{i=0}^{n-1} w_i c_i. \tag{5.1}$$

To sample a C-space path between two configurations c_1 and c_2, the vector v_{step} is calculated (Equation 5.2). For a C-space displacement of v_{step} it is guaranteed that the maximum workspace displacement is 1 mm.

$$v_{step}(c_1, c_2) = \frac{(c_2 - c_1)}{d_{WS}(c_2 - c_1)}. \tag{5.2}$$

The maximum workspace step size ε_{ws} can be specified in millimeters, which allows one to control the granularity of the planning algorithms in an easy way. The step size ε_{ws} is used to generate $n = \lceil \frac{d_{WS}}{\varepsilon_{ws}} \rceil - 1$ intermediate sampling configurations c_k on the path between two configurations c_1 and c_2:

$$c_k = c_1 + k\varepsilon_{ws} v_{step}(c_1, c_2), \;\; k = (1, .., n). \tag{5.3}$$

This sampling of the C-space path $(c_2 - c_1)$ guarantees that the workspace movements of the robot for two successive intermediate configurations is smaller than the upper limit of ε_{ws} mm.

Table 5.1 Worst case workspace movement for ARMAR-III

DoF	mm	DoF	mm	DoF	mm	DoF	mm
Platform x	1	Head pitch	300	Elbow	390	Hand index 1	70
Platform y	1	Head roll	300	Wrist 1	150	Hand index 2	70
Platform α	1,176	Head yaw	300	Wrist 2	150	Hand middle 1	70
Torso pitch	1,176	Shoulder 1	700	Wrist 3	150	Hand middle 2	70
Torso roll	1,176	Shoulder 2	700	Hand thumb 1	70	Hand ring	70
Torso yaw	1,176	Shoulder 3	390	Hand thumb 2	70	Hand pinkie	70

Since the collision status of a path is gained by checking all discrete samples for collisions, it is not guaranteed that intermediate configurations don't result in a collision. This guarantee can be given by Quinlan's *Free Bubble* approach [23] or an extension presented in [30] where enlarged robot models and lazy path validation approaches are used to efficiently validate complete path segments.

5.4 Planning Grasping Motions

Grasping an object is a typical and often needed operation for a service robot that operates in a human-centered environment. In some cases the task of grasping an object includes a suitable grasping pose, which implicitly defines the target configuration of the grasping motion, but in most cases there is no need for limiting the grasping to one specific grasp pose. In this section we want to show how the search for collision-free grasping configurations can be included in the planning process. In [7] a goal distance heuristic is used to implicitly compute reachable goal configurations during the planning process. Instead of using heuristic workspace goal functions, we are using predefined grasping poses in order to be able to use non-symmetric objects. In [5] a subset of feasible grasps is pre-calculated via a scoring function and the planning is done with the so determined target poses. The BiSpace approach, presented in [11], builds up two search trees, one in the C-space and one in the goal space and a global solution is searched by connecting the search spaces. The motion planning algorithms described in [6] work on continuous workspace goal regions instead of single target configurations. The JT-RRT approach, presented in [32], avoids the explicit search for IK solutions by directing the RRT extensions toward a 3D workspace goal position. Therefore the transposed Jacobian is used to generate C-space steps out of a workspace goal direction. The JT-RRT approach can be useful when no IK-solver is present for a robot system and only a grasping pose in workspace is known. Since there is no explicit C-space target configuration defined, the approach could not be implemented as a bi-directional RRT and the advantages of the Bi-RRT algorithms cannot be applied.

Our proposed algorithms unite the search for collision-free motions with the search for solutions of the IK problem to one planning scheme. The planners are initialized with a set of grasping poses which are used to calculate feasible target configurations. The computation of feasible grasping poses is done during the planning process and thus the planning is not limited to a potentially incomplete set of target configurations.

5.4.1 Predefined Grasps

If an object is to be grasped with an end effector of the robot, a collision-free trajectory has to be planned in order to bring the hand to a pose P_{grasp} which allows application of a feasible grasp. This grasping pose is defined with respect to the pose of the target object and could be derived by applying the grasp-specific transformation T_i.

For each object to be grasped or manipulated by the robot, a collection of feasible grasps is stored in a database. This collection of feasible grasps holds information about the transformations between the end effector and the final grasping position, the type of grasp, the preposition of the end effector and some grasp quality descriptions. These database entries can be generated automatically (e.g., with GraspIt! [22] or GraspStudio [29]) or, like in the following examples, by hand. A wok with 15 feasible grasps for each hand of the humanoid robot ARMAR-III can be seen in Figure 5.4(a).

To grasp an object o (located at position P_o) with the end effector e by applying the grasp g_k of the feasible grasp collection gc_o^e, the IK problem for the pose P_k^o has to be solved:

$$P_k^o = T_k^{-1} * P_o. \tag{5.4}$$

5.4.2 Randomized IK-solver

To grasp a fixed object with one hand, the IK-problem for one arm has to be solved. In case of the humanoid robot ARMAR-III, the operational workspace can be increased by additionally considering the three hip joints of the robot. This leads to a 10 DoF IK problem. Typically, an arm of a humanoid robot consists of six to eight DoF and is part of a more complex kinematic structure. If an analytical method exists for solving the IK problem for one arm, a randomized algorithm can be constructed which randomly samples the preceding joints and uses the analytical IK-solver for determining the final arm configuration. This probabilistic approach increases the operational workspace of the robot arm and is suitable for randomized planning algorithms.

For ARMAR-III we use a specialized analytic approach for solving the 7 DoF IK problem for one arm where all possible elbow positions are computed and, depending on the parameterization, the best one is chosen [1]. If there are multiple solutions, the behavior can be adjusted. Either the one with the lowest accumulated joint movement or a random solution out of the set of possible results is chosen.

In most cases it is desirable to consider the joints of the hip since the reachable workspace increases significantly when using additional degrees of freedom. In this case the three hip joints of ARMAR-III are randomly sampled until an IK query is successfully answered. If a configuration was found which brings the end effector to the desired pose, the IK solution has to be checked against self-collisions and collisions with obstacles in order to avoid invalid configurations. If the collision checker reports any collisions, the solution is rejected and the search is continued.

The approach is probabilistically complete, which means if time goes to infinity the algorithm will find a solution if at least one exists. To avoid endless runtime, the search for an IK solution is stopped after a specific number of tries and it is assumed that there is no valid result. Since this IK-solver is used within a probabilistic planning algorithm, this approach fits well in the planning concept.

5.4.2.1 Reachability Space

The use of a reachability space can speed up the randomized IK-solver. The reachability space represents the voxelized 6D-pose space where each voxel holds information about the probability that an IK query can be answered successfully [4, 11]. It can be used to quickly decide if a target pose is too far away from the reachable configurations and therefor if a (costly) IK-solver call makes sense.

The reachability space of the two arms of ARMAR-III is shown in Figure 5.4(b). Here the size of the 3D projections of the 6D voxels is proportional to the probability that an IK query within the extent of the voxel can be answered successfully. The reachability space is computed for each arm and the base system is linked to the corresponding shoulder.

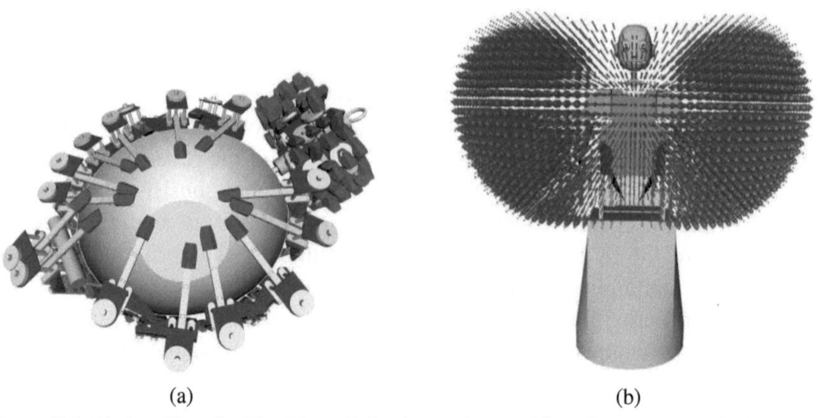

(a) (b)

Figure 5.4 (a) An object (wok) with predefined grasping positions for two arms of ARMAR-III (b) The 3D projection of the reachability spaces for both arms of ARMAR-III

The reachability spaces can be computed by solving a large number of IK requests and counting the number of successful queries for each voxel. Another way of generating the reachability space is to randomly sample the joint values while using the forward kinematics to determine the pose of the end effector and thus the corresponding 6D voxel [4]. An analytic approach generating a representation of the reachability is presented in [15].

Since the reachability space is linked to the shoulder, it moves when setting the three hip joints randomly in the search loop of the probabilistic IK-solver. For this reason, the target pose P_k, which is given in the global coordinate system, is transformed to the shoulder coordinate system and the corresponding voxel of the resulting pose P'_k is determined. The analytical IK-solver is only called if the entry in this voxel is greater than zero or a given threshold.

5.4.2.2 A 10 DoF IK-solver for Armar-III

The most convenient kinematic chain for reaching or grasping an object with ARMAR-III consists of the three hip joints followed by seven arm joints. This 10 DoF structure leads to a large reachable workspace and thus enables the robot to perform grasping and manipulation operations without moving the base platform.

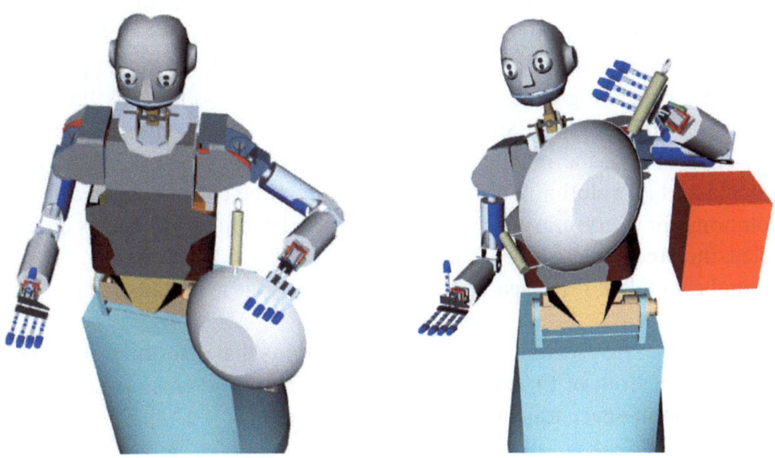

Figure 5.5 Solutions of the 10 DoF IK-solvers

To measure the performance of the 10 DoF IK-solver, the wok with 15 associated grasping poses is set to a random pose in front of the robot. Then the IK-solvers with and without reachability space are called in order to find a valid configuration for bringing the end effector to one of the 15 grasping poses. Two exemplary results of the IK-solver in an empty and a partly blocked scene are shown in Figure 5.5. The results shown in Table 5.2 are determined by building the averages of 100 IK queries with different object poses. [1] The average runtime and the number of calls of the analytical 7 DoF IK-solver are given for setups with/without reachability space and in scenes with/without obstacles. It turns out that the use of the reachability space speeds up the IK-solver enormously and it allows the use of these approaches in real-world applications.

Table 5.2 Performance of the 10 DoF IK-solvers

	Without obstacle		With obstacle	
	Avg. runtime	# IK calls	Avg. runtime	# IK calls
Without reach. space	1,404 ms	101.9	2,880 ms	217.3
With reach. space	60 ms	6.1	144 ms	13.6

5.4.3 RRT-based Planning of Grasping Motions with a Set of Grasps

In most cases the task of grasping an object should not be limited to one specific grasp. Instead of configuring the motion planner with one explicit target configuration, the framework can choose a set of grasp candidates and configure the motion planning services accordingly. It is possible to calculate an IK solution for each grasp and to use this set of configurations as targets for the planning process. This will lead to a planning scheme where the search for a solution is limited to the pre-calculated IK solutions. Since in general there are infinite numbers of solutions for the IK problem, the planner could fail although there is a valid motion for a not considered IK solution. Furthermore, it can be time consuming to calculate the IK solutions for every grasping pose in advance. If the feasible grasps are densely sampled, the precalculation has to be done for a lot of workspace poses. These problems can be avoided, if the search for valid IK solutions is included in the planning process.

The following two sections present two algorithms that determine an IK solution while planning. Both of these algorithms take as input the grasp set for the object and output a joint-space trajectory to reach the object.

5.4.3.1 J^+-RRT

The J^+-RRT planner can be used to search a collision-free grasping trajectory without the explicit implementation of an IK-solver. During RRT-buildup the planner tries to connect the existing tree to predefined 6D grasping poses defined in workspace. If a grasping pose is successfully connected to the RRT, a solution is

[1] These performance evaluations have been carried out on a DualCore system with 2.0 GHz.

found and the resulting trajectory is optimized with standard path pruning techniques [12].

Algorithm 5.1 J^+-RRT(q_{start},p_{obj},gc)

```
RRT.AddConfiguration(q_start);
while (!TimeOut()) do
    ExtendRandomly(RRT);
    if (rand() < p_ExtendToGoal) then
        Solution ← ExtendToGoal(RRT, p_obj, gc);
        if (Solution ≠ NULL) then
            return PrunePath(Solution);
    end
end
```

Algorithm 5.2 $ExtendToGoal(RRT, p_{obj}, gc)$

```
grasp ← GetRandomGrasp(gc);
p_target ← ComputeTargetPose(grasp);
q_near ← GetNearestNeighbor(RRT, p_target);
repeat
    p_near ← ForwardKinematics(q_near);
    Δ_p ← p_target − p_near;
    Δ_q ← J⁺(q_near) * LimitCartesianStepSize(Δ_p);
    q_near ← q_near + Δ_q;
    if (Collision(q_near) || !InJointLimits(q_near)) then
        return NULL;
    RRT.AddConfiguration(q_near);
until (Length(Δ_p) > Threshold_Cartesean) ;
return BuildSolutionPath(q_near);
```

In Algorithm 5.1 the main part of the J^+-RRT planner is shown. The planner is initialized with the starting configuration q_{start}, the workspace pose p_{obj} of the object and a set of feasible grasps ($gc = \{g_0, .., g_k\}$) defined relative to the object. The RRT algorithm is used to build up a tree of reachable and collision-free configurations. When a new configuration is added to the tree, the corresponding workspace pose of the hand is stored with the configuration data in order to use it later for calculating the approach movements. Approach movements are generated by the *ExtendToGoal* method which is called with some probability at each iteration of the planner. In Algorithm 5.2 a random grasp is chosen and, combined with the pose of the manipulation object, the workspace target pose p_{target} of the end effector is determined. The workspace metric described in Equation 5.8 is used to determine the nearest neighbor q_{near} out of all RRT-entries. q_{near} is marked, so that in the case of a failed extension loop, it will never be chosen twice as a nearest neighbor. Then the Jacobian $J(q_{near})$ is calculated and it's pseudoinverse $J^+(q_{near})$ is built by single value decomposition. The workspace pose difference Δ_p between the nearest node and p_{target} is used to calculate a delta in C-space which biases the extension toward the target. If the new configuration is collision-free and does not violate joint limits, the tree is extended by q_{new}. The workspace directed C-space extension is performed until the new configuration is not valid or p_{target} is reached, which means that a solution was found.

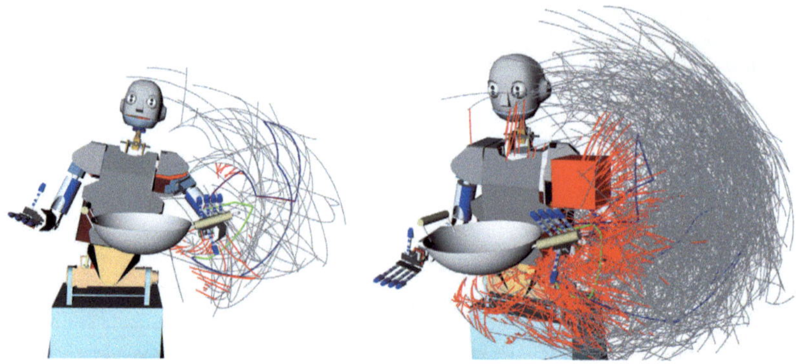

Figure 5.6 Results of the J^+-RRT algorithm. The red parts are generated by the *ExtendToGoal* method of the approach

In Figure 5.6 a resulting RRT of a J^+-RRT planner in an empty scene and in an environment with an obstacle is depicted. The resulting grasping trajectory and its optimized version are shown in blue and green. The optimized version was generated by standard path pruning techniques [12]. The red parts of the search tree have been generated by the *ExtendToGoal* part of the approach, where the pseudoinverse Jacobian is used to bias the extension to a grasping pose. The figure shows that the search is focused around the grasping object, but in most cases the joint limits and collisions between hand and wok prevent the generation of a valid solution trajectory. The performance of the J^+-RRT planner can be seen in Table 5.3. It turns out that the usability of the approach is limited in cluttered scenes because of the long runtime.

5.4.3.2 A Workspace Metric for the Nearest Neighbor Search

The metric used for determining the nearest workspace neighbor combines the Cartesian distance $d_{position}$ with the rotational distance $d_{orientation}$ of two poses:

$$d_{position}(p_0, p_1) = \sqrt{(x_1 - x_0)^2 + (y_1 - y_0)^2 + (z_1 - z_0)^2}. \tag{5.5}$$

To compute the orientational distance, the angle between the two orientations is computed. Since the orientational component of the pose is represented with quaternions, the angle between two quaternions q_0 and q_1 has to be computed. This can be done by computing q_{dif}, the quaternion representing the rotation between q_0 and q_1:

$$q_{dif}(q_0, q_1) = q_0^{-1} * q_1. \tag{5.6}$$

The rotational distance is set to the angle of q_{dif}, which can be retrieved using:

$$d_{orientation}(q_0, q_1) = 2 * acos(q_{dif}(q_0, q_1).w). \tag{5.7}$$

The final metric is the weighted sum of the Cartesian and the rotational distance. In our experiments we set the factor α to a value so that a difference in the orientation of 1 degree has the same effect as a translation of 3 mm.

$$d(p_0, p_1) = \alpha * d_{position} + (1 - \alpha) * d_{orientation}. \tag{5.8}$$

5.4.3.3 IK-RRT

To speed up the planning, an IK-solver can be used in order to generate goal configurations during the planning process. The planner is configured with a set of feasible grasping poses, which, combined with the pose of a manipulation object, defines a set of workspace target poses. These poses are used as input for the IK-solver calls. As shown in Algorithm 5.3 the IK-RRT algorithm is initialized by adding the start configuration to the first RRT while the second RRT is empty until an IK solution is found. During the planning loop, both trees are extended with standard RRT-based methods and attempts made to connect them via an intermediate configuration. With some probability, a random grasp from the set of feasible grasps is chosen and a call to the randomized IK-solver is performed. When a feasible IK configuration q_{IK} is found, it is added to the second tree and the new node is marked as a solution node.

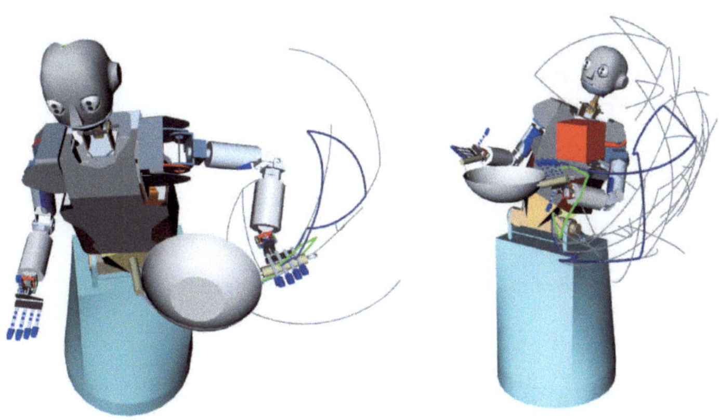

Figure 5.7 Results of the IK-RRT planner: the solution is marked blue, the optimized solution is shown in green

Since the IK search is probabilistically complete for the set of grasps and the RRT-Connect algorithm is known to be probabilistically complete [16], the IK-RRT approach is probabilistically complete. This means, that as time goes to infinity the algorithm will find a solution if at least one exists.

In Figure 5.7 a result of the IK-RRT approach is shown. The original and the optimized solution path are depicted in blue and green. Due to the bi-directional approach of the IK-RRT algorithm the search tree is much smaller compared with the results of the J^+-RRT approach (Figure 5.6).

A comparison of the average performance of the two proposed planners J^+-RRT and IK-RRT can be seen in Table 5.3. Both planners are queried 100 times with a random object position in front of the robot. The task was to find a collision-free trajectory to one of the 15 associated grasps in two scenes, one with and one without an obstacle. Furthermore both planners take care of self-collisions and collisions with the target object. The results in Table 5.3 point out that the IK-RRT approach is much faster than the J^+-RRT algorithm.

Algorithm 5.3 IK-RRT(q_{start},p_{obj},gc)

```
RRT1.AddConfiguration(q_start);
RRT2.Clear();
while (!TimeOut()) do
    ExtendRandomly(RRT1);
    ExtendRandomly(RRT2);
    if (#IKSolutions == 0 || rand() < p_IK) then
        grasp ← GetRandomGrasp(gc);
        p_target ← ComputeTargetPose(p_obj, grasp);
        q_IK ← ComputeIK(p_target);
        if (q_IK != NULL & !Collision(q_IK)) then
            RRT2.AddConfiguration(q_IK);
    end
    q_r ← GetRandomConfiguration();
    if (RRT1.Connect(q_r) & RRT2.Connect.(q_r)) then
        Solution ← BuildSolutionPath(q_r);
        return PrunePath(Solution);
    end
end
```

Table 5.3 The performance of the proposed planning approaches

	Without obstacle Average runtime	With obstacle Average runtime
J^+-RRT	2,032 ms	18,390 ms
IK-RRT	140 ms	480 ms

5.5 Dual Arm Motion Planning for Re-grasping

To plan a re-grasping motion with two arms, two problems have to be solved. The configuration for handing off the object from one hand to the other hand must be determined. This configuration must bring the object, which is grasped with one hand, to a position where the other hand can apply a feasible grasp. This search also includes choosing which grasp should be applied with the second hand. The configuration is only valid if there are no collisions between the arms, the environment, the object and the robot. Furthermore there must exist a collision-free trajectory which brings the arm with the attached object and the other arm to the re-grasping position.

5.5.1 Dual Arm IK-solver

If the robot is to re-grasp or hand-off an object, the search for a valid re-grasping configuration includes a collision-free object pose and a valid and collision-free IK-solution for both arms. This leads to a 23D IK problem, where the combination of the 6D object pose, three hip joints and 7 DoF for each arm results in a 23D solution vector.

To find a reachable object pose in the workspace of the robot, the 6D pose of the object and the configuration of the three hip joints can be sampled randomly until a call of the IK-solver is successful for one of the poses P_k^o. Therefore the Cartesian position of the object is limited to the extent of the reachable workspace and the orientation part does not have any restrictions.

5.5.2 Reachability Space

Since the computational costs of IK-solver calls could be high, the search for feasible object poses can be sped up by the use of reachability spaces. During the IK search loop, the analytic 7-DoF IK-solvers are called only, if the IK-probability of at least one left and at least one right grasping pose in the corresponding reachability space is above a threshold. If the IK-probability is below that threshold, the random generated hip configuration and object pose are discarded and a new sample is generated. If the the IK-probability is high enough it is likely that the costly IK-solver calls will succeed and that the pose is valid.

5.5.3 Gradient Descent in Reachability Space

For further speedup we propose a gradient descent approach which can be used to optimize the search for a graspable object pose. If an object pose was found, where the corresponding reachability space entry lies above a threshold, we apply a search for a local maximum. This is done by checking the neighbor voxels of the reachability space. If there is a voxel with a higher reachability space entry and the new pose is collision-free, the object 6D position is moved toward this voxel by the extent of the corresponding dimensions of a voxel. The new position then lies inside the voxel with the higher reachability entry. This is repeated until there are no neighbors with higher entries which means the position is at a local maximum of the discretized reachability distribution.

If a feasible object pose for re-grasping is needed, the gradient descent approach can be used with two reachability spaces by searching a local maximum for the left and the right grasping pose. Therefore the sum of both reachability space entries is optimized by moving the object pose until a local maximum is reached.

To avoid losing probabilistic completeness by applying the discretized reachability space and the gradient descent approach, these extensions to the original algorithm are only used with some probability during the search loop. Thus, the theoretical behavior of the IK-solvers remain untouched while the performance can be considerably increased. The resulting run time of the the dual arm IK-solvers

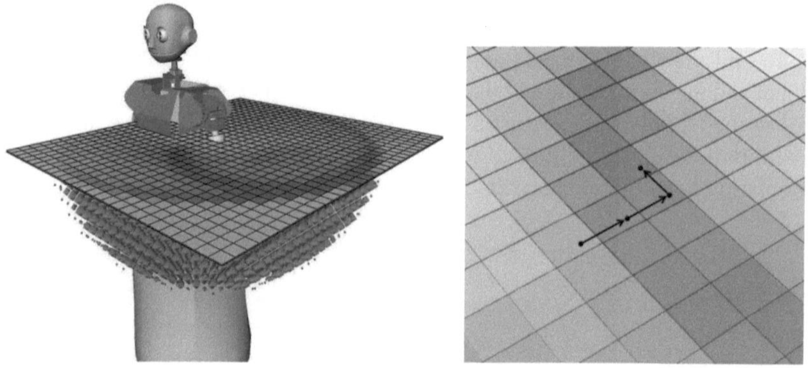

(a) (b)

Figure 5.8 (a) A 2D view of the reachability space of ARMAR-III (b) The 2D projection of a gradient descent optimization. The color intensity is proportional to the probability that a pose inside the voxel is reachable

are shown in Table 5.4. The IK-solver returns a valid object position and the corresponding joint configuration for the hip and both arms. In this configuration the object and the robot are in a collision-free state and a grasp can be applied for the left and the right hand (row 1). The second row shows the performance of the IK-solver when the object is already grasped with one hand.

Table 5.4 Performance of the dual arm IK-solvers

	Without obstacle		With obstacle	
	Average runtime	# IK calls	Average runtime	# IK calls
Flexible grasp selection	47 ms	3.3	161 ms	6.5
Object grasped with left hand	162 ms	3.2	220 ms	4.3

5.5.4 Dual Arm J^+-RRT

The dual arm J^+-RRT is an extension of the J^+-RRT approach presented in Section 5.4.3.1.

Instead of defining the target by a fixed workspace pose and a set of grasps, the object is attached to a hand and thus the target is implicitly defined by the set of grasps. These grasping poses lead to a set of transformations between both hands, defining all dual arm configurations for a re-grasping procedure. The *ExtendToGoal* part of the J^+-RRT approach (see Algorithm 5.1) has to be adapted for the dual arm algorithm. Instead of moving one arm toward a fixed goal pose, the two end effectors are moved toward each other in order to produce a configuration where the object can be grasped with both hands. The *DualArmExtendToGoal* part of the algorithm selects a random grasp and the configuration with the smallest distance between the two end effector poses and tries to move both arms toward a re-grasping pose. This is done by alternately moving the arms toward the corresponding goal poses in workspace. Thus the pseudoinverse Jacobians are calculated for every step and sample configurations are generated. These samples are tested for collision and violations of joint limits and added to the RRT. If a re-grasping pose can be reached a solution to the planning problem is found, otherwise the chosen RRT nodes are marked in order to exclude them for further goal extension steps.

Algorithm 5.4 $DualArmExtendToGoal(RRT, gc)$

```
grasp ← GetRandomGrasp(gc);
n ← GetNodeMinDistanceTCPs(RRT);
while (!Timeout()) do
    n ← MoveLeftArm(n, grasp);
    if (!n) then
        return NULL;
    n ← MoveRightArm(n, grasp);
    if (!n) then
        return NULL;
    if (HandOffPoseReached(n, grasp)) then
        return BuildSolutionPath(n);
end
```

Algorithm 5.5 MoveLeftArm$(n, grasp)$

$p_{left} \leftarrow TCPLeft(n);$
$p'_{left} \leftarrow TargetPoseLeft(grasp);$
$\Delta_p \leftarrow p'_{left} - p_{left};$
$\Delta_q \leftarrow J^+(q_{left}) * LimitCartesianStepSize(\Delta_p);$
$q_{left} \leftarrow q_{left} + \Delta_q;$
if $(Collision(q_{left}) \,||\, !InJointLimits(q_{left}))$ **then**
  return $NULL$;
return $BuildNewConfigurationLeft(n, q_{left})$;

5.5.5 Dual Arm IK-RRT

With the IK-solver methods of Section 5.5.1 it is possible to generate feasible configurations for a re-grasping procedure. The search for these configurations can be included in a RRT-based planner as described in Section 5.4.3.3. The dual arm IK-solver is used to generate IK solutions during the planning process. These IK solutions include a valid pose of the object with the corresponding joint configuration of hip and both arms for grasping the object with both hands. The Algorithm 5.3 has to be adapted slightly to include the dual arm IK-solver. Instead of a predefined object pose, the object is attached to the kinematic structure of one arm and thus the IK-solver just operates on the set of feasible grasps. The resulting dual arm IK-RRT planner can be used for building collision-free re-grasping trajectories in cluttered environments.

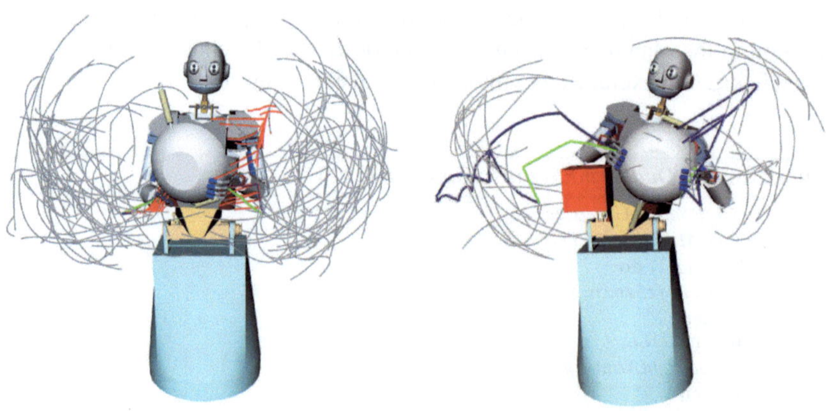

(a) (b)

Figure 5.9 The results of the dual arm J^+ and the dual arm IK-RRT planner. The solution is marked blue, the optimized solution is shown in green. (a) The re-grasping motion is planned with the dual arm J^+-RRT. The red parts are generated by the *ExtendToGoal* part of the algorithm. (b) Dual arm IK-RRT: the wok is grasped with the left hand and the collision-free solution trajectory results in a re-grasping configuration

The results of the dual arm re-grasp planners are shown in Table 5.5. The planners were queried 100 times and the average planning time was measured in scenes with and without an obstacle. In this evaluation the wok is grasped with the left hand and a re-grasping trajectory is searched. Again, the dual arm IK-RRT planner is much faster than the dual arm J^+ approach because of the possibility to take advantage of the bi-planning approach.

Table 5.5 Performance of the dual arm re-grasping planners

	Without obstacle Average runtime	With obstacle Average runtime
J^+-RRT	1,662 ms	5,192 ms
IK-RRT	278 ms	469 ms

5.5.6 Planning Hand-off Motions for Two Robots

The proposed algorithms can be used to generate collision-free re-grasping motions for two robots. Instead of considering two arms of one robot, two arms of two different robot systems can be used as input for the planning algorithms.

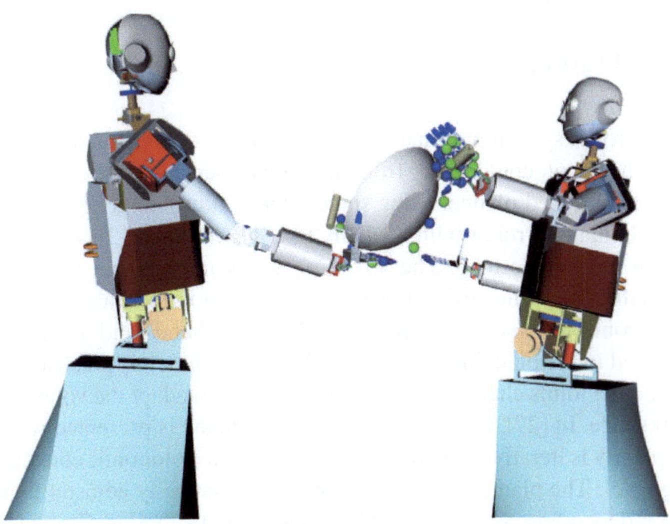

Figure 5.10 A hand-off configuration for two robots

A result of such a re-grasping motion can be seen in Figure 5.10. The performance of the two arm hand-off planning algorithms is similar to the one-robot case. From the algorithmic point of view the only difference between the one-robot and the two-robot problem are the additional hip joints of the second robot.

5.5.7 Experiment on ARMAR-III

In this experiment ARMAR-III is operating in a kitchen environment where the partly opened cabinet and a box are limiting the operational workspace of the robot. A planner for dual arm re-grasping is used to find a hand-off configuration and to plan a collision-free hand-off motion for both arms. The resulting trajectory moves both hands and the plate that is already grasped with the right hand, to the hand-off position and after re-grasping the arms are moved to a standard pose. This real-world experiment shows how the dual arm re-grasping planners enable the humanoid robot ARMAR-III to hand-off objects from one hand to the other.

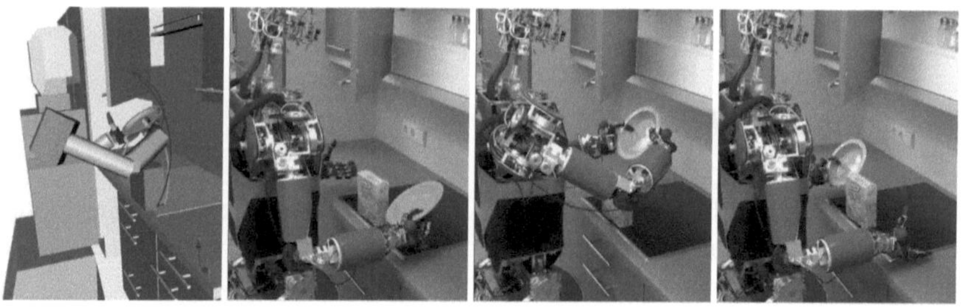

Figure 5.11 The humanoid robot ARMAR-III is re-grasping a plate in the kitchen

5.6 Adaptive Planning

Since a complex robot system has many degrees of freedom, a planner considering all the joints of the robot, could suffer from the high dimensionality of the configuration space. A RRT-based planner using standard techniques to generate a motion trajectory for a humanoid robot with 43 DoF, like ARMAR-III [2], is not able to find solutions in reasonable time. The 43D C-space is not suitable for searching collision-free paths, even when using large step sizes for approximation.

A humanoid robot has several subsystems, like arms, hands and a head, which should be involved in the planning process. In [28], an approach is presented where the number of active joints changes dynamically in order to adapt the volume of the reachable workspace. In [27], a multi-level planning scheme is presented where the planning complexity is iteratively increased by adding non-holonomic constraints at each planning level. The planner in [33] is able to automatically adjust 4 DoF of a humanoid robot depending on the detected environmental situation. This planning scheme strongly depends on the situation detecting, which can be difficult if many joints are used for planning. In [10], a multi-level planner is presented, which starts with a low C-space resolution and increases the sampling resolution to finer levels when a coarse solution is found.

The idea of changing the number of DoF during the planning process is picked up for the adaptive planning approach [31]. The proposed planner benefits from the partly reduced dimensionality of the configuration space since free space that has to be covered by the RRT is limited. The general planning approach can be used to build a planner that unites coarse and fine planning tasks, e.g., navigation planning usually considers a low-dimensional C-space for searching a collision-free trajectory for positioning the robot, but reaching or grasping tasks need a more complex selection of joints for planning. The adaptive planner for ARMAR-III combines the coarse search for positioning the platform with the finer levels of planning reaching motions for an arm and the dexterous motions for planning the movements of the five finger hand. As shown in the experiments, the planning times were noticeable decreased and the resulting planner is fast enough for use on a real robot platform.

5.6.1 Adaptively Changing the Complexity for Planning

To accelerate the planning process, we want to introduce an adaptive RRT-based planning scheme that is able to change the dimensionality of the C-space adaptively. This concept is implemented for unidirectional and bi-directional planners. To explain the algorithm, first the unidirectional method is described, the bi-directional planner is introduced in Section 5.6.4.

To configure the planner, the workspace is divided into *Planning Areas*, which represent an implicit knowledge of the planning problem. The definition in workspace allows it to easily define and adjust the structure of the these areas. For each area the corresponding set of joints, the *Planning Levels*, are defined.

5.6.2 A 3D Example

In Figure 5.12(a)–(e), a simple 3D example of the adaptive planning algorithm is shown. In this example the workspace is equal to the C-space, which means that a configuration $\{x, y, z\}$ leads to a similar workspace position. The effect of increasing and decreasing the *Planning Areas* and *Planning Levels* is visualized. The complete workspace except the red box represents the planning area 1 with the corresponding planning level 1, which includes the two C-space dimensions x and y. The red box represents planning area 2 with the planning level 2, covering the complete C-space.

In Figure 5.12(a) the search tree grows in planning level 1 until a new configuration falls in planning area 2 (Figure 5.12(b)). When a new configuration lies in planning area 2, the planning level is changed to planning level 2, which means that from now on the search tree is extended in all three C-space dimensions (Figure 5.12(c)). In Figure 5.12(d) the new configuration lies outside the current planning area and thus the planning level should be decreased. For this reason, a path to a configuration in the next lower planning level is searched and the planning goes on in planning level 1 (Figure 5.12(e)).

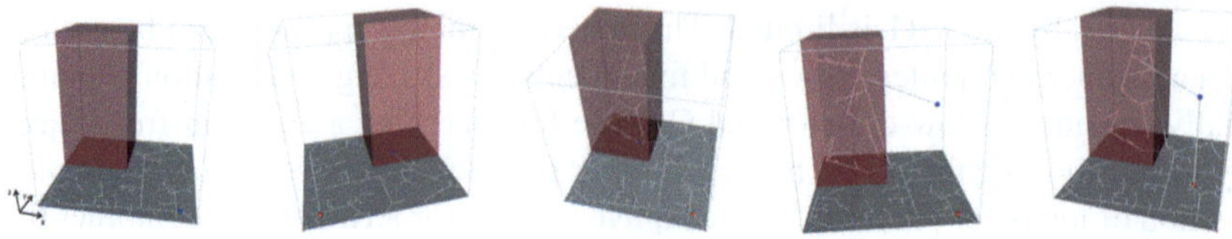

(a) (b) (c) (d) (e)

Figure 5.12 Adaptive planning: a 3D example

5.6.3 Adaptive Planning for ARMAR-III

To use the adaptive planning scheme for the humanoid robot ARMAR-III, the general approach is adjusted. The use of kinematic subsystems of ARMAR-III allows an easy and straightforward definition of planning areas with corresponding sets of joints.

5.6.3.1 Kinematic Subsystems

To divide the planning problem into smaller subsets, where the use of a RRT-based planning algorithm is more promising, we define different subsystems of the robot. These subsystems are robot specific and like the kinematic structure they have to be defined once for a system.

Table 5.6 Subsystems of ARMAR-III

Subsystem	Involved joints	# Joints
Platform	Translation x,y, rotation	3
Torso	Pitch, roll, yaw	3
Right arm	Shoulder 1,2,3, elbow	4
Right wrist	Wrist 1,2,3	3
Right hand	Thumb 1,2, index 1,2, middle 1,2, ring, pinkie	8
Left arm	Shoulder 1,2,3, elbow	4
Left wrist	Wrist 1,2,3	3
Left hand	Thumb 1,2, index 1,2, middle 1,2, ring, pinkie	8
Head neck	Tilt, pitch, roll, yaw	4
Head eyes	Eyes tilt, eye pan right, eye pan left	3

With these subsystems for the robot, we are able to reduce the complexity of the planning process by considering only the systems that are needed for a given planning task. The planning framework decides which systems are included in the planning process and configures the planning algorithms automatically. For example, the task of grasping an object out of the cupboard, may need the subsystems *Platform, Torso, Right Arm, Right Wrist* and *Right Hand* to get involved for planning. The chosen subsystems result in a 21D configuration space, which is used for searching a path in C_{free}. The joints that are not considered, remain in their standard pose and can be adopted in a post-processing step, e.g., the head can be adjusted to see the target.

5.6.3.2 The Approach

The planner starts with a low dimensional C-space in order to move the robot in the environment to a position that is near to the target object. For this positioning in the environment the detailed kinematic structures of the robot (e.g., the finger joints) are not considered, since they do not support the planning process in this rough planning step. If the planner has found a path in C-space that brings the robot near to the target object, or if the reduced planning failed, more joints are used to allow a more detailed planning. Which joints or subsystems are chosen to increase the complexity depends on the planning task. This planning scheme is performed until a solution is found or the full complexity of the robot is reached.

A parameter that directly affects the planning time is $d_{PlanningArea}$, the minimum workspace distance of the tool center point (TCP) to the target configuration. When the TCP distance falls below this value the planner changes to the next level and increases the number of involved subsystems. For this reason the TCP workspace distance to the goal configuration is calculated for each new configuration that is added to the RRT.

To avoid a manual definition of $d_{PlanningArea}$ for each planning level, the minimum TCP distance for the first level is set to the doubled maximum reaching distance (in case of AMRAR III, this is 1176 mm) and the following values are calculated by iteratively bisecting $d_{PlanningArea}$. The extent of the planning levels are shown in Figure 5.13 for the subsystems *Platform, Torso, Right Arm, Right Wrist* and *Right Hand.*

Table 5.7 shows the extent of the planning levels for five subsystems of ARMAR-III. The value for the *Right Hand* subsystem is set to zero, since the number of joints could not increase any more and the planner should go on until a global solution is found.

Table 5.7 Extent of the planning levels for ARMAR-III

Subsystem	$d_{PlanningArea}$ (mm)
Platform	2,352
Torso	1,176
Right arm	588
Right wrist	294
Right hand	0

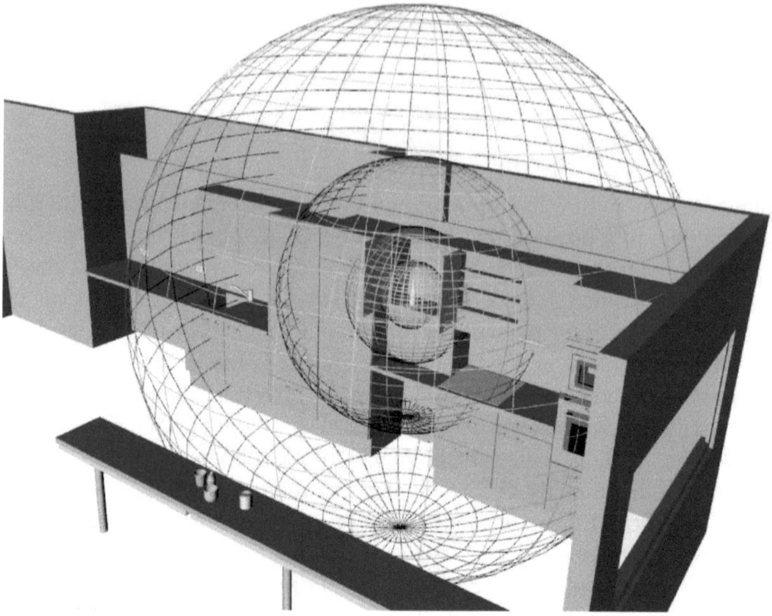

Figure 5.13 The extent of the planning levels around a target object for five subsystems of ARMAR-III

Algorithm 5.6 AdaptivePlanning(c_{start}, c_{goal})

$PlanningArea \leftarrow 1$;
$RRT.addConfig(c_{start})$;
while *(!TimeOut())* **do**
  $c_r \leftarrow RandomConfig(PlanningArea)$;
  $c_{nn} \leftarrow GetNearestNeighbor(RRT, c_r)$;
  if $(RRT.Connect(c_{nn}, c_r)$ && $!IsInPlanningArea(c_r, PlanningArea))$ **then**
    $PlanningArea \leftarrow ChangePlanningArea(c_r)$;
  if $(IsMaximalPlanningArea(PlanningArea)$ && $rand() < p_{goal})$ **then**
    $c_{nn} \leftarrow GetNearestNeighbor(RRT, c_{goal})$;
    if $(RRT.Connect(c_{nn}, c_{goal})$ **then**
      return BuildSolution();
  end
end

Algorithm 5.7 RandomConfig($PlanningArea$)

```
for (i ← 0; i < RRT.dimension; i ← i+1) do
    if (IsDimensionInPlanningArea(i, PlanningArea)) then
        c[i] ← RandomValue(DimBoundaryLo[i], DimBoundaryHi[i]);
    else
        c[i] ← StandardValue(i);
end
return c;
```

5.6.4 Extensions to Improve the Planning Performance

5.6.4.1 Randomly Extending Good Ranked Configurations

As described in [7] each node in the RRT can hold a ranking of its configuration, which can be used to support the planning. The ranking is calculated as the workspace distance of the TCP from the current to the goal configuration. To improve the planning performance, the planner sometimes chooses one of the last k best ranked nodes and does an extension step in an arbitrary direction. To avoid trapped situations, failures are counted and configurations with many failed extension steps are removed from the ranking.

5.6.4.2 Bi-planning

The planning algorithm should find a trajectory for a given start and goal configuration of the robot. In most cases the target configuration is more critical than the start configuration, since typical planning tasks will generate grasping or reaching trajectories. Hence the target configurations often result in low obstacle distances and thus in limited free space to operate. These situations are difficult for sampling-based planners, since only short paths in C_{free} can be found [13, 8]. To support the RRT-based planner, a bi-directional planning approach can be used, which builds up trees from the start and goal configuration and iteratively tries to connect them [16, 25].

The adaptive planner has to be adapted slightly to support the bi-directional search. The forward tree which starts the search from the start configuration is built as in the unidirectional case. The second tree which starts at the goal configuration needs some changes in the algorithm.

The planner starts with full resolution from the goal configuration. Each new configuration c_n that is added to the RRT, is checked whether $d_{tcp-Target}$, the TCP workspace distance of the TCP between the new and the goal configuration, is greater than the current extent of the planning level. In this case the planning level is decreased and further planning is done in a C-space with less dimensions.

With this extension of the adaptive planning algorithm, the bi-planner builds up two trees, one starting at the start configuration and one inverted tree starting at the goal configuration. The bi-planning loop generates random configurations and tries to connect them to both trees. If this connection succeeds for both trees, a global solution is found. If the connection fails, the trees are extended as long as possible until a collision occurs. This behavior supports the covering of the free configuration space and leads to a fast and robust planning algorithm.

5.6.4.3 Focusing the Search on the Area of Interest

By defining the planning levels, areas of interests with different extent are defined around the target object. The planning process can be accelerated by focusing the search on these areas. Since the global goal trajectory is unknown, the search should not be limited to one area, otherwise a solution can be missed. To achieve a focus on a target area, an adaption of the classical RRT-Connect and RRT-Extend algorithms is proposed. The standard extension of the C-space tree will connect a random configuration c_r to c_{nn}, the nearest neighbor of the existing tree. This behavior guarantees a uniform coverage of the C-space, which is a helpful property for a global planning problem, but in a locally bounded planning task the planning time will increase, since many areas of the C-space that are not important for planning are unnecessarily investigated. To emphasize a specific area in workspace an adaption of the *GoalZoom* [19] algorithm is used. Sometimes, instead of an arbitrary configuration c_r, a more promising configuration c_{zoom} next to the goal configuration c_{goal} is used to extend the tree. The maximum distance d_{zoom} between c_{goal} and the randomly chosen c_{zoom} depends on the current planning level. To avoid the introduction of another parameter, d_{zoom} is defined in workspace and set to the current planning extent value $d_{PlanningArea}$. This means that a random position c_{zoom}, used to bias the search toward the goal, has to meet the constraint that the maximum workspace distance d_{WS} is smaller than $d_{PlanningArea}$:

$$d_{WS}(c_{goal}, c_{zoom}) < d_{PlanningArea}. \tag{5.9}$$

In the bi-directional case the enhancement works as follows: the randomly chosen configuration, for which a connection to both trees is tested, sometimes is chosen surrounding the start or goal configuration, whereby the distance depends on the planning level.

5.6.5 Experiments

For evaluation, the simulation environment of the humanoid robot ARMAR-III is used. The robot model has 43 DoF and for each limb there are two 3D models, one for visualization and one simplified model for collision checking purposes. The robot is operating in a kitchen environment, which is also modeled with full and reduced resolution for visualization and collision checking as described in Section 5.2.

The starting position of the robot is located outside the kitchen and a trajectory for grasping an object is searched. In this experiment the target object is placed inside the fridge (Figure 5.14(a)). For this task, the planner uses the subsystems *Platform, Torso, Right arm, Right Wrist* and *Right Hand*. In our test setup the subsystem for the right hand consists of six instead of eight joints because the two middle and the two index finger joints are coupled and thus are counted as one DoF. The overall number of joints used for planning, and therefore the dimensionality of the C-space, is 19.

We make the assumption that a higher level task planning module has already calculated a goal position for grasping the object, thus c_{goal}, the target in C-space, is known.

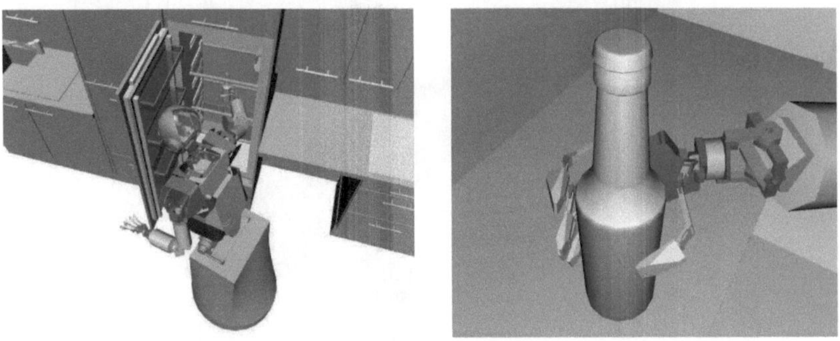

(a) (b)

Figure 5.14 Target position of the planning task: (a) distance view and (b) narrow view

The maximum workspace stepsize ε_{ws} was set to 30 mm for collision checking on path segments. Since ε_{ws} is a worst-case approximation, the real step sizes are significant lower. When adding a path segment to the RRT, intermediate nodes are generated in order to support the nearest neighbor search. These nodes are generated with a maximum workspace distance of 90 mm in all tested planners. [2]

5.6.5.1 Unidirectional Planning

A first evaluation of the planning problem described above, has been done by using a RRT-Connect one-way planner (A). The planner builds up a tree, covering the free C-space, and randomly tries to connect this tree to c_{goal}. The results point out that the planner often has difficulties escaping from local minima and finding a correct solution path to the goal configuration. These local minima may occur from situations when the right hand moves under the cupboard or when the position of

[2] These tests were carried out on an Intel Core 2 Duo Linux System with 2.16 GHz processor.

the finger joints is disadvantageous. In more than 60% of the test cases, planning was stopped because a time limit of 10 minutes or a limit on RRT nodes (100,000) was exceeded. If the planner can not find a solution within these limitations, it is not sufficient for use in a real-world scenario.

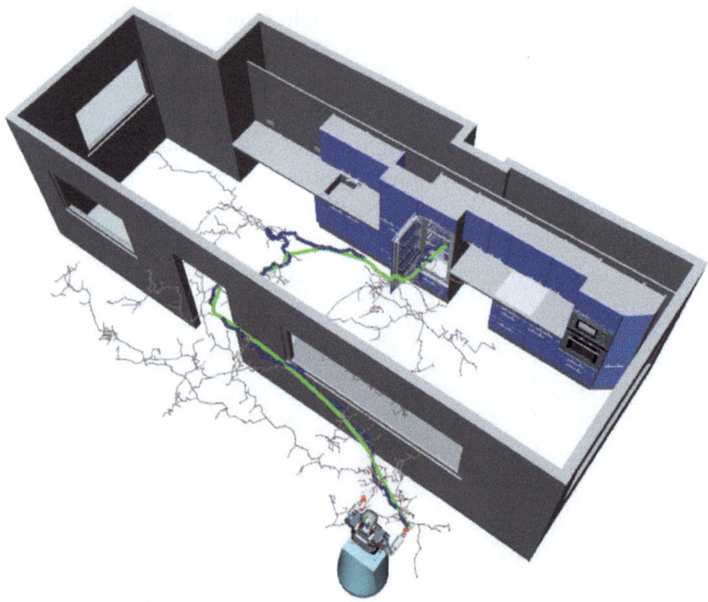

Figure 5.15 RRT bi-planning: the search tree with original (blue) and optimized (green) TCP paths

The adaptive planning (B) performs better than the RRT-Connect method. More test runs succeeded and the planning time can be reduced by over 60 %. Nevertheless a lot of planning cycles failed.

For further improvement of the adaptive planning, the *GoalZoom* enhancement and the random extend steps of Section 5.6.4 have been implemented and tested. The planner (C) benefits from these improvements and therefore the planning time and the number of failed planning runs can be decreased.

Table 5.8 Unidirectional planning

	Planning succeeded	Avg. planning time (success)	Avg. number of RRT nodes	Avg. number of collision checks
RRT-connect (A)	37.5 %	98.6 s	30,907	252,605
Adaptive planning (B)	43.5 %	31.3 s	10,089	83,017
Extended adapt. planning (C)	57.5 %	14.2 s	5,123	40,522

5.6.5.2 Bi-directional Planning

The RRT-Connect planner (A) often fails due to local minima in which the search frequently gets stuck. To support the planning, the bi-planning approach was implemented for the RRT-Connect algorithm (D). The planning succeeded in every test run and the average runtime was 3 s (Table 5.9 row D).

A planned RRT with original (blue) and optimized (green) solution paths is depicted in Figure 5.15.

Although the adaptive planner (B) achieves better results than the RRT-Connect planner (A), there are also settings in which the planning fails. The results of the bi-directional RRT-Connect planner (D) point out that planning can benefit a lot from building up two search trees. The adaptive bi-planner (E) was implemented and tested as described in Section 5.6.4. The adaptive reduction of the subsystems combined with a bi-planner results in an average planning time of 477 ms (Table 5.9 row E).

The *GoalZoom* enhancement, described in Section 5.6.4, which noticeable increased the planning performance for unidirectional planners, just decreases the planning time slightly for the adaptive bi-directional planner (F). As shown in Table 5.9 the average planning time decreases from 477 ms to 423 ms. Figure 5.16 shows a typical RRT with original and reduced solution paths for this configuration of the planning algorithm.

Table 5.9 Bi-directional planning

	Planning succeeded	Avg. planning time	Avg. number of RRT nodes	Avg. number of collision checks
RRT-connect (D)	100.0 %	3,037 ms	784	4,802
Adaptive planning (E)	100.0 %	477 ms	477	1,443
Extended adapt. planning (F)	100.0 %	423 ms	388	1,181

5.7 Conclusion

A service or humanoid robot can benefit from a planning framework that offers specialized approaches for motion planning. The concept presented here includes multiple planning algorithms, which can be selected and parametrized by the framework. The diversity of applications and use in a human-centered environment necessitates fast and robust algorithms for a wide range of planning problems. A generalized approach, covering all requested needs, is desirable, but with current techniques it is not realistic to achieve an implementation that is fast enough for real-world applications. For this reason we presented and evaluated several specialized algorithms that can be used within the planning framework and thus a fast system for a wide range of planning problems can be constructed.

The presented approaches can be used to include the search for inverse kinematics solutions in the planning process. For this reason a set of feasible grasps are stored with each manipulation object. The proposed J^+-RRT and IK-RRT planners

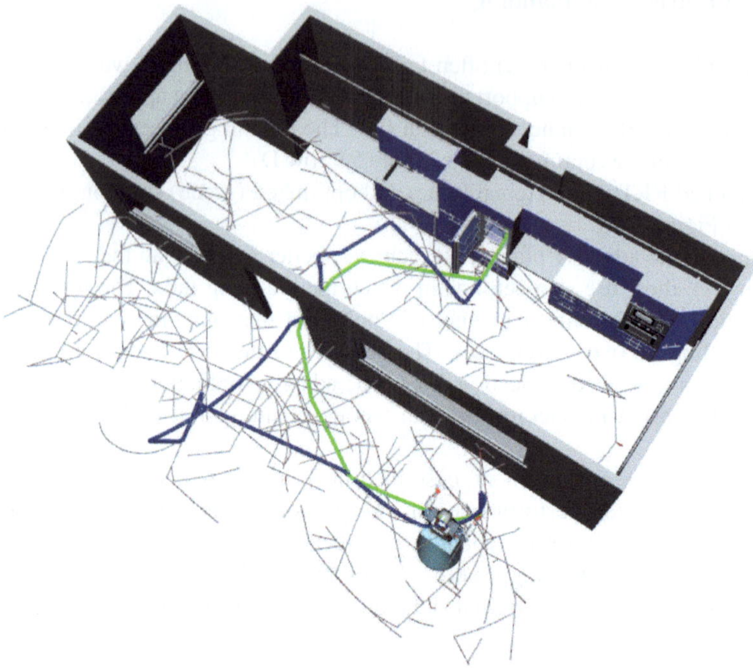

Figure 5.16 Adaptive bi-planning: the search tree with original (blue) and optimized (green) TCP paths

are not limited to one specific goal configuration and thus trajectories for the often needed task of grasping an object can be generated in a fast and robust manner. The dual arm extensions of the J^+-RRT and the IK-RRT algorithms offer a possibility for efficiently planning dual arm motions for grasping, re-grasping and hand-off procedures. As shown in the experiments, the algorithms can be applied for two arms of one or two humanoid robots. The performance evaluations pointed out the possibility of using these planners on real robot platforms.

The adaptive planning approach can be used to efficiently unite different planning problems like navigation, reach and grasp planning. The algorithms have been evaluated with a common planning problem for a humanoid robot. As test case a grasping task in a kitchen environment was chosen where the planners had to find trajectories for 19 DoF of the robot ARMAR-III. As a reference a standard RRT planner was used for which the poor performance and the high number of unsuccessful planning cycles have been overcome by using a bi-planning approach. The results of the adaptive planner point out that the planning time can be noticeabley decreased if the planning task and the subsystems of the robot used are known. The use of several extensions combined with the adaptive approach leads to a planner that is able to find solutions for a 19 DoF grasping task in about half a second on average. The planning performance is sufficient for real-world applications and use on a hardware platform.

Acknowledgments The work described in this paper was partially conducted within the the German Humanoid Research project SFB588 funded by the German Research Foundation (DFG: Deutsche Forschungsgemeinschaft) and the EU Cognitive Systems projects GRASP (FP7-215821).

References

[1] Asfour T, Dillmann R (2003) Human-like motion of a humanoid robot arm based on a closed-form solution of the inverse kinematics problem. In: IEEE/RSJ international conference on intelligent robots and systems (IROS), Las Vegas, USA, pp 407–1412
[2] Asfour T, Regenstein K, Azad P, Schröder J, Bierbaum A, Vahrenkamp N, Dillmann R (2006) Armar-III: an integrated humanoid platform for sensory-motor control. In: IEEE-RAS international conference on humanoid robots (Humanoids06), Genova, Italy, pp 169–175
[3] Asfour T, Azad P, Vahrenkamp N, Regenstein K, Bierbaum A, Welke K, Schröder J, Dillmann R (2008) Toward humanoid manipulation in human-centred environments. Robotics and Autonomous Systems, vol 56(1), pp 54–65
[4] Badler NI, Phillips CB, Webber BL (1993) Simulating humans: computer graphics animation and control. Oxford University Press, New York
[5] Berenson D, Diankov R, Nishiwaki K, Kagami S, Kuffner J (2007) Grasp planning in complex scenes. In: IEEE-RAS international conference on humanoid robots (Humanoids07), Pittsburgh, USA, pp 42–47
[6] Berenson D, Srinivasa S, Ferguson D, Collet A, Kuffner J (2009) Manipulation planning with workspace goal regions. In: IEEE international conference on robotics and automation (ICRA), Kobe, Japan, pp 618–624
[7] Bertram D, Kuffner J, Dillmann R, Asfour T (2006) An integrated approach to inverse kinematics and path planning for redundant manipulators. In: IEEE international conference on robotics and automation (ICRA), Orlando, USA, pp 1874–1879
[8] Boor V, Overmars M, Stappen A (1999) The Gaussian sampling strategy for probabilistic roadmap planners. In: IEEE international conference on robotics and automation (ICRA), Detroit, USA, pp 1018–1023
[9] Canny JF (1988) The complexity of robot motion planning. MIT Press, Cambridge, MA, USA
[10] Chen PC, Hwang YK (1998) SANDROS: a dynamic search graph algorithm for motion planning. IEEE Transactions on Robotics & Automation, vol 14(3), pp 390–403
[11] Diankov R, Ratliff N, Ferguson D, Srinivasa S, Kuffner J (2008) Bispace planning: concurrent multi-space exploration. In: international conference on Robotics: Science and Systems (RSS), Zurich, Switzerland
[12] Geraerts R, Overmars MH (2005) On improving the clearance for robots in high-dimensional configuration spaces. In: IEEE/RSJ international conference on intelligent robots and systems (IROS), Alberta, Canada, pp 4074–4079
[13] Hsu D, Latombe JC, Motwani R (1997) Path planning in expansive configuration spaces. In: IEEE international conference on robotics and automation (ICRA), Albuquerque, USA, vol 3, pp 2719–2726
[14] Jim A, Ninez P, Thomas F, Torras C (2001) 3D collision detection: a survey. In: Computers and Graphics, pp 269–285

[15] Kee D, Karwowski W (2002) Analytically derived three-dimensional reach volumes based on multijoint movements. In: Human factors: the journal of the human factors and ergonomics society, pp 530–544
[16] Kuffner J, LaValle S (2000) Rrt-connect: an efficient approach to single-query path planning. In: IEEE international conference on robotics and automation (ICRA), San Francisco, USA, pp 995–1001
[17] Larsen E, Gottschalk S, Lin MC, Manocha D (2000) Fast proximity queries with swept sphere volumes. In: Technical report, Department of Computer Science, University of North Carolina
[18] LaValle S, Kuffner J (2000) Rapidly-exploring random trees: progress and prospects. In: Workshop on the Algorithmic Foundations of Robotics
[19] LaValle SM (2006) Planning algorithms. Cambridge University Press, Cambridge, UK, available at http://planning.cs.uiuc.edu/
[20] LaValle SM, Kuffner JJ (2001) Randomized kinodynamic planning. In: Int J Robotics Res, vol 20(5), pp 378–400
[21] Lin M, Gottschalk S (1998) Collision detection between geometric models: a survey. In: IMA conference on mathematics of surfaces
[22] Miller AT (2001) Graspit!: a versatile simulator for robotic grasping. PhD thesis, Department of Computer Science, Columbia University
[23] Quinlan S (1994) Real-time modification of collision-free paths. PhD thesis, Stanford University
[24] Reif JH (1979) Complexity of the mover's problem and generalizations. In: SFCS '79: Proceedings of the 20th annual symposium on foundations of computer science (sfcs 1979), IEEE Computer Society, Washington, DC, USA, pp 421–427
[25] Sanchez G, Latombe J (2001) A single-query bi-directional probabilistic roadmap planner with lazy collision checking. In: International symposium on robotics research, Lorne, Victoria, Australia
[26] Schwartz JT, Sharir M (1983) On the piano movers problem: coordinating the motion of several independent bodies. In: Int J Robotics Res, pp 97–140
[27] Sekhavat S, Laumond J, Overmars MH (1998) Multilevel path planning for nonholonomic robots using semiholonomic subsystems. In: Int J Robotics Res, vol 17, pp 840–857
[28] Sian NE, Yokoi K, Kajita S, Tanie K (2004) A framework for remote execution of whole body motions for humanoid robots. In: IEEE-RAS international conference on humanoid robots (Humanoids04), Los Angeles, USA, vol 2, pp 608–626
[29] Vahrenkamp N, Asfour T, Dillmann R (2010) Simox: a simulation and motion planning toolbox. In: Technical report, Institute for Anthropomatics, Karlsruhe Institute of Technology (KIT), Karlsruhe, Germany
[30] Vahrenkamp N, Asfour T, Dillmann R (2007) Efficient motion planning for humanoid robots using lazy collision checking and enlarged robot models. In: IEEE/RSJ international conference on intelligent robots and systems (IROS), San Diego, USA, pp 3062–3067
[31] Vahrenkamp N, Scheurer C, Asfour T, Kuffner JJ, Dillmann R (2008) Adaptive motion planning for humanoid robots. In: IEEE/RSJ international conference on intelligent robots and systems (IROS), Nice, France, pp 2127–2132
[32] Weghe MV, Ferguson D, Srinivasa S (2007) Randomized path planning for redundant manipulators without inverse kinematics. In: IEEE-RAS international conference on humanoid robots (Humanoids07), Pittsburgh, USA, pp 477–482
[33] Yoshida E (2005) Humanoid motion planning using multi-level DoF exploitation based on randomized method. In: IEEE/RSJ international conference on intelligent robots and systems (IROS), Edmonton, Canada, pp 3378–3383

Chapter 6
Multi-contact Acyclic Motion Planning and Experiments on HRP-2 Humanoid

Adrien Escande and Abderrahmane Kheddar

Abstract We investigate the problem of planning sequences of contacts that support acyclic motion for a humanoid robot on uneven floors or in highly constrained environments. We propose a method that combines two basic techniques for motion generation: planning and optimization. The planning part is guided by a potential field function that grows a tree whose nodes are sets of possible contacts; the optimization part consists in a posture generator that checks the feasibility of contacts-set candidates under constraints such as static stability, joint limits, non-desirable collisions, etc. The difference between successive nodes in the tree is exactly one contact (creation or removal). Our planner allows motion-supporting contacts to be made with any parts of each robot's links on any part of the environment: it can be seen as the generalization of motion in which walking is a particular case. Therefore, the meaning of obstacles is different from the classical usage in planning: any part of the environment is at the same time a potential support (for some parts of the robot) and an obstacle (for the remaining parts). We demonstrated our planner in various scenarios involving the HRP-2 humanoid robot and conducted several experiments.

6.1 Introduction

Humanoids are anthropomorphic robotic systems. Their particular design raises various challenging problems. Some of these problems are fundamental and traditionally tackled in robotics research, whereas others are specific. Near human-size humanoids such as Honda' ASIMO or Toyota's Partner-robots are aimed to

Adrien Escande
CEA-LIST, Fontenay-aux-Roses, France, e-mail: adrien.escande@cea.fr

Abderrahmane Kheddar
CNRS-UM2 LIRMM, Montpellier, France and CNRS-AIST JRL, UMI3218/CRT, Tsukuba, Japan
e-mail: kheddar@lirmm.fr

be the missing component of the actual information and communication technology systems, working and serving humans in co-located spaces. Other human-size humanoids, such as Kawada's recent HRP-3, are clearly designed for industrial applications such as large building and construction sites, yards, nuclear power-plant maintenance, etc. Small size humanoids such as Fujitsu's HOAP or Sony's Qrio are targeted for entertainment and robot companion. Humanoids are conceived to be advanced manipulators performing a large variety of tasks in a standalone or collaborative way. They are shaped as human-kind not only for the very reason that our environment and its infrastructures fits our physical, mobility, motion, and cognitive capabilities, but also because anthropomorphism allows any person not familiar with a robot to guess easily what task capabilities and dexterity s/he might expect from these robots.

Achieving whole-body support mobility is an important function of both systems when it comes to human assistance and service. A considerable amount of research has tackled this problem from the planning and the control viewpoints. Efficient and robust walking algorithms have been implemented on humanoid platforms. But for humanoids to walk, footprints' planning is necessary. Footprint planning has received dedicated attention in robotics [1] and computer graphics [2]. Combining footprint planning and walking control strategy allows humanoid robots to walk on horizontal flat soil, slightly sloped ground or climb stairs. In most cases, the robot uses only its feet, which reduces the number of possible motions. Walking is realized by sequencing different contacts between the robot body and its environment. We humans often use other parts of the body to either help biped motion (e.g., by increasing stability) or to perform motions that are not possible with the usual upright biped posture.

Our aim is to devise solutions to accomplish similar functionalities on humanoids and to increase their motion capabilities in non-structured environments or structured but highly cluttered ones. Therefore, we are addressing the problem of planning non-gaited acyclic motions allowing robots to take support on any part of the environment with any part of their body.

Several methods were inspired by fundamental robotic motion planning [3, 4], to plan footprints of multi-legged robots [5], robotic humanoids [6][1] or virtual avatars [2]. Extending footprint planning to other terminal points taking support on several holds (predefined) was first addressed in the context of simulation. In [7], non-gaited motion planning for a humanoid avatar is done in several steps. First, a precursory planner finds a route using a descent gradient method combined with backtracking to escape eventual local minima (randomize path planning). Then based on a finite state machine and a heuristic dictated by the current state, holds are selected or contacts removed from predefined holds on the environment. Because the target application is virtual animation, constraints such as torque limits, balance, contact stability and its unilateral nature have not been the core concern.

Our problem can find several similarities with the so called *multi-step planning* for free-climbing robots, which has been thoroughly studied in [8, 9], and with ma-

[1] See also dedicated workshop in the 2007th edidtion of the IEEE-RAS Humanoids conference `http://staff.aist.go.jp/kensuke.harada/Humanoids07.htm`

nipulation planning as described in [10]. Indeed, a similar fundamental basis was applied to humanoid non-gaited motion in [11]. An improved version of this work using motion primitives is presented in [12]. In the former work, holds are predefined on both the robots and on the environment; in the latter, the motion primitives help to find holds on the environment; a contact is defined as a pair (hold, robot terminal point). Each contact stance (possible contact) forms a constraint manifold. Additional constraints on a given manifold reduce its feasible space. Possible transitions between two successive contact stances have direct mappings in the non-empty intersection between neighboring manifolds' feasible spaces. A stance transition and component graphs are built, pruning of which allows one to find possible contact transitions. This method is a contact before motion planning and is clearly close to what we adopted in our approach.

Our planner allows contact supports to occur on any part of the humanoid with any part of the environment, but to ease the computation, and input some knowledge about safety (avoiding breakable parts of the robot or parts of the environment, favoring parts of the robot or the environment enabling stable contacts), we predefine some contact spots (surfaces) on the robot, and choose a subset of the surfaces of the environment). We, however, still have non-discrete sets of possible contacts. Contrarily to what has been claimed in [8], the problem is not simpler relative to considering a numerable set of holds that can be contacted by a numerable set of terminal-points. Planning for contact points is a complex interleaved mix of continuous and discrete choices, yielding a tremendously high combinatorial complexity: there is the discrete choice of which sequence of contacting parts is to be used by the robot, together with the continuous choice of where on the environment each contact should happen. Both types of choice impact the possibilities of the other. Lastly, choices have to be made under the constraint that a continuous path must exist between two elements of the contact sequence. Our approach uses motion before contact at a first stage, similarly to [7], but with a different approach, to find a rough path for the robot. We then use this path to drive an incremental building of a contact configurations tree by combining a potential-like best first planning with a posture generator that checks the feasibility of each contact configuration.

6.2 Overview of the Planner

In [13] we presented a planner based on the following principle: planning is made in the space of sets of contacts *SC*, by building incrementally a tree of such sets. The difference between a father and its son is exactly one contact. To drive the planning, a potential function f is given. At each iteration, the best leaf of the tree (according to f) is selected and its sons are built by adding or removing a contact. If some of the new leafs are too close to existing nodes or leafs, they are discarded. This mechanism is inspired by the potential-field-based planner best first planning [14]. However, we are planning here in *SC*. This allows a substantial reduction of the search space compared with the usual configuration space, but it does not allow taking into account the geometrical and physical limitations of the robot: two contacts of a set may be too far from one another, a contact may force collisions or instability of the robot, etc. Feasibility of a set must be checked, and this is done with a posture generator.

For a given set of contacts as input, the posture generator writes and attempts to solve an optimization problem with (non-linear) constraints, whose variables are the robot's degrees of freedom. Constraints include the matches of points and frames of the contacts, collisions with environment and self-collisions, joint limits and stability. A successful output is a configuration of the robot which respects these constraints. Upon failure of the posture generator, the set of contacts is discarded. An illustration of the planner is given in Figure 6.1.

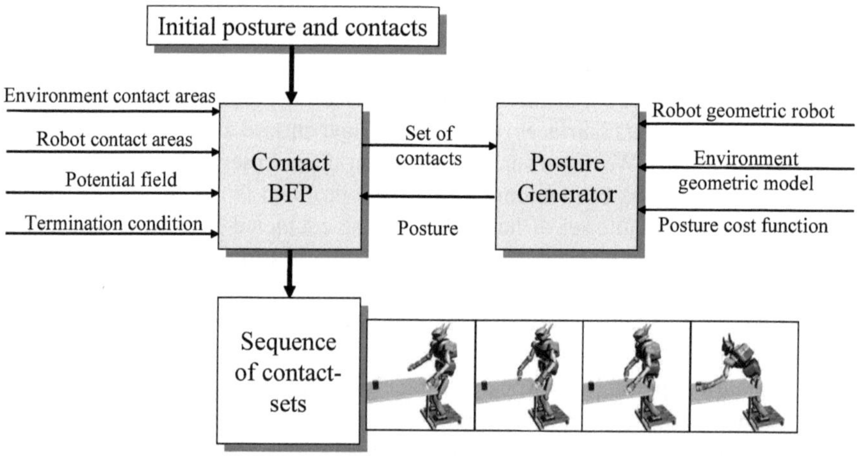

Figure 6.1 Planner overview with its different modules and requests/information exchange flow

Planning is thus made in the sets of contacts space, but with a constant link to the configuration space. The inputs of our planner are the data of the environment, the data of the robot, a feasible starting set of contacts and some end conditions. Output is a sequence of sets of contacts along with their associated postures. It must be noted that associated postures are only feasibility witnesses of the sets and can be changed by post-processing.

6.3 Posture Generator

For a given set of contacts $\{C_i\}$ as input, the posture generator writes and attempts to solve an optimization problem with (non-linear) constraints, whose variables are the n degrees of freedom of the robot q:

$$\min_{q \in \mathcal{Q}} o(q), \tag{6.1}$$

where Q is the set of admissible postures for these contacts:

$$\mathcal{Q} = \begin{cases} q_i^- \leq q_i \leq q_i^+, \quad \forall i \in [|1,n|], & (6.2) \\ \varepsilon_{ij} \leq d(r_i(q), r_j(q)), \quad \forall (i,j) \in \mathcal{I}_{\text{auto}}, & (6.3) \\ \varepsilon_{ik} \leq d(r_i(q), O_k), \quad \forall (i,k) \in \mathcal{I}_{\text{coll}}, & (6.4) \\ s(q) \leq 0, & (6.5) \\ g_i(q) = 0, \quad \forall C_i, & (6.6) \\ h_i(q) \leq 0, \quad \forall C_i. & (6.7) \end{cases}$$

Inequalities 6.2 describe the joint limits. We consider 1D joints (*rotational* and *prismatic* joints), bounds for more complex joints could be written in a more general (and sometimes approached) form $b(q) \leq 0$.

Equations 6.3 and 6.4 translate the auto-collision and collision avoidance constraints. $d(A,B)$ is the (signed) distance between two objects. ε_{ij} and ε_{ik} are security distances, thus usually positive. $\mathcal{I}_{\text{auto}}$ is a subset $[|1,n+1|]^2$: some pairs of bodies can never collide because of joint limits or other collisions that will always occur before, and it is not always possible to write collision avoidance constraint as a positive distance requirement. Denoting N_O the number of objects in the environment, $\mathcal{I}_{\text{coll}}$ is a subset of $[|1,n+1|] \times [|1,N_O|]$: pairs (r_i, O_k) involved in a contact we want not to be part of it, for example.

s is a stability criterion written only on the configuration q (a simple example would be the projection of the CoM inside the supporting hull in the case of horizontal, coplanar contacts – an extension of this criterion can be used as proposed in [15]). g_i and h_i are, respectively, the equality and inequality constraints describing the i-th contact – basically, they force a point and a frame of a link to correspond to a point and a frame of the environment.

Writing the problem only on q prevents taking torque limitations into account. This could be done by generalizing the problem to the product of the configuration and torques spaces. We are not considering any constraints on the torque at this stage (planning).

The optimization criterion o is optional. It lets the user control the overall shape-style of the obtained posture. For example, the user may want to have human-like postures, e.g., [16]. In [13], we used the distance to a reference posture, whereas in [17], the potential function f of the planner level is used as the optimization criterion.

To solve such an optimization problem we use FSQP (see [18]), which allows the use of any function provided it is at least twice continuous[2]. In particular, it copes with non-linear criteria and constraints.

The posture generator is a powerful module of our planner. It is used for projections on sub-manifolds of the configuration space as several other methods (see [19]

[2] This solver was suggested after discussion with Dr P-B. Wieber from INRIA, since he used it successfully in his framework HUMANS.

and [20]), but in a very meaningful way, since the user has great control of the projection method, by defining both the constraints and the optimization criteria.

It must be noted that functions g_i and h_i defined in Equations 6.6 and 6.7 can be of any kind, since our solver only requires them to be twice continuous. We can thus write many other types of constraints than those of contacts C_i.

Equality constraints $g_j(q) = 0, j \in [1, j_i]$ define a sub-manifold on which the posture we are looking for must be. Such a set is called *task* in [19]. It is indeed really close to the concept of task function defined by Samson *et al.* in [21], which relates to regulating the distance to a sub-manifold. This notion of regulation is the only difference between both definitions. This difference encompasses and allows taking into account the discrepancy between the planning in a perfect world with supposedly flawless models and the execution of the plan in a real environment (still, this is not that simple).

Yet we are also using inequalities $h_j(q) \leq 0$ that restrict this sub-manifold to one of its subsets. We will call *task*, as a generalization of [19], a set of equalities and inequalities. Contact is a particular kind of task. Having the parallel in mind with Samson's tasks is interesting because we can translates easily our tasks to be executed by a stack-of-tasks [22, 23].

With these remarks, we can rewrite Equations 6.6 and 6.7 in a more general way:

$$\begin{cases} g_i(q) = 0, & \forall \mathscr{T}_i, \\ h_i(q) \leq 0, & \forall \mathscr{T}_i. \end{cases} \qquad \begin{matrix}(6.8)\\(6.9)\end{matrix}$$

the $\mathscr{T}_i$ being tasks.

Through the addition of tasks to the posture generator, inclusion of tasks may be then done in our planner in a very natural and easy way that does not need any particular coding effort, but the interface to insert these tasks in the posture generation problem related to a new node.

Here are some tasks that may be of great interest in motion planning:

- maintaining the orientation of a carried object, as we will show below;
- keep looking at a target, or maintaining visual features in the field-of-view. The latter can be useful, for example, for self-localization;
- looking at the contact that is being created, so that, once the plan is executed on the real robot, there will be visual feedback, especially in the absence of contact sensors, etc.

Let us take as an example the following mission: the robot must carry a glass containing a liquid. If the robot does not have the glass in the gripper, it must first reach it. Separating the problem into sub-missions as "go to the glass", "grasp the glass", "carry it", and so on, would be the purpose of a higher-level task planner and is beyond the scope of this work. Here we assume the robot with the glass grasped. In order not to spill a drop of liquid, the robot must keep the glass in a vertical position all the time and in particular at each witness posture associated to a node of our planner. This task can be described with:

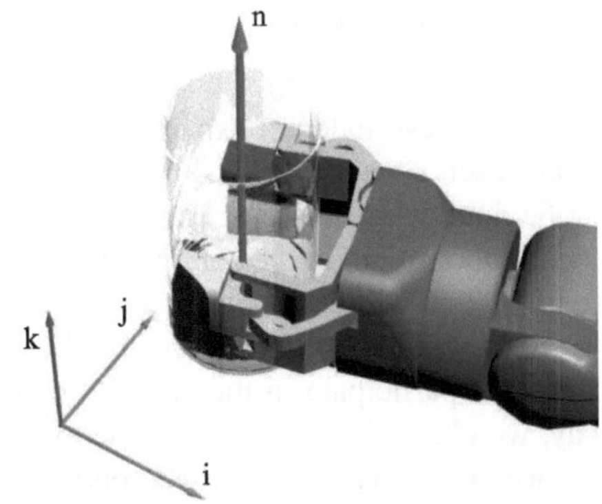

Figure 6.2 Vectors linked to the definition of task $\mathscr{T}_{\text{glass}}$

$$\mathscr{T}_{\text{glass}} = \begin{cases} n(q).i = 0 \\ n(q).j = 0, \\ -n(q).k < 0 \end{cases} \tag{6.10}$$

where $n(q)$ is the axis of the hand carrying the glass, and (i, j, k) is a frame of the world (cf. Figure 6.2). This task is added to every posture generation. Every posture of the output plan will then cope with these constraints. Adding a task in the planning decreases the number of degree of freedom of the robot, meaning a witness posture will be found for fewer sets of contacts. It has thus a direct impact on the plan, since it reduces the possibilities of the planner [24].

6.4 Contact Planning

We name *contact* the match between one point and frame of the robot's body with one point and frame of the environment such as the distance between them is 0. We are mostly using plane/plane contacts to encompass stability. In this case, points to be matched are taken on the planes and frames are defined by a normal vector to the planes (inner for the body, outer for the environment) and a vector of the plane. For a given set of contacts, sons are built by either adding a new contact or deleting an existing one.

In this section, we will introduce the two most important modules for generating the tree of sets of contacts: the generation of new nodes, used to expand the tree, and how to build the potential function to guide this expansion in an efficient way. We will then propose a heuristic to speed up the expansion process.

6.4.1 Set of Contacts Generation

¿From a selected node (the best node not expanded yet), growing the tree is done by building new contacts sets obtained from this best node by removing or adding a contact. This must be done while ensuring each new set of contacts is reachable from its father by a continuous path in the configuration space. This last constraint is dealt with by a heuristic that will be detailed below.

When deleting a contact, we keep it as geometrical constraints for the posture generation, but it does not participate in the stability of the robot any longer. In this conservative way, we check for feasibility at the exact moment the contact is broken, in particular, we check that the removed contact is in the stability area of the remaining one(s). This makes it very likely to find a feasible path between the postures of the old and new sets of contacts. When building sons, the planner tries to delete every existing contact in turn.

Generating new contacts is one of the most difficult tasks of contact planners and an efficient solution is the major theoretical contribution of this paper. We first discuss the basic challenges and solutions. New contacts must be as few as possible to avoid a combinatorial explosion (i), yet numerous enough to correctly cover the possibilities of the robot (ii). Moreover, contact candidates (i.e., before the feasibility check by the posture generator) should lead to a failure of the posture generation as seldom as possible (iii), since failure is most of the time much more time consuming than success. For new contacts, we also use a similar heuristic to that with removed contacts: a set with a new contact is checked without this contact being taken into account in the stability constraints. The new contact must be in the feasibility area of the old one(s).

A pair (R, E) made of a contact spot and a surface of the environment defines only a floating contact (for example a hand on the table) in the sense that it does not fix the relative position (x, y) and orientation (θ) between the two surfaces. A particular fixed contact between R and E is a choice of these three parameters. The possible contacts for a given pair (R,E) is thus a bounded subset C of a 3D space. If the surface is convex, so is C.

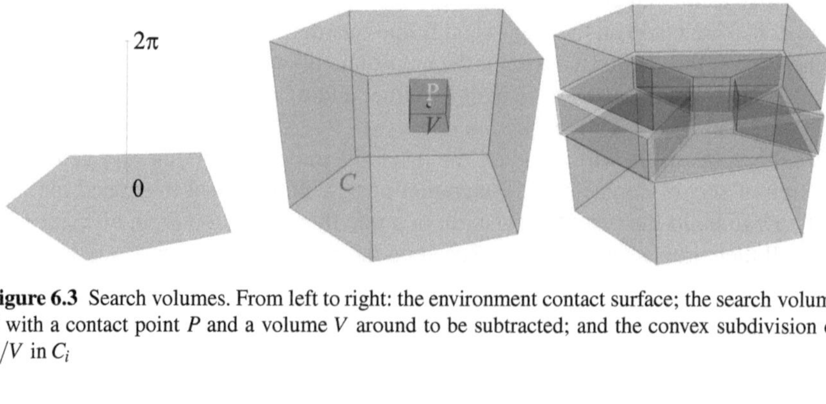

Figure 6.3 Search volumes. From left to right: the environment contact surface; the search volume C with a contact point P and a volume V around to be subtracted; and the convex subdivision of C/V in C_i

Generating new contacts for a pair (R, E) becomes choosing points in C with the following constraints:

- points must not be too close;
- must ensure a good coverage of C;
- must correspond to feasible contacts.

The first two points correspond to (i) and (ii), the last one ensures (iii). The contacts feasibility is the bottleneck issue because it is checked by a call to the posture generator, which takes time when it fails. Rather than predefining contact pairs by some choice of (x, y, θ), we let the posture generator decide the contact parameters so that feasibility is directly verified if such a point exists under given constraints and optimization criteria. Once a point is chosen, a small volume around it is subtracted to C and a new posture generation is done within this updated C. The process is repeated until either no point is found or C is empty. Practically, the surface is chosen convex so that C is convex (or cut into convex parts otherwise). When a volume V is subtracted to C, C/V is cut into convex parts C_i on which the search process is repeated, see Figure 6.3. Keeping convex search volumes is important to avoid local minima in the posture generation process.

Furthermore, for each C_i, we obtain the best node in it. This will prove to be a useful property for further improvement of the new contacts generation. Contact generation is thus made according to Algorithm 6.1.

Algorithm 6.1 Contact Generation

foreach *contact spot R* **do**
    foreach *surface of the environment E* **do**
        Construct the search volume $C = E \times]-\pi, \pi]$
        $L \leftarrow C$
        while *L is not empty, take one search volume* C_{i0} *in L* **do**
            if *a contact is successfully generated* **then**
                build a volume V around this contact
                cut C_{i0}/V into convex parts and put them in L
            end
        end
    end
end

6.4.2 Rough Trajectory

Simple potential functions f as the distance to a goal are enough in simple environments or when there is no big obstacles between the initial position and the goal. However, it gives poor results in other cases, because it leads the planner into local minima from which the time to escape is huge. Moreover, it may drive the planner along complicated and unexpected path. For example, in the early attempts to

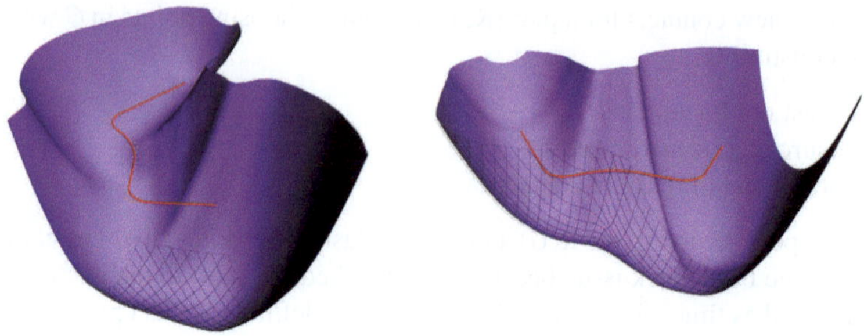

Figure 6.4 A rough trajectory and the potential field based on it. Based on a trajectory in the configuration space (red) acting as reference postures, we build a descending valley-like potential by combining the distance to the trajectory and the distance traveled along the trajectory

solve the scenario presented in Section 6.5, the robot was climbing on the table. To overcome the problems of a too simple potential function, we build our potential function on a very rough trajectory T in the configuration space. This trajectory is defined by a few key postures and can either be given manually if the user wants to impose the overall shape of the movement, or be computed automatically by adapting some classical planning approaches [25]. Path between key postures is obtained by linear interpolation so that we have a piecewise linear curve in the configuration space. This trajectory might be colliding with the environment, even at key postures. As long as the penetration depth is not too big, it will not cause any problem. Its purpose is to give a rough approximation of the intended path in the workspace, together with an idea of the postures along it. We designed the potential function as a descending valley-like potential field around the trajectory T (see Figure 6.4) whose minimum is at the end of T.

We denote F the 6D subspace of the configuration space corresponding to the translation and rotation of the base frame of the robot (free-flyer). Let P be a point of the configuration space, P' and T' are the projection of P and T onto F. We suppose that the projection p from T to F is bijective, that is to say no points of T have the same translation and rotation coordinates. This is a reasonable hypothesis in our case. We denote Q' the closest point of T' to P'. Let $Q = p^{-1}(Q')$ and s' the curvilinear abscissa of Q' along T'. We then define the potential function f_T base on T as:

$$f_T(P) = \|P - Q\|_2^2 - k \cdot s', \tag{6.11}$$

where $\|.\|_2$ is the Euclidean norm and k is a positive coefficient. The first part of this function aims at keeping the planner as close as possible to the rough trajectory and gives the function its valley-like shape, the second part introduces a slope that will drive the planner towards the goal by measuring the path made along T. k is a weight that controls the relative effects of these two parts. Because we will need f_T to be a two times continuous function of P for the heuristic of the next section, we needed to slightly reshape T. Roughly speaking this potential field embeds two terms, one being the distance d from the current configuration to the trajectory T, the other measuring the distance l traveled along the trajectory:

$$f_T(q) = d(q,T)^2 - k \cdot l(q). \tag{6.12}$$

T thus acts as a trajectory of reference postures. While this copes with relatively big local minima and proved to be effective, we do not tackle some smaller ones, which are linked to the intrinsic discrete nature of contact planning: advancing along the rough trajectory may require moving lightly and/or shortly away from the trajectory: moving a foot forward while walking, for example, requires the robot to shift its weight onto the other foot, moving away from the symmetrical standing posture the trajectory T classically indicates. In this example, walking implies regularly climbing up the potential field, as if the robot was trapped in a local minimum. Anticipating that moving forward a foot would help to move forward the robot, even if it briefly implies the robot moves "backward": anticipation would push going over potential barriers. We therefore use similar potential fields as our global field f_T but acting only on some bodies of the robot: from a global trajectory T we can derive the trajectories T_i of each robot's body and define a potential field $f_{T_i}(q) = d_i(q,T_i)^2 - k_i \cdot l_i(q)$ that is solely based on the absolute 6D position of the body. We then build a total potential field

$$f'_T(q) = \alpha f_T(q) + \sum \lambda_i f_{T_i}(q), \tag{6.13}$$

where the λ_i are parameters whose values are 1 or 0 depending on whether we take into account the i-th individual potential field or not. Typically we take $\lambda_i = 1$ for the feet and hands and sometimes for the knees. This method allows the selected bodies to anticipate by being ahead of the overall motion: since advancing along a trajectory gives a bonus, this compensate for a biggest yet meaningful gap between the posture found and the reference posture given by T [26].

6.4.3 Using Global Potential Field as Local Optimization Criterion

Up to now, when adding a contact to a set S, the planner generates all possibilities, ending with a set of leafs L_S, that also includes leafs obtained by removing a contact from S. However, we know it will first consider only the best leaf l_0 in L_S. Let's denote $\mathscr{L}$ the set of all existing leafs outside L_S. If l_0 is better than any leaf of $\mathscr{L}$, then the planner will select it and build its sons. These sons will then be compared with the elements of $\mathscr{L}$ and of L_S/l_0, but if we know which leaf l_1 is the second best one in L_S, then we only need to compare them with the elements of $\mathscr{L}$ and with l_1. Let's go further: we only need to know the best son of l_0 to decide which leaf is the best among all existing leafs (in $\mathscr{L}$ and L_S).

We thus remark that if we are able to generate the sons of a set in the decreasing order of their score with respect to the potential function f_T, then we only need to maintain one leaf among its sons at all times (when a leaf is selected, it becomes a node). That is, we could only generate one son at first, generate a second one when the first is selected, and so on... We would end up with a tree for which each node has a single son that is a leaf.

The advantage behind this is that we need to generate far fewer leafs, and since we do not generate until it is not possible, we have far less costly failures of the posture generator. We can roughly expect to be 10 to 20 times faster. But the remark is based on a strong hypothesis: the capacity to generate sons in the good order. We do not fully have this capacity: it is really difficult to predict for which pair (R,E) or which removed contact we will have the best leaf. However for a specific pair (R,E), we already know that we can generate sons in the decreasing order of their optimization criteria value. By using the potential field f_T as optimization criterion for the posture generation, we can thus generate sons in the good order with respect to f_T, for each pair (R,E). We needed to regularize f_T because the posture generator is using two times continuous functions. We end up with a hybrid solution where we still have to generate every leaf obtained by deleting a contact, and have to maintain only one leaf per pair (R,E).

With this solution, computation is sped up by a factor 3 to 5 for the kind of scenarios we tackle. It must be noted that this does not affect the complexity of the tree, but it avoids a large number of posture generations.

6.5 Simulation Scenarios

Preliminary simulations of the proposed method, obtained with a basic implementation of the planner, have been realized for two scenarios: grasping a can, and walking on stairs. The first one requires the humanoid HRP-2 to bring a can placed on a table. Only part of the mission is experimented. Indeed, the robot is put near the table with a given initial posture. Then it has to reach the can, grasp it and return to its initial posture. The second scenario consists in highlighting that the approach is in fact a generalization of walking, the results and experiments for both scenarios are thoroughly described in [13].

Since then, more advanced scenarios have been simulated and then experimented, and we present a few

In the first one, the HRP-2 is sitting at a table on a fixed chair and is asked to go to a position on the side of the table. It can use its feet, thighs, knees, forearms and hands to do so and can make contact on the table's top surface, the horizontal of the chair and the floor. The goal is set as a coordinate the waist must exceed. The main challenge of this scenario consists in the really narrow space around the chair, and especially between the chair and the table, making it really difficult for the robot to stand on its sole feet. There are numerous potential collisions involving the legs, from self-collisions to collisions with the legs of the chair. The fact the chair cannot move is both a limitation of our planner, which does not handle moving objects, and an additional challenge for it, since it cannot widen the moving space of the robot as a human would do.

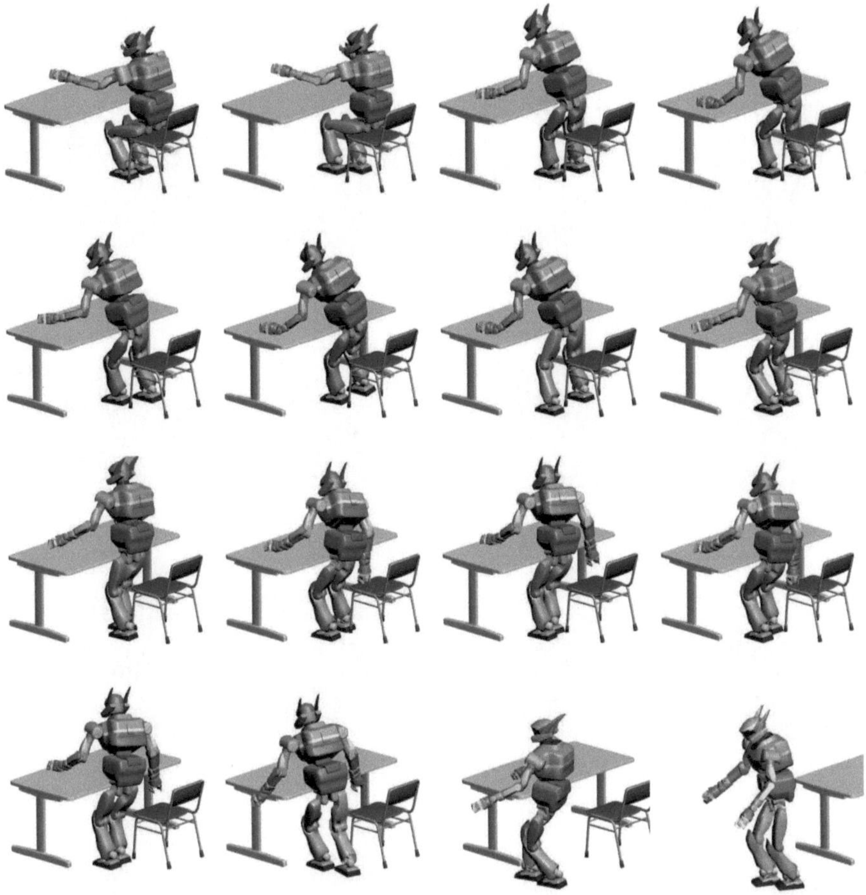

Figure 6.5 Some witness postures of the output sequence. The robot is sitting on a chair and first moves its feet slightly before getting up, and then puts its left wrist on the table (first line). It then shifts the wrist contact, which allows it to move the left foot outside of the chair (second line). It uses then twice its right hand to help it place its right foot outside of the chair (third line). Finally it can walk toward the goal (fourth line)

The output of our planner for this scenario is a plan consisting of 81 sets of contacts together with their associated witness postures; it took around 5 hours, without the use of the individual potential field described by Equation 6.13. Figure 6.5 displays a chosen subset of the witness postures that illustrates how the robot makes use of its hand to help him to go outside the chair. The robot needs 19 nodes to put a first foot on the chair side, and 22 more to have both feet outside the chair. To do so, it needs to help with its left wrist on the table, then its right hand on the chair. The remaining part of the movement is a static walk. The total movement corresponds to a sequence of 30 steps which, while being small, do not differ much from what a human would do.

Figure 6.6 Passing under the table using individual trajectories for hands, feet and knees. The total movement is made in 141 nodes (from 0 to 140). Between two pictures, there is a "time" of 10 nodes (additional in-between picture is added last line, column 2). Note that the longest part of the plan is to go from the upright posture to the four-legged one

Our second scenario emphasizes the gain due to individual potential fields as explained by Equation 6.13. The HRP-2 is put initially in a collision-free non-cumbersome space and is asked to reach, in the same posture, another position where the only way to go is to pass under a table. Passing under a table requires coping with both the very small steps of four-legged motion and above all the change of posture types, which happens twice: the robot first stands in front of the table, then must pass under it, and finishes in an upright posture on the other side.

In this scenario, when no individual potential field is used, i.e., using only Equation 6.12 as a criterion and reference trajectories, the planner has great difficulty entering and moving under the table. 20 000 nodes are generated before the planner stops unsuccessfully. In contrast, with individual fields the planner ends successfully in 1.3 hours and 2916 nodes. The result is a path of 141 nodes depicted in Figure 6.6. Half of the planning time is spent before the table, while passing under the table is done with only 30 nodes. We emphasize the fact that we did not provide any specific knowledge for the robot to evolve from a standing biped posture to a four-legged posture.

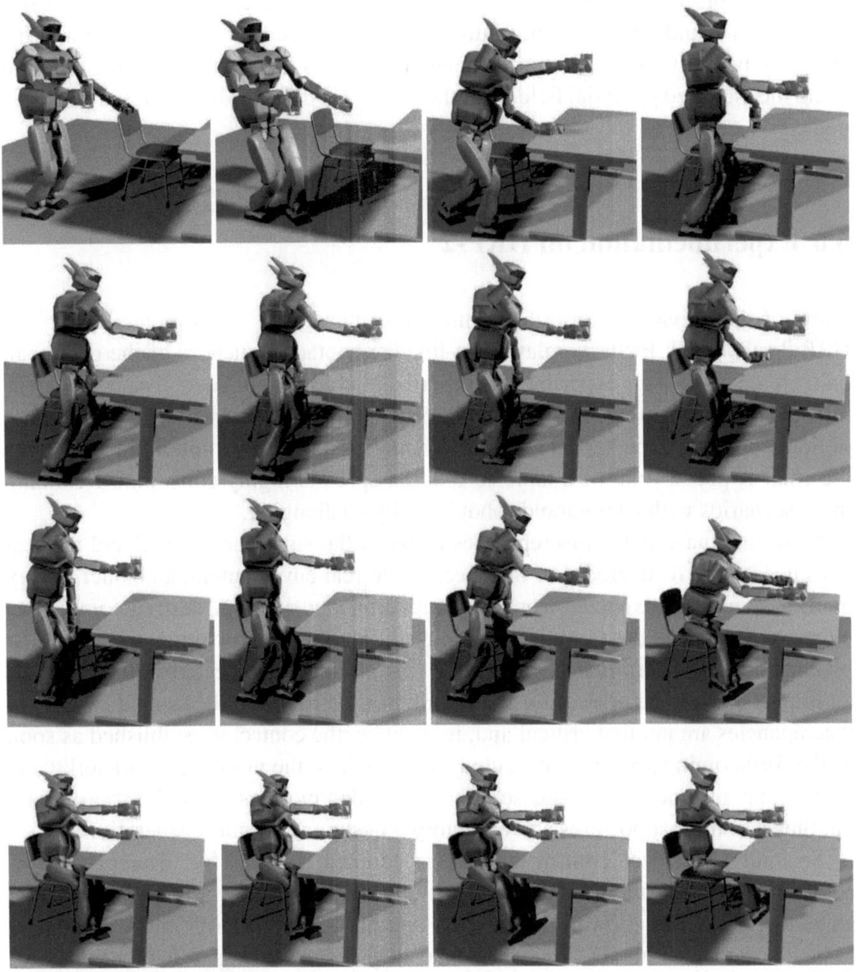

Figure 6.7 Plan for sitting on a chair while holding a filled glass (69 nodes are generated). Plan is displayed every five nodes (additional in-between picture at line 3, row 4)

A third scenario is the 'reverse' of the first one: the HRP-2 initial position is in an obstacle-free non-cumbersome space, in a stand-up position, holding a filled glass of liquid. It is asked to sit on a chair near a table. This scenario aims at demonstrating the inclusion of an additional task in the planning. To do so we take the scenario described in Section 6.3, namely carrying a filled glass without spilling a drop of liquid, and we merge it with the planning of sitting on a chair at a table. This planning is difficult since it forces the robot to enter a narrow space.

Planning is successfully achieved in about 4 hours, during which 5400 nodes are generated. The output plan consists in 69 nodes, some of which are depicted in Figure 6.7. The robot is first walking, then by helping with his left hand it finally manages to place its left hand in front on the chair (around the 50th node). At this stage it has found an entry point into the narrow space between the table and the chair and begins to move on the chair with the help of its thighs. The same scenario using the modified potential field of Equation 6.13 is done in about 3 hours and 3800 nodes.

6.6 Experimentation on HRP-2

Some of the previously described simulations have been experimented using the HRP-2 robot [27]. First, we played on-line the contact sequences of the grasp can on table scenario. This experiment is described in [13]. In the second step we experimented the getting up from chair scenario and sitting on chair while holding a glass of liquid needing to be kept at constant orientation during any motion. In both cases the chair is put in front of a table so that the space is highly constraint. Realizing these scenarios with a humanoid robot is really challenging.

First, the unavoidable discrepancies between the simulation model and the real ones need to be recovered. For instance, in the real environment, a contact is very likely not made as expected from the simulation's position. A desired contact can either occur before the motion is completed or not occur (or removed) when the planned motion is terminated. In the first case, sudden unexpected contacts induce an impact that can be critical for the stability of the robot. In the second case, light discrepancies are not that critical and, in practice, the contact is established as soon as the forthcoming posture is executed. Nevertheless, the motion appears jerky and impact may also occur in this case. Ideally guarded motions would be appropriate; but they call for fast postures and trajectory adjustments, if not regeneration.

Secondly, whatever control strategy is adopted, solving for discrepancies or achieving dynamic motion between postures require haptic sensing; especially; since the contact is allowed to occur on any part of any robot's links, they need to be detected and characterized as soon as they occur. Force guarded motions cannot be performed without haptic sensing. Flexibility of the cover allows having more *robust* contacts by adjusting and complying with the contacting surface local shape. Nowadays, humanoid robots lack of the sensors to detect contacts on their whole surface.

Last, but not least, a challenging problem comes from a specificity of the HRP-2 humanoid: the flexibility of the ankles is not controllable and induces non-desirable motions that are reduced or even compensated by a dedicated stabilizer. This stabilizer works in conjunction with the ZMP reference trajectory and is designed to work well only with planar contacts. Hence it must be switched off when this is not the case. Some multi-contact configurations are clearly not coplanar in our scenario cases; therefore the stabilizer needs to be switched off when HRP-2 takes support on the table or on the chair. The stabilizer on/off transitions result in discontinuities in the posture configurations. For proper use, one needs to switch it off just before a non-coplanar contact is created and this is difficult to guarantee in the presence of geometric discrepancies. Switching on/off the stabilizer has also the undesirable effect of suddenly changing the posture, because the stabilizer modifies the ideally desired joint posture.

All the experiments have been realized by taking into account previously described limitations: we took extra care to reduce as much as possible the geometric discrepancies by *a priori* good initial calibration of the initial positions of the robot, chairs and tables; we found manually the right time to switch on/off the stabilizer; we also reduced the stability margins at the feet, filtered some postures, etc.

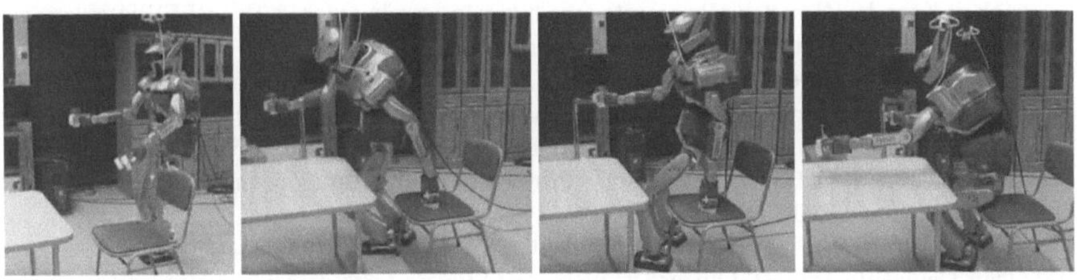

Figure 6.8 The glass scenario played on HRP-2. The glass remains perfectly vertical during the whole motion

The simulation of the Figure 6.5 has been realized in a repeatable way (more then five trials). It is thoroughly described in [17]. Figure 6.8 illustrates snapshots of the execution of the glass scenario using again the HRP-2 robot. This experiment corresponds to the simulated scenario of Figure 6.7. Since some postures were really demanding for the robot and some steps redundant, we manually removed some nodes.

6.7 Conclusion

Planning for non-gaited motions extends humanoid whole-body motion capabilities. Our approach to this problem proved to be efficient and allowed performing complex experiments using the HRP-2 platform. However, several technical bottlenecks are still to be resolved before reaching a sophisticated implementation. Some of them deal with how to guarantee the robustness of the supporting contacts through guarded motion, haptic sensitized flexible cover, and dynamic motion generation between successive contact configurations. Some possible challenging extensions of the problem might deal with contact support on deformable objects, sliding contacts, and motion with movable objects. We have already started to deal with these issues, future work is dedicated to solve these technical problems and realize the challenging extensions.

Acknowledgements This work is partially supported by grants from the European Commission FP6 ImmerSence IP project, Contract No. 27141 `www.immersence.info` and experiment on a robot are in the frame of the STREP ROBOT@CWE, Contract No. 34002 `www.robot-at-cwe.eu`. Both authors are very thankful to Dr Kazuhito Yokoi, and for early discussions of this work: Dr Pierre-Brice Wieber (also for the usage of FSQP) and Dr Slyvain Miossec and all the JRL members.

References

[1] Joel Chestnutt, Manfred Lau, Kong Man Cheung, James Kuffner, Jessica K Hodgins, and Takeo Kanade. Footstep planning for the honda asimo humanoid. In: IEEE international conference on robotics and automation, April 2005.

[2] Min Gyu Choi, Jehee Lee, and Sung Yong Shin. Planning biped locomotion using motion capture data and probabilistic roadmaps. ACM Trans Graphics, 22(2), pp 182–203, 2003.

[3] Howie Choset, Kevin M. Lynch, Seth Hutchinson, George Kantor, Wolfram Burgard, Lydia E. Kavraki, and Sebastian Thrun. Principles of robot motion: theory, algorithms, and implementation. MIT Press, 2005.

[4] Steven M. LaValle. Planning Algorithms. Cambrige University Press, 2006.

[5] Jean-Daniel Boissonnat, Olivier Devillers, and Sylvain Lazard. Motion planning of legged robots. SIAM J Computing, 30:2000, 2000.

[6] James J. Kuffner, Koichi Nishiwaki, Satoshi Kagami, Masayuki Inaba, and Hirochika Inoue. Motion planning for humanoid robots. In: international symposium of robotics research, Siena, Italy, 2003.

[7] Maciej Kalisiak and Michiel van de Panne. A grasp-based motion planning algorithm for character animation. J Visualization Computer Anim, 12(3), pp 117–129, 2001.

[8] Tim Bretl. Multi-Step Motion Planning: Application to Free-Climbing Robots. PhD thesis, Stanford University, 2005.

[9] Timothy Wolfe Bretl. Motion planning of multi-limbed robots subject to equilibrium constraints: the free-climbing robot problem. Int J Robotics Res, 25(4), pp 317–342, April 2006.

[10] Thierry Siméon, Juan Cortès, Jean-Paul Laumond, and Anis Sahbani. Manipulation planning with probabilistic roadmaps. Int J Robotics Res, 23(7–8), pp 729–746, July-August 2004.

[11] Kris Hauser, Tim Bretl, and Jean-Claude Latombe. Non-gaited humanoid locomotion planning. In: IEEE/RSJ international conference on humanoid robots, pp 7–12, December 5-7 2005.

[12] Kris Hauser, Tim Bretl, Kensuke Harada, and Jean-Claude Latombe. Using motion primitives in probabilistic sample-based planning for humanoid robots. In: workshop on the algorithmic foundations of robotics, 2006.

[13] Adrien Escande, Abderrahmane Kheddar, and Sylvain Miossec. Planning support contact-points for humanoid robots and experiments on HRP-2. In: IEEE/RSJ international conference on robots and intelligent systems, pp 2974–2979, Beijing, China, 9–15 October 2006.

[14] Jean-Claude Latombe. Robot motion planning. Kluwer Academic Publishers, Boston-Dordrecht-London, 1991.
[15] Tim Bretl and Sanjay Lall. Testing static equilibrium for legged robots. IEEE Trans Robotics, 24, pp 794–807, 2008.
[16] Kensuke Harada, Kris Hauser, Tim Bretl, and Jean-Claude Latombe. Natural motion generation for humanoid robots. In: IEEE/RSJ international conference on robots and intelligent systems, 2006.
[17] Adrien Escande, Abderrahmane Kheddar, Sylvain Miossec, and Sylvain Garsault. Planning support contact-points for acyclic motions and experiments on HRP-2. In: international symposium on experimental robotics, Athens, Greece, 14–17 July 2008.
[18] Craig Lawrence, Jian L. Zhou, and André L. Tits. User's guide for CFSQP version 2.5: A C code for solving (large scale) constrained nonlinear (minimax) optimization problems, generating iterates satisfying all inequality constraints, 1997.
[19] Mike Stilman. Task constrained motion planning in robot joint space. In: IEEE/RSJ international conference on robots and intelligent systems, 2007.
[20] Juan Cortés. Motion Planning Algorithms for General Closed-Chain Mechanisms. PhD thesis, 2003.
[21] Claude Samson, Michel Le Borgne, and Bernard Espiau. Robot Control: the Task Function Approach. Clarendon Press, Oxford, United Kingdom, 1991.
[22] Nicolas Mansard and François Chaumette. Task sequencing for sensor-based control. IEEE Trans Robotics, 23(1), pp 60–72, February 2007.
[23] Nicolas Mansard, Oussama Khatib, and Abderrahmane Kheddar. Task sequencing for sensor-based control. IEEE Trans Robotics, 25(3), pp 670–685, June 2009.
[24] Adrien Escande and Abderrahmane Kheddar. Planning contact supports for acyclic motion with task constraints and experiment on hrp-2. In: IFAC 9th international symposium on robot control (SYROCO'09), pp 259–264, Gifu, Japan, September 9–12 2009.
[25] Karim Bouyarmane, Adrien Escande, Florent Lamiraux, and Abderrahmane Kheddar. Collision-free contacts guide planning prior to non-gaited motion planning for humanoid robots. In: IEEE international conference on robotics and automation, 2009.
[26] Adrien Escande and Abderrahmane Kheddar. Contact planning for acyclic motion with tasks constraints. In: IEEE/RSJ international conference on intelligent robots and systems, pp 435–440, St. Louis, USA, October 11–15 2009.
[27] Kenji Kaneko, Fumio Kanehiro, Shuuji Kajita, Hirohisa Hirukawa, Toshikazu Kawasaki, Masaru Hirata, Kazuhiko Akachi, and Takakatsu Isozumi. Humanoid robot HRP-2. In: IEEE international conference on robotics and automation, pp 1083–1090, New Orleans, LA, April 2004.

Chapter 7
Motion Planning for a Humanoid Robot Based on a Biped Walking Pattern Generator

Kensuke Harada

Abstract In this chapter, we address the motion planning problem of humanoid robots dynamically walking on the ground. We first explain the analytical-solution-based online walking pattern generator. Then, we explain several motion planning problems based on the walking pattern generator. Section 7.3 explains a two-stage time-parametrized approach for the whole-body motion planning where, in the first phase, the constraint condition is generated as a function of time by using the biped walking pattern generator, and in the second phase, collision-free whole-body motion is planned by using this constraint condition. Section 7.4 explains the simultaneous planning method of both the foot-place and the upper-body motion taking the dynamics of the robot into consideration. To realize this simultaneous planning, we use the online walking pattern generator for the offline motion planning. In Section 7.5, we explain two methods for the whole-body manipulation problem. We first explain the method for modifying the motion by considering the hand reaction force, and then explain the online gait planning approach based on the hand reaction force in order to realize the stable pushing manipulation. We will confirm the effectiveness of the approach by several simulation and experimental results.

7.1 Introduction

In recent years, many humanoid robots capable of performing complex and human-like whole-body motion have been developed [17]. By observing the motion of humanoid robots, it is difficult for us to realize that human beings typically select parameters carefully in order to avoid unnecessary collisions of links and to keep the robot's balance. This chapter presents some methods for planning the whole-body motion of a humanoid robot. Especially, we focus on the situation where a

Kensuke Harada
National Institute of Advanced Industrial Science and Technology (AIST), Umezono 1-1-1, Tsukuba, Ibaraki, 305-8568, Japan, e-mail: kensuke.harada@aist.go.jp

humanoid robot dynamically walks. Although the biped gait for humanoid robots has been extensively researched [1, 2, 3, 4, 6, 7], there has not been much research to collision-free motion planning.

Let us consider generating the walking pattern of a humanoid robot. By using most of the existing walking pattern generator, walking motion has been generated for a given trajectory of the ZMP (zero moment point). Once the desired ZMP trajectory is given, the position/orientation of the feet are determined such that the ZMP is included in the support polygon. Then, by solving an ordinary differential equation, the trajectory of the CoG (center of gravity) is calculated. In this chapter, we consider how we can extend this walking pattern generator to the whole-body motion planning problems.

This chapter presents three methods. First, we present the whole-body motion planning framework. In this framework, to keep the robot balanced, the CoG trajectory works as time-parametrized constraint conditions imposed on the robot's configuration. When planning the collision-free whole-body motion of a humanoid robot, we consider these time-parametrized constraint conditions by adding a time parameter to each milestone of the planner. In addition, we present the method for making the path a smooth one by using b-spline interpolation.

Second, we present a method for simultaneously planning both the foot-place and the upper-body motion, taking the dynamics of the robot into consideration. To realize this function, we focus on the differential equation expressing the relation between the ZMP and the CoG. To solve this differential equation, we use the online walking pattern generator. We randomly sample the configuration space to search for the path connecting the start and the goal configurations. When sampling the configuration space, three milestones are sequentially connected to the parent milestone.

Third, we present the methods for whole-body manipulation. To realize whole-body manipulation, we have to consider the hand reaction force in the motion planner. We presents two methods; first we consider modifying the motion of a humanoid robot using the hand reaction force, and second we present a method for planning the foot place in real time in order to keep the robot balanced during the pushing manipulation.

We show the effectiveness of the method through several simulation and experimental results by using the humanoid robot HRP-2.

7.2 Gait Generation Method

This section introduces the analytical-solution-based online gait generation method [4, 14]. For example, when a biped humanoid robot moves in an environment including many obstacles, the robot has to detect the obstacles and change the walking direction in real time to avoid the obstacle. Although several methods have been proposed, two distinctive features of the method explained in this section are (1) by using the analytical solution, the gait can be generated in real time, and (2) by simultaneously planning the ZMP and the CoG, the continuity between two CoG trajectories can be guaranteed. Among the motion planning methods explained in this chapter, Sections 7.4 and 7.5.2 use the online gait generation method while Sections 7.3 and 7.5.2 use the offline one.

7.2.1 Analytical-solution-based Approach

Let us consider the humanoid robot walking on the flat floor. While we focus on the motion of a humanoid robot within the sagittal $(x-z)$ plane, the motion in the lateral $(y-z)$ plane can be treated in the same fashion. By splitting the gait of a humanoid robot into multiple walking phases, we assume that the ZMP trajectory of a gait is given by the spline function. For example, a gait can be split into single and double support phases. Let $p_{zmp}^{(j)} = [x_{zmp}^{(j)}\ y_{zmp}^{(j)}\ z_{zmp}^{(j)}]^T$ be the ZMP trajectory belonging to the jth segment of time. By using the spline function, the ZMP trajectory within the sagittal plane can be expressed as

$$x_{zmp}^{(j)} = \sum_{i=0}^{n} a_i^{(j)} (t - \bar{t}_{j-1})^i,$$
$$\bar{t}_{j-1} \leq t \leq \bar{t}_(j), \quad j = 1, \cdots, m, \tag{7.1}$$

where $a_i^{(j)}$ $(i = 1, \cdots, m,\ j = 1, \cdots, m)$ denote scalar coefficients.

Let $p_G^{(j)} = [x_G^{(j)}\ y_G^{(j)}\ z_G^{(j)}]^T$ be the trajectory of the CoG corresponding to the ZMP trajectory of the jth segment of time. Also, let $\mathscr{L}^{(j)} = [\mathscr{L}_x^{(j)}\ \mathscr{L}_y^{(j)}\ \mathscr{L}_z^{(j)}]^T$ be the angular momentum of the robot about the CoG. The relation between the ZMP and the CoG can be expressed by the following ordinary differential equation:

$$x_{zmp}^{(j)} = \frac{-\dot{L}_y^{(j)} + M x_G^{(j)} (\ddot{\bar{z}}_G^{(j)} + g) - (\bar{z}_G^{(j)} - z_{zmp}^{(j)}) \ddot{\bar{x}}_G^{(j)}}{M(\ddot{\bar{z}}_G^{(j)} + g)}. \tag{7.2}$$

Next, by setting $\bar{x}_G^{(j)} = x_G^{(j)} + \Delta x_G^{(j)}$, Equation 7.2 can be split into the following two equations:

$$x_{zmp}^{(j)} = x_G^{(j)} - \frac{z_G^{(j)} - z_{zmp}^{(j)}}{g} \ddot{x}_G^{(j)}, \tag{7.3}$$

$$\frac{\dot{\mathscr{L}}_y^{(j)}}{Mg} = \Delta x_G^{(j)} - \frac{z_G^{(j)} - z_{zmp}^{(j)}}{g} \Delta \ddot{x}_G^{(j)}. \tag{7.4}$$

In this research, we assume that the effect of Equation 7.4 is small and can be compensated by the stabilizing controller installed in our simulation/experimental setup. However, it is possible to consider the effect of Equation 7.4 by using the fourier series[5]. Substituting Equation 7.1 into Equation 7.3 and solving with respect to $x_G^{(j)}$, we can obtain the analytical solution of the CoG position as

$$x_G^{(j)} = V^{(j)} \cosh(w_c(t - \bar{t}_{j-1})) + W^{(j)} \sinh(w_c(t - \bar{t}_{j-1})) + \sum_{i=0}^{n} A_i^{(j)} (t - \bar{t}_{j-1})^i, \quad j = 1, \cdots, m, \tag{7.5}$$

$$a_i^{(j)} = A_i^{(j)} - \frac{1}{w_c^2}(i+1)(i+2)A_{i+2}, \quad i = 0, \cdots, n-2, \tag{7.6}$$

$$a_i^{(j)} = A_i^{(j)}, \quad i = n-1, n, \tag{7.7}$$

where $w_c = \sqrt{g/(z_G - z_{zmp})}$, and $V^{(j)}$ and $W^{(j)}$ $(j = 1, \cdots, m)$ denote the scalar coefficients.

Here, Equation 7.5 includes sinh and cosh in its homogenous part. To prevent the solution of Equation 7.5 diverging as time goes by, $V^{(j)}$ and $W^{(j)}$ $(j = 1, \cdots, m)$ should be determined by using the two-point boundary-value problem where both the initial and final position of the COG are specified. On the other hand, since the initial velocity of the COG cannot be specified, it is difficult to ensure the continuity of the CoG when connecting two gait trajectories.

7.2.2 Online Gait Generation

Before showing the detail of the online gait generation method [14], we show an overview (Figure 7.1). Given the ZMP trajectory for a humanoid robot walking straight forward as shown in Figure 7.1(a), we calculate the horizontal CoG trajectory. Then, if we want the robot to change the walking direction to the left, we recalculate the trajectory of the horizontal CoG by using the ZMP trajectory turning left. As shown in Figure 7.1(b), the change of walking direction can be realized by smoothly connecting the new CoG trajectory to the currently executing one.

In the online walking pattern generaror, we periodically calculate the new CoG trajectory while the robot walks. Let us assume that the configuration at the beginning of the next single-support phase is given. Let us also assume that the trajectories of both $\hat{q}$ and p_{zmp} for a few footsteps from the beginning of the next single-support phase are given. The online walking pattern generator calculates the trajectory of x_G and y_G for a few footsteps starting from the next single-support phase using Equation 7.3.

As is described in [4, 14], the difficulty of the online walking pattern generator is the smooth connection of two CoG trajectories. When solving Equation 7.3, we use the two-point boundary-value problem such that the robot does not finally fall down. We ensure the continuity of the CoG trajectory by simultaneously planning the CoG and the ZMP trajectories. Here, the coefficients of the spline function of the ZMP trajectory belonging to the first segment of time is also set to be unknown

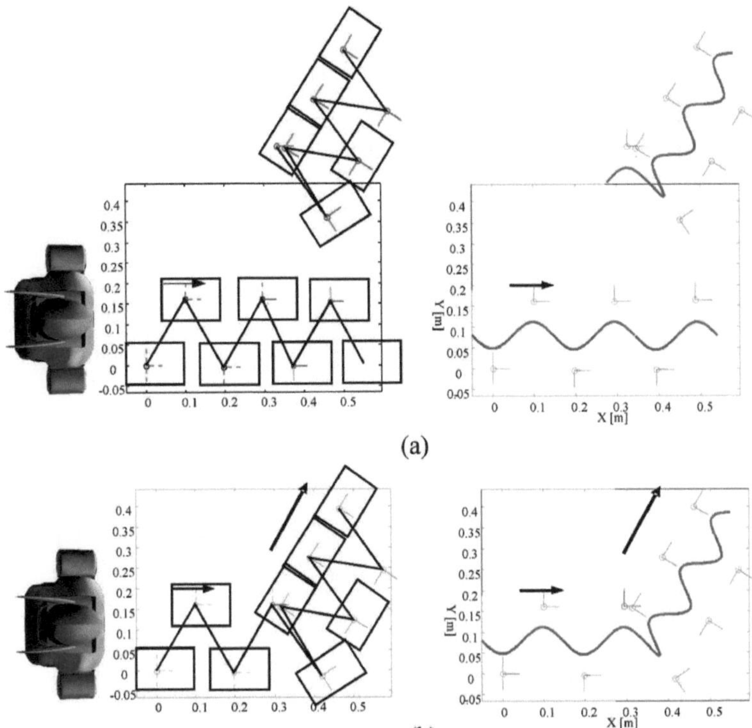

Figure 7.1 Overview of online walking pattern generation: (a) calculation of new trajectory (left: ZMP, right: CoG); (b) connection of new trajectory to current one (left: ZMP, right: CoG)

and is calculated such that both the new ZMP and CoG trajectories are smoothly connected to the currently executing ones. In addition to the $2m$ unknowns $V^{(j)}$ and $W^{(j)}$ $(j = 1, \cdots, m)$ in Equation 7.5, we consider setting $A_i^{(1)}$ $(i = 0, \cdots, n)$ as unknown constants. When $n = 2$ for the first section, the $2m + n + 1$ unknowns $(A_0^{(1)}, \cdots, A_n^{(1)},\ V^{(1)}, W^{(1)},\ \ldots,\ V^{(m)}, W^{(m)})$ can be determined by considering the boundary conditions as follows:

$$\tilde{y} = \tilde{Z}^{-1}\tilde{w}, \tag{7.8}$$

where

$$\tilde{y} = [A_0^{(1)}\ A_1^{(1)}\ A_2^{(1)}\ V^{(1)}\ W^{(1)}\ \ldots\ V^{(m)}\ W^{(m)}]^T$$

$$\tilde{Z} = \begin{bmatrix} Z_0 & 0 & 0 & \cdots & & 0 \\ Z_{11} & Z_{12} & 0 & \cdots & & 0 \\ 0 & 0 & Z_2 & 0 \cdots & & 0 \\ & 0 & & \ddots & & \vdots \\ \vdots & \vdots & & & Z_j & \\ & & & & \ddots & 0 \\ 0 & 0 & \cdots & & 0 & Z_{m-1} \\ 0 & 0 & \cdots & & 0 & z_{m-1} \end{bmatrix}$$

$$Z_0 = \begin{bmatrix} 1 & 0 & -2/w_c^2 \\ 1 & \bar{t}_1 - \bar{t}_0 & (\bar{t}_1 - \bar{t}_0)^2 - 2/w_c^2 \end{bmatrix}$$

$$Z_{11} = \begin{bmatrix} 1 & 0 & 0 \\ 0 & 1 & 0 \\ 1 & (\bar{t}_1 - \bar{t}_0) & (\bar{t}_1 - \bar{t}_0)^2 \\ 0 & 1 & 2(\bar{t}_1 - \bar{t}_0) \end{bmatrix}$$

$$Z_{12} = \begin{bmatrix} 1 & 0 & 0 & 0 \\ 0 & w_c & 0 & 0 \\ \cosh(w_c(\bar{t}_1 - \bar{t}_0)) & \sinh(w_c(\bar{t}_1 - \bar{t}_0)) & -1 & 0 \\ w_c \sinh(w_c(\bar{t}_1 - \bar{t}_0)) & w_c \cosh(w_c(\bar{t}_1 - \bar{t}_0)) & 0 & -w_c \end{bmatrix}$$

$$\tilde{w} = [x_{zmp}^{(1)}(\bar{t}_0)\ x_{zmp}^{(2)}(\bar{t}_1)\ x_G^{(1)}(\bar{t}_0)\ \dot{x}_G^{(1)}(\bar{t}_0)$$
$$A_0^{(2)}\ A_1^{(2)}\ \cdots A_0^{(j+1)} - \sum_{i=0}^{n} A_i^{(j)}(\bar{t}_j - \bar{t}_{j-1})^i$$
$$A_1^{(j+1)} - \sum_{i=1}^{n} i A_i^{(j)}(\bar{t}_j - \bar{t}_{j-1})^{i-1} \cdots x_G^{(m)}(\bar{t}_f) - \sum_{i=0}^{n} A_i^{(m)}(\bar{t}_f - \bar{t}_{m-1})^i]^T.$$

Here, we confirmed numerically that the matrix $\tilde{Z}$ in Equation 7.8 is invertible.

7.2.3 Experiment

We performed an experiment using the humanoid robot HRP-2 [4]. In the experiment, the human operator holds one hand of HRP-2. The force/torque sensor is attached at the wrist of the HRP-2, and the position of the footstep changes depending on the information from the force/torque sensor. The experimental result is shown in Figure 7.2. We can see that the humanoid robot walks in the direction that the human operator pushes or pulls the hand of HRP-2. It is just like navigating a person with closed eyes by holding his/her hand.

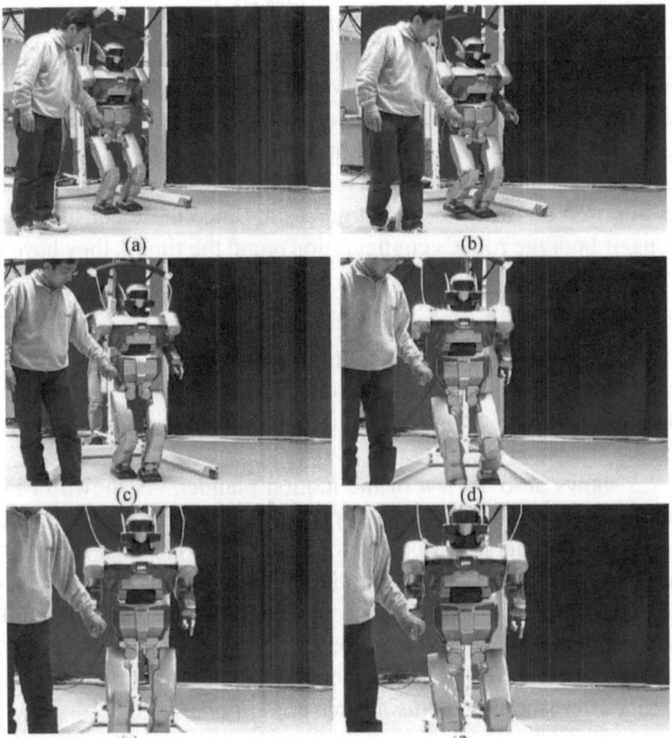

Figure 7.2 Experiment navigating the humanoid robot HRP-2 based on force sensor information at the wrist: (a) $t = 0.0$ s; (b) $t = 2.0$ s; (c) $t = 4.0$ s; (d) $t = 6.0$ s; (e) $t = 8.0$ s; (f) $t = 10.0$ s; (g) $t = 12.0$ s; (h) $t = 14.0$ s

7.3 Whole-body Motion Planning

In this section we explain the offline planning method of the collision-free whole-body motion for a humanoid robot dynamically walking on the floor [16]. For this purpose, we supply a two-stage time-parametrized framework for the whole-body motion planning of a humanoid robot dynamically walking on the floor while maintaining its balance.

7.3.1 Definitions

A configuration $q \in \mathscr{C}$ of the humanoid robot is composed of the position/orientation of the waist (p_B/ϕ_B) and all the joint angles (θ). We plan the robot's motion such that the feet make contact with the environment at the desired position within the specified period of time. At the same time, the robot has to avoid all other collisions of the links. Let $\mathscr{C}_{free} \subset \mathscr{C}$ be the set of configurations where such unnecessary collisions do not occur. Moreover, we impose the desired trajectories to some parts of the robot such as the feet, hands, and horizontal position of the CoG. Here, the horizontal trajectory of the CoG is obtained by using the walking pattern generator and is used to plan the motion of the robot while maintaining its balance.

We regard the condition for the robot to follow the desired trajectory as constraint conditions imposed on the robot's configuration. Since these constraint conditions are functions of both the robot's configuration q and the time t, they have the form

$$f(q,t) = 0. \tag{7.9}$$

Let $\mathscr{C}_{cons}(t) \subset \mathscr{C}$ be a subset of the configuration such that the robot follows the desired trajectory at a specified time. In our motion planning problem, we search for the configuration $q(t)$ $(t_0 \leq t \leq t_n)$ of the robot from the start q_{st} to the goal q_{ed} included in the set $\mathscr{C}_{free} \cap \mathscr{C}_{cons}(t)$.

Figure 7.3 shows an overview of the motion planner. As shown in Figure 7.3, we first run the walking pattern generator described in Section 7.2. While running the walking pattern generator, we monitor whether or not unnecessary collisions occur. If collisions do not occur, the motion planner returns the joint trajectory of the robot. On the other hand, if unexpected collisions occur, we plan a collision-free path of the configuration within the period of time where collisions occur by using the method described in Section 7.3.3. Then, the planner returns the collision-free joint trajectory after smoothing it using spline interpolation.

7.3.2 Walking Pattern Generation

Let the sum of the gravity and the inertia force applied to the robot be f_G and the sum of the moments about the COG of the robot be τ_G with respect to the reference coordinates. Although we explained the online walking pattern generation method in the previous section, this section uses the offline one [7] where the CoG trajectory is generated for a given trajectory of f_G/τ_G. Here, the ZMP can be expressed as a function of f_G/τ_G. While executing the walking pattern generator, we record the time trajectories of the horizontal COG trajectory, the foot trajectory with respect to the inertia coodinate system, the hand trajectory with respect to the chest coordinate system and the neck joint trajectory. These trajectories are used when planning the collision-free whole-body motion.

7.3.3 Collision-free Motion Planner

This planner incrementally constructs a network of milestones m composed of two trees T_{st} and T_{ed} rooted at m_{st} and m_{ed}, respectively. The planner grows trees of

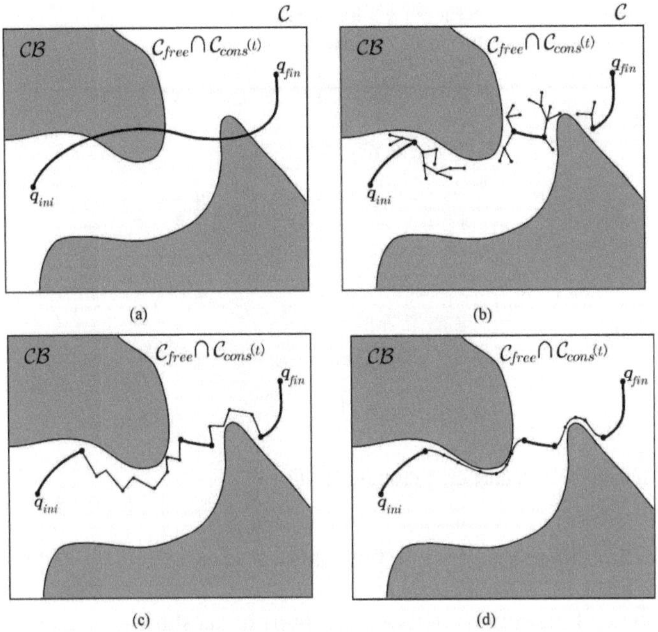

Figure 7.3 Overview of the whole-body motion planning method: (a) walking pattern generation and collision detection; (b) motion planning using PRM; (c) path generation; (d) shortcut and smoothing

collision-free milestones until a connection is found between two trees. Once a connection is found, the planner checks for collisions of the path between the two collision-free milestones included in the path.

Figure 7.4 shows an overview of the proposed planner. As shown in Figure 7.4(a), while running the walking pattern generator, we record the period of time in which unnecessary collisions occur. Then, before starting the motion planner, we determine the time of the start/goal of the planner, t_0 and t_n, such that the period of collision is included between t_0 and t_n. We further consider discretizing the time span as $t = t_0, t_1, \cdots, t_{n-1}, t_n$. In our proposed planner, each milestone m is composed of the configuration of the robot q and the time parameter t. If time t_i is associated with milestone m_i, we have

$$t_i = \mathrm{time}(m_i). \tag{7.10}$$

In this case, the milestone m_i has to satisfy the constraint condition $f(q, t_i) = 0$.

We assume that the root configurations satisfy $q_{st} \in \mathscr{C}_{free} \cap \mathscr{C}_{cons}(t_0)$ and $q_{ed} \in \mathscr{C}_{free} \cap \mathscr{C}_{cons}(t_n)$. If the milestone m_j is a child of the milestone m_i and is included in T_{st}, we set $\mathrm{time}(m_j) = t_{i+1}$, $q(t_{i+1}) \in \mathscr{C}_{free} \cap \mathscr{C}_{cons}(t_{i+1})$. $\mathrm{time}(m_i) = t_i$ and $q(t_i) \in \mathscr{C}_{free} \cap \mathscr{C}_{cons}(t_i)$. On the other hand, if m_i is included in T_{ed}, we set $\mathrm{time}(m_j) = t_{i-1}$.

The path obtained by using this method may not be a smooth one; it may include detours and unexpected discontinuity of velocity. Our smoothing algorithm comprises two steps: the shortcut path is found in the first step and spline interpola-

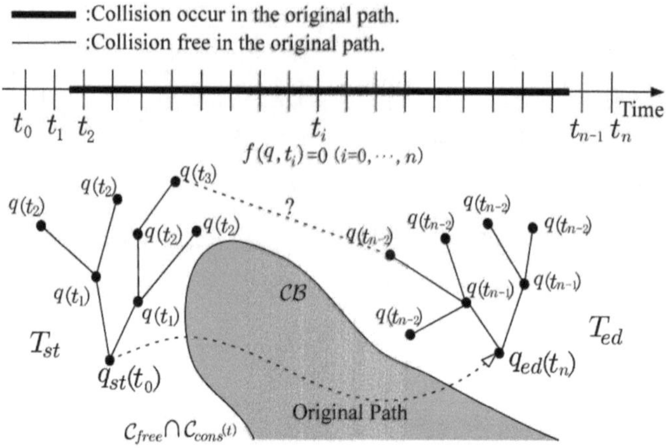

Figure 7.4 Method for including the constraint condition

tion is applied in the second step. In both steps, it is essential to avoid unnecessary collisions of links.

An overview of the spline interpolation algorithm is shown in Figure 7.5. Figure 7.5(a) shows the path obtained by the shortcut operation. Then, as shown in Figure 7.5(b), we split the trajectory between $q(t_i)$ and $q(t_j)$ into n segments $S_1, S_2, \cdots, S_n$. By applying b-spline interpolaton, we can obtain a set of curved trajetories. Figure 7.5(b) shows the case in which the trajectory is split into n segments.

We check for collisions of the curved segment from S_1 to S_n. If the collision occurs in S_i, we consider adding an additional node to S_i and splitting S_i into S_{i1} and S_{i2}. Figure 7.5(b) shows the case where collision occurs at S_2. Then, Figure 7.5(c) shows that an additional node is inserted to S_2 and shows that S_2 is split into S_{21} and S_{22}. After checking for collisions of S_i, we check for collisions of S_{i+1}. If we finish checking for collisions of S_n, we return to S_1. We iterate this operation until we obtain a collision-free trajectory connecting $q(t_i)$ to $q(t_j)$.

If the difference between the shortcut trajectory and the splined trajectory reduces as the number of node increases and if the minimum distance between the shortcut trajectory and the obstacle is greater than 0, we can obtain a smooth and collision-free path by using this spline interpolation algorithm. As far as we tried, as the number of nodes increased, the difference between the shortcut trajectory and the splined trajectory decreased. However, the same may not necessarily be true if we use spline interpolation.

7.3.4 Results

We used OpenHRP to generate the biped walking pattern combined with the motion planner MPK (motion planning kit) [18].

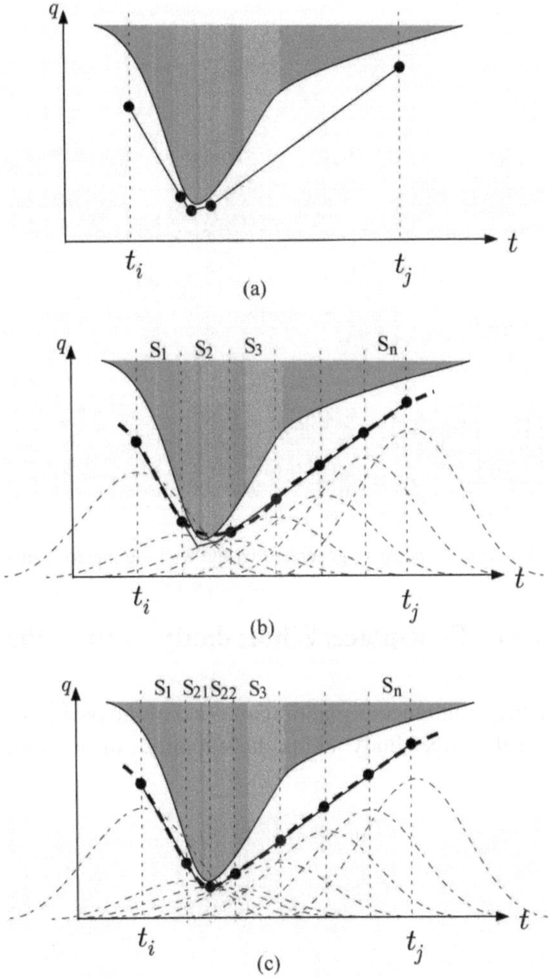

Figure 7.5 Spline interpolation: (a) input path generated by shortcut operation; (b) initial spline interpolation; (c) spline interpolation inserting an additional node

By using a 2.4 GHz PC, the robot's motion of approximately 15 sec is calculated within 3 min. We planned the motion of a humanoid robot walking through a gate. Figure 7.6 shows the output of the walking pattern generator. As we can see from Figure 7.6 (d) and (e), a collision occurs between the gate and the robot. Then, between 5 sec before and 3.5 sec after the collision, we plan the collision-free motion of the robot. The result of motion planning is shown in Figure 7.7; the robot avoids the collision between itself and the gate. As shown in this result, the humanoid robot dynamically walks through the gate while maintaining its balance. Figure 7.8 shows another example where the humanoid robot passes through the narrow space.

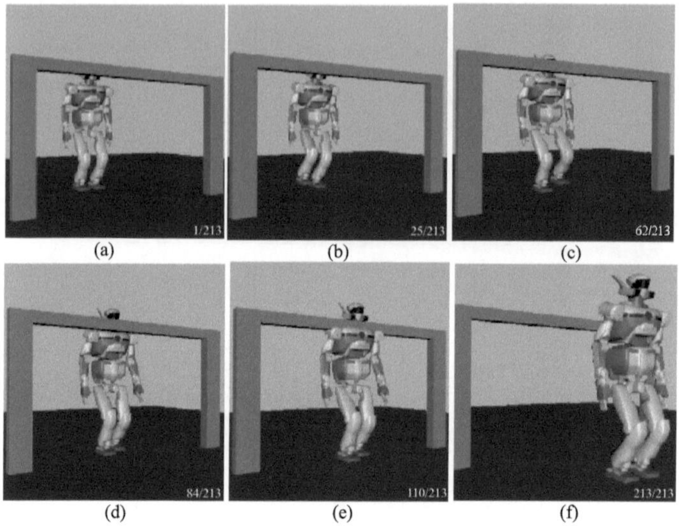

Figure 7.6 Original motion of the robot generated by the walking pattern generator

7.4 Simultaneous Foot-place/Whole-body Motion Planning

Next, we explain the simultaneous planning method of the position of the foot-place and the posture of the upper-body taking the dynamics of the robot into consider-

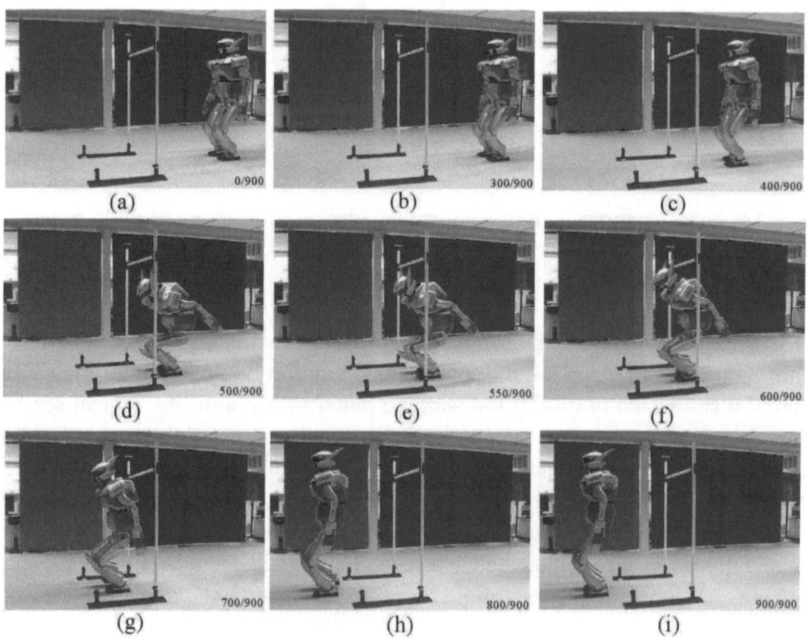

Figure 7.7 Experimental result for walking through gate

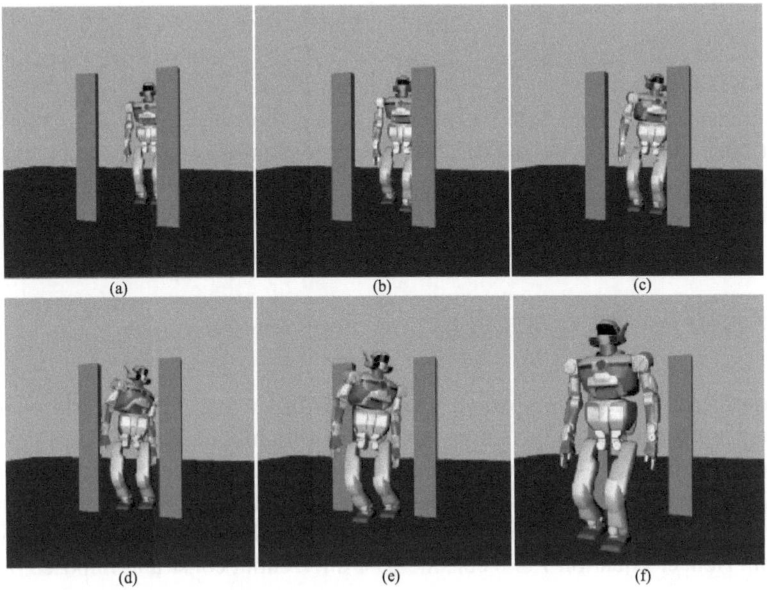

Figure 7.8 Calculation result of passing through a narrow space

ation [15]. In the whole-body motion planning method explained in the previous section, the position of the feet is determined before running the walking pattern generator and is not changed when planning the collision-free motion. The key idea to realize such function is to use the online walking pattern generator explained in the previous section for the offline motion planning problem.

7.4.1 Definitions

Let us consider the following coordinate transformation:

$$(p_G, \phi_B, \xi_{Fr}, \xi_{Fl}, \xi_{Hr}, \xi_{Hl}, \theta_r) = f(\xi_B, \theta), \qquad (7.11)$$

where
$p_G = [x_G\ y_G\ z_G]^T$ position of the COG,
ϕ_B orientation of the waist,
$\xi_{Fj}\ \ j = l, r$ 6 dimensional position/orientation of the left/right foot,
$\xi_{Hj}\ \ j = l, r$ 6 dimensional position/orientation of the left/right hand,
θ_r Joint angle vector of the chest, the neck and the finger.
Since we specify the desired trajectory for the position/orientation of the hands, the dimension of the configuration space can be reduced and we can obtain:

$$q = (p_G, \phi_B, \xi_{Fr}, \xi_{Fl}, \theta_r) \in \mathscr{C}, \quad (7.12)$$
$$= (x_G, y_G, \hat{q}) \quad (7.13)$$

We simultaneously plan both the position of the footstep (ξ_{Fr} and ξ_{Fl}) and the upper-body motion (z_G, ϕ_B and θ_r). Our proposed planner can be applied to general humanoid robots, its configuration having the form of Equation 7.12 and both legs having at least 6 DOF.

7.4.2 Gait Pattern Generation

We use the online walking pattern generator in order to simultaneously plan the footstep and the upper body motion. Among the variables defined in Equation 7.12, the horizontal CoG position has to be determined such that the robot can maintain its balance. We obtain the horizotal CoG position by solving Equation 7.3. Here, in order to plan the footstep, we solve Equation 7.3 while planning the collision-free motion. Hence, Equation 7.3 is considered a differential constraint imposed on the motion planner.

The online walking pattern generator calculates the trajectory of x_G and y_G for a few footsteps starting from the next single-support phase by using Equation 7.3. This trajectory is connected to the currently executing trajectory. The trajectory composed of a few foot-steps is calculated since we can determine the position/velocity of the CoG if we arrange that the robot stops at the last footstep.

7.4.3 Overall Algorithm

Algorithm H1 shows the planning algorithm. Given the start (q_{init}) and the goal (q_{goal}) configurations, the planner builds a tree T rooted at the start configuration (EXPAND-TREE) and tries to connect the tree to the goal configuration (CONNECT-GOAL). If the tree can be connected to the goal, the planner returns the path τ from the start to the goal. After the path is obtained, we check for collisions on the path.

Algorithm H1 PLANNER(q_{init}, q_{goal})

```
Install q_init as the root of T
Repeat s times
    EXPAND-TREE
    τ ← CONNECT-GOAL
    If τ ≠ nil, then return tau
Return failure
```

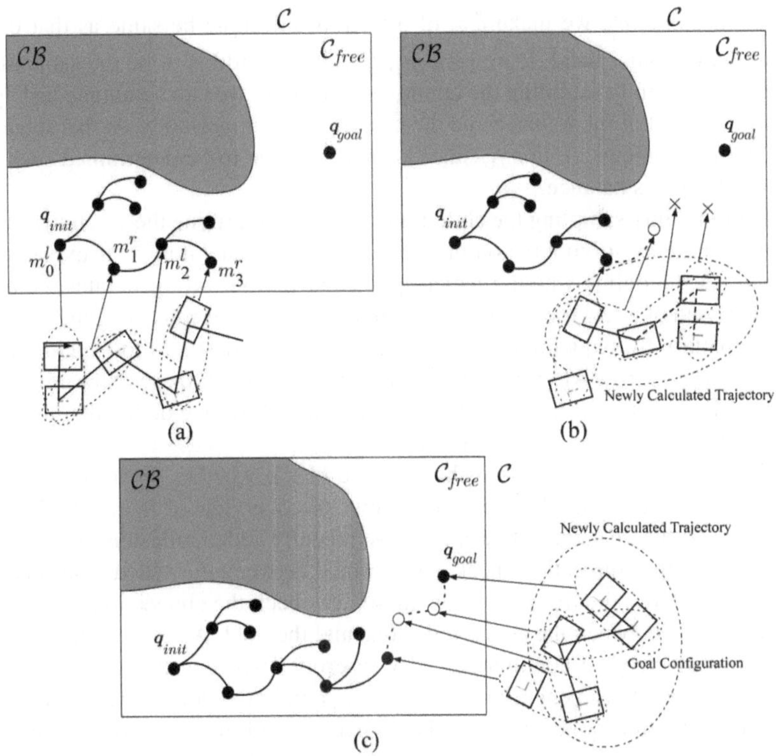

Figure 7.9 Overview of motion planning method: (a) definition of tree; (b) expansion of tree; (c) connection of tree

Figure 7.9 shows an overview of our planning algorithm. Figure 7.9(a) shows the definition of the tree of milestones used in our planner. We define each milestone by using the configuration of the robot just before the single-support phase begins. Here, we can define two kinds of milestones depending on the forthcoming single-support phase. The milestones m^r and m^l denote that the forthcoming single-support phase is the right-support and the left-support phase, respectively. Let m_i^r be the ith milestone with the forthcoming right-support phase. The parent of this milestone can be defined by

$$m_j^l = \text{parent}(m_i^r), \tag{7.14}$$

where the function parent($*$) is used to define the parent of the milestone $*$. Equation 7.14 means that the left-support phase comes after the right-support phase and vice versa.

An overview of the tree expansion algorithm is shown in Figure 7.9(b). First, we randomly pick a (parent) milestone from T. Then, by randomly sampling the configuration space, we generate a new (child) milestone. Here, if we picked the parent milestone m_j^r, we make ξ_{Fr} of the child milestone be same as that of the parent. On the other hand, if we pick m_j^l, ξ_{Fl} of the child is to be the same as the parent. By randomly sampling the configuration space, we can simultaneously plan the displacement of the footstep and the posture of the upper-body. At this stage, we can determine neither x_G nor y_G since x_G and y_G have to be determined such that the robot keeps its balance.

After randomly sampling the child milestone, we determine the x_G and y_G of the chile milestone by solving the online walking pattern generator. As is explained in Section 7.2, the online walking pattern generator periodically generates the biped gait composed of a few footsteps. Hence, in our planner, we consider generating the walking pattern by sequentially connecting two extra milestones to the child milestone (Figure 7.9(b)). After generating the child milestone, these extra milestones will be deleted and will not be used by the planner. Also, we define that the robot stops at the second added extra milestone.

Once a milestone is added to the tree, we check whether or not the tree can be connected to the goal. Figure 7.9(c) shows the overview of the method of tree connection. If the distance between the most recently added milestone and the goal is less than ρ, the connection check is performed between the added milestone and the goal. Similar to the tree expansion phase, we check the connection by inserting two milestones between the added milestone and the goal. Also, we define that the robot stops at the goal configuration. By inserting two milestones, we solve the online walking pattern generator. Then, we can obtain the horizotal position of the CoG for all the milestones between the added one and the goal. By inserting two milestones, the robot can smoothly stop at the goal configuration.

If all the milestones between the added one and the goal are found to be collision-free, we can obtain a path τ between the start and the goal. After a candidate path τ is generated, we check for collisions on all the curved segments included in the path.

7.4.4 Experiment

We calculated the motion of the humanoid robot HRP-2 walking to avoid an obstacle. We simultaneously planned the 2D waist joint angles and x-y position of the footstep. When planning, we descretized the displacement of the footstep. The experimental result is shown in Figures 7.10 and 7.11. As we can see from the figure, the robot avoids the obstacle walking to the right. At the same time, the robot changes the joint angles of the waist.

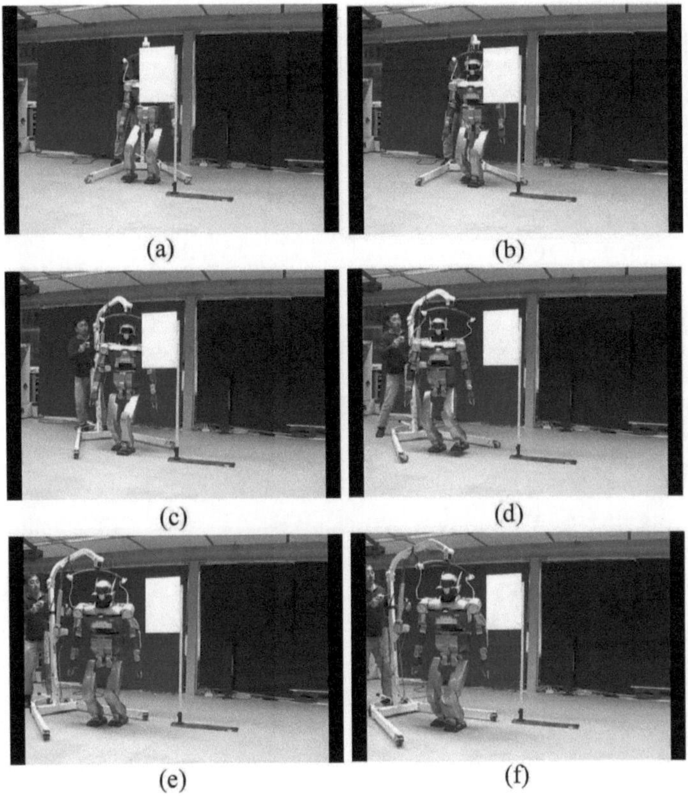

Figure 7.10 Walking motion of HRP-2 avoiding an obstacle (side view)

7.5 Whole-body Manipulation

When a humanoid robot performs a task, the hands contact the environment and it becomes important to consider how we can include the hand reaction force into the motion planner. This section discusses two methods.

As an example of whole-body manipulation of a humanoid robot, let us consider dynamically walking while pushing a large object placed on the floor. In such a case we can consider two styles of pushing manipulation, *i.e.*, position-control-based pushing manipulation and force-control-based pushing. As for the position-control-based method, the gait pattern is determined before the robot actually moves and is not changed,depending on the hand reaction force. Although the position-control-based method can be easily implemented, the robot may not keep balance if the weight of the object or the friction coefficient between the object and the floor is different from that predicted. On the other hand, as for the force-control-based method, the gait pattern is adaptively changed depending on the force sensor information at the tip of the arms. By using the force-control-based method, we expect that the

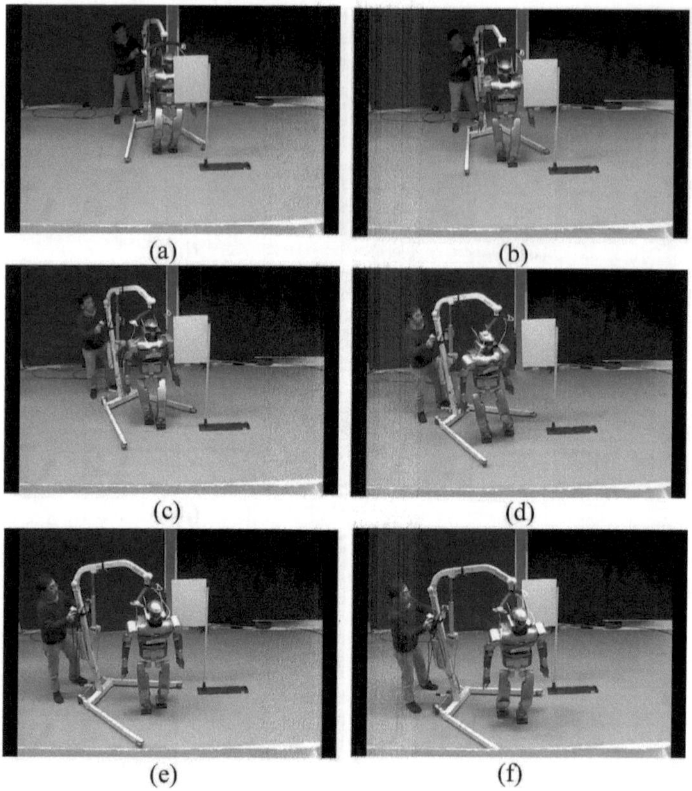

Figure 7.11 Walking motion of HRP-2 avoiding an obstacle (top view)

robot will not lose balance even if the weight of the object or the friction coefficient between the object and the floor is changed during robot movement. To realize the pushing manipulation, we use the online walking pattern generator. In this section, we do not consider collisions among the links of the robot.

7.5.1 Motion Modification

We first consider the position-control-based method. In this case, whole-body manipulation can be realized by modifying the motion of the robot based on the predicted hand reaction force [10, 12].

Let $\bar{p}_{zmp}$ be the ZMP position without considering the hand reaction forces. From Equation 7.2, the change of the ZMP position due to the hand reaction force in the sagittal plane is given by

$$x_{zmp} - \bar{x}_{zmp} = \sum_{j=1}^{2} \frac{(z_{Hj} - z_{zmp}) f_{Hj}^{(x)} + (x_{zmp} - x_{Hj}) f_{Hj}^{(z)}}{M(\ddot{z}_G + g)}, \tag{7.15}$$

where $p_{Hj} = [x_{Hj}\ y_{Hj}\ z_{Hj}]$ and $f_{Hj} = [f_{Hj}^{(x)}\ f_{Hj}^{(y)}\ f_{Hj}^{(z)}]$ $(j = 1, 2)$ denote the position of the hand and the hand reaction force, respectively.

Let us explain the physical interpretation of Equation 7.15. Figure 7.12 shows the effect of the hand reaction forces onto the position of the ZMP. As shown in Figure 7.12(a), when the hand does not contact the environment, $p_Z = \bar{p}_Z$ is satisfied. On the other hand, when pushing the wall as shown in Figure 7.12(b), the position of the ZMP will shift to the back of the robot. Also, when carrying an object as shown in Figure 7.12(c), the position of the ZMP will shift to the front of the robot.

Figure 7.13 shows the result of the pushing manipulation where the humanoid robot walks while pushing a large object placed on the floor [10]. In this case, by using the weight of the object and the friction coefficient between the floor and the object, the hand reaction force in the horizontal direction can be predicted. By substituting the predicted hand reaction force into Equation 7.15, the modification of the ZMP position can be calculated. The horizontal position of the waist is modified using this value. The picture shows that the humanoid robot can push a large object without falling down.

Figure 7.14 shows the results of an experiment with humanoid robot carrying an object of weight 8 kg[12]. Using the measured hand reaction force, the horizotanl position of the waist is modified based on Equation 7.15.

7.5.2 Force-controlled Pushing Manipulation

Next, we consider force-control-based pushing manipulation [11, 13]. Let us consider the case where a humanoid robot pushes the object while walking forward. Let us also consider the motion of the robot within the sagittal plane. In this subsection,force-control-based pushing manipulation is realized by separating the pushing phase from the stepping phase. The timing chart of the force-control-based pushing manipulation is shown in Figure 7.15.

During the pushing phase, the robot pushes the object,controlling the reaction force applied at the hands using impedance control. Here, the desired impedance is given by

$$m\ddot{x}_H + c\dot{x}_H = f_{xd} - f_x. \tag{7.16}$$

An overview of the impedance control used in the pushing manipulation is shown in Figure 7.16. When $f_{xd} \le \mu m_O g$, the object will not move since f_x balances f_{xd}. On the other hand, when $f_{xd} > \mu m_O g$, the object moves,the acceleration of the object being dependent on the difference between f_{xd} and f_x. When the target impedance is realized, an object with light weight moves with high acceleration. The heavier

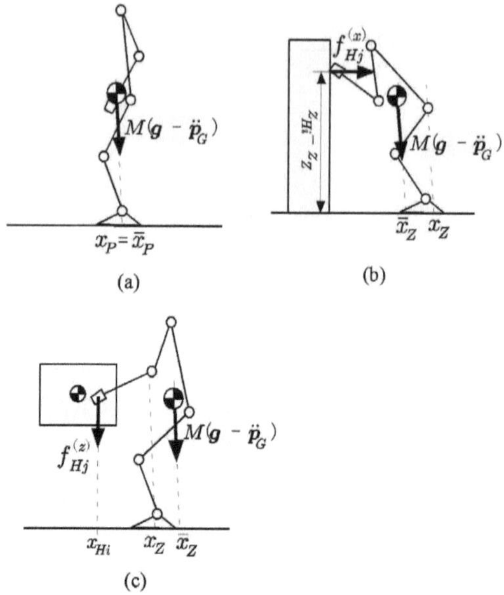

Figure 7.12 Modification of the ZMP position: (a) humanoid robot without manipulation; (b) humanoid robot pushing wall; (c) humanoid robot carrying an object; (d) definition of heel and toe

the weight of the object, the lower the acceleration of the object. Finally the object will not move if it is very heavy. Therefore, by using this algorithm, we can expect the robot to keep dynamical balance without taking the mass of the object into consideration.

During the stepping phase, the step length is set the same as the amount of pushing the object in the pushing phase. By using the amount of pushing the object, the desired ZMP trajectory is recalculated, and is smoothly connected to the current ZMP trajectory. To realize the desired ZMP trajectory, the CoG trajectory is also recalculated using the online walking pattern generator.

Figure 7.17 shows an experimental result [11]. In the experiment, we used a table weighing about 10kg. While the motion of the table is disturbed between $t = 12sand$t=20s, we can see that the robot keeps balance by adaptively changing the gait pattern depending on the pushing force information.

7.6 Conclusion

In this chapter we explained the method of planning the motion of a humanoid robot using the walking pattern generator. We explained the method of planning the whole-body motion, the method of simultaneously planning the foot-place and the upper-body motion, and the whole-body manipulation problem. In a future research

Figure 7.13 Experimental Result of Pushing Manipulation: (a) $t = 2.0$ s; (b) $t = 6.0$ s; (c) $t = 10.0$ s; (d) $t = 14.0$ s

we plan to construct the motion planner of a humanoid robot taking all the functions explained in this chapter into account.

References

[1] A. Takanishi, H. Lim, M. Tsuda, and I. Kato, Realization of Dynamic Biped Walking Stabilized by Trunk Motion on a Sagittally Uneven Surface, In: proceedings of IEEE international workshop on intelligent robots and systems (IROS '90), pp 323–330, 1990.

[2] S. Kagami, K. Nishiwaki, T. Kitagawa, T. Sugihiara, M. Inaba, and H. Inoue, A Fast Generation Method of a Dynamically Stable Humanoid Robot Trajectory with Enhanced ZMP Constraint, In: proceedings of IEEE international conference on Humanoid Robots, 2000.

[3] S. Kajita, F. Kanehiro, K. Kaneko, K. Fujiwara, K. Harada, K. Yokoi, and H. Hirukawa, Biped Walking Pattern Generation by using Preview Control of Zero-Moment Point, In: proceedings of IEEE international conference on robotics and automation, pp 1620–1626, 2003.

[4] K. Harada, S. Kajita, K. Kaneko, and H. Hirukawa, An Analytical Method on Real-time Gait Planning for a Humanoid Robot, In: proceedings of 2004 IEEE-RAS/RSJ international conference on Humanoid Robots (Humanoids2004), 2004

[5] K. Harada, S. Kajita, K. Kaneko, and H. Hirukawa, An Analytical Method on Real-time Gait Planning for a Humanoid Robot, J. of Humanoid Robotics, 3-1, pp 1–19, 2006

[6] T. Sugihara and Y. Nakamura, A Fast Online Gait Planning with Boundary Condition Relaxation for Humanoid Robots, In: proceedings of IEEE international conference on robotics and automation(ICRA'05), 2005.

[7] H.Hirukawa, S.Hattori, S.Kajita, K.Harada, K.Kaneko, F.Kanehiro, M.Morisawa, and S.Nakaoka, A Pattern Generator of Humanoid Robots Walking on a Rough Terrain, In: pro-

Figure 7.14 Experiment of lifting an object: (a) $t = 0.0$ s; (b)$t = 3.0$ s; (c) $t = 6.0$ s; (d) $t = 9.0$ s; (e) $t = 12.0$ s; (f) $t = 15.0$ s; (g) $t = 18.0$ s; (h) $t = 21.0$

ceedings of 2007 IEEE international conference on robotics and automation, pp 2181–2187, 2007.

[8] T. Takubo, K. Inoue, K. Sakata, Y. Mae, T. Arai, Mobile Manipulation of Humanoid Robots - Control Method for CoM Position with External Force -, In: proceedings of IEEE/RSJ international conference on intelligent robots and systems (2004) 1180–1185

[9] Y. Hwang, A. Konno, and M. Uchiyama, *Whole Body Cooperative Tasks and Static Stability Evaluations for a Humanoid Robot*, In: proceedings of IEEE/RSJ international conference on intelligent robots and systems, pp 1901–1906, 2003.

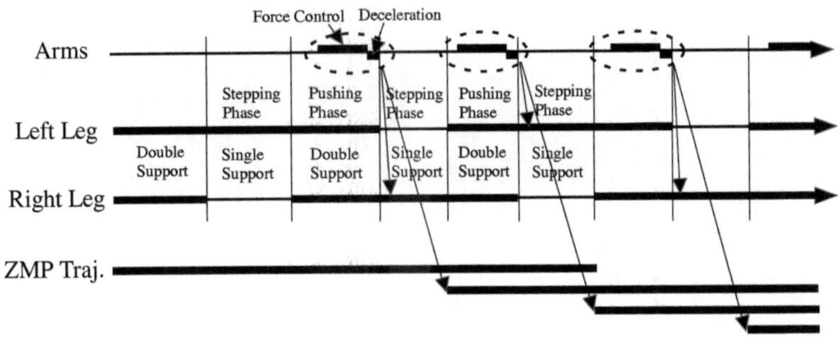

Figure 7.15 Time chart of force controlled pushing manipulation

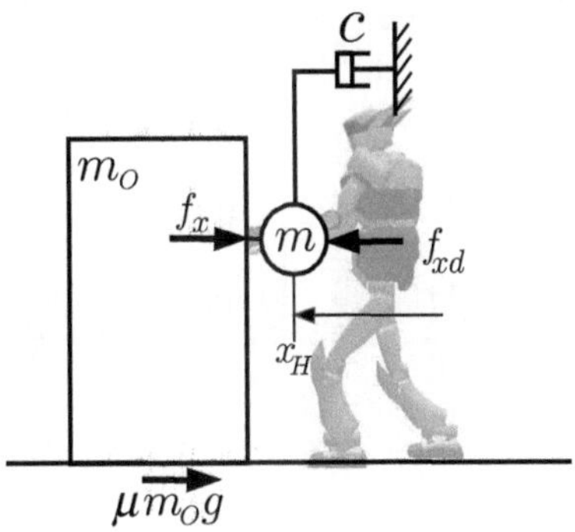

Figure 7.16 Arm Impedance Control

[10] K. Harada, S. Kajita, K. Kaneko, and H. Hirukawa, *Pushing Manipulation by Humanoid considering Two-Kinds of ZMPs*, In: proceedings of IEEE international conference on robotics and automation, pp 1627–1632, 2003.

[11] K. Harada, S. Kajita, F. Kanehiro, K. Fujiwara, K. Kaneko, K. Yokoi, H. Hirukawa: Real-Time Planning of Humanoid Robot's Gait for Force Controlled Manipulation, In: proceedings of IEEE international conference on robotics and automation (2004) 616–622

[12] K. Harada, S. Kajita, H. Saito, M. Morisawa, F. Kanehiro, K. Fujiwara, K. Kaneko, H. Hirukawa, A Humanoid Robot Carrying a Heavy Object, In: proceedings of IEEE international conference on robotics and automation (2005) 1724–1729

[13] K. Harada, S. Kajita, F. Kanehiro, K.Fujiwara, K. Kaneko, K.Yokoi, and H. Hirukawa, Real-Time Planning of Humanoid Robot's Gait for Force Controlled Manipulation, IEEE/ASME Trans on Mechatronics, 12-1, pp 53-62, 2007.

[14] M.Morisawa, K.Harada, S.Kajita, S.Nakaoka, K.Fujiwara, F.Kanehiro, K.Kaneko, H.Hirukawa, Experimentation of Humanoid Walking Allowing Immediate Modification of Foot Place Based on Analytical Solution, In: proceedings of 2007 IEEE international conference on robotics and automation, pp 3989–3994, 2007.

[15] K. Harada, M. Morisawa, K. Miura, S. Nakaoka, K. Fujiwara, K. Kaneko, and S. Kajita, Kinodynamic Gait Planning for Full-Body Humanoid Robots, In: proceedings of 2008 IEEE/RSJ international Conference on intelligent robots and systems, pp 1544–1550, 2008.

[16] K. Harada, S. Hattori, H. Hirukawa, M. Morisawa, S. Kajita, and E. Yoshida, Motion Planning for Walking Pattern Generation of Humanoid Robots, In: proceedingsIEEE/RSJ international conference on intelligent robots and systems, 2007.

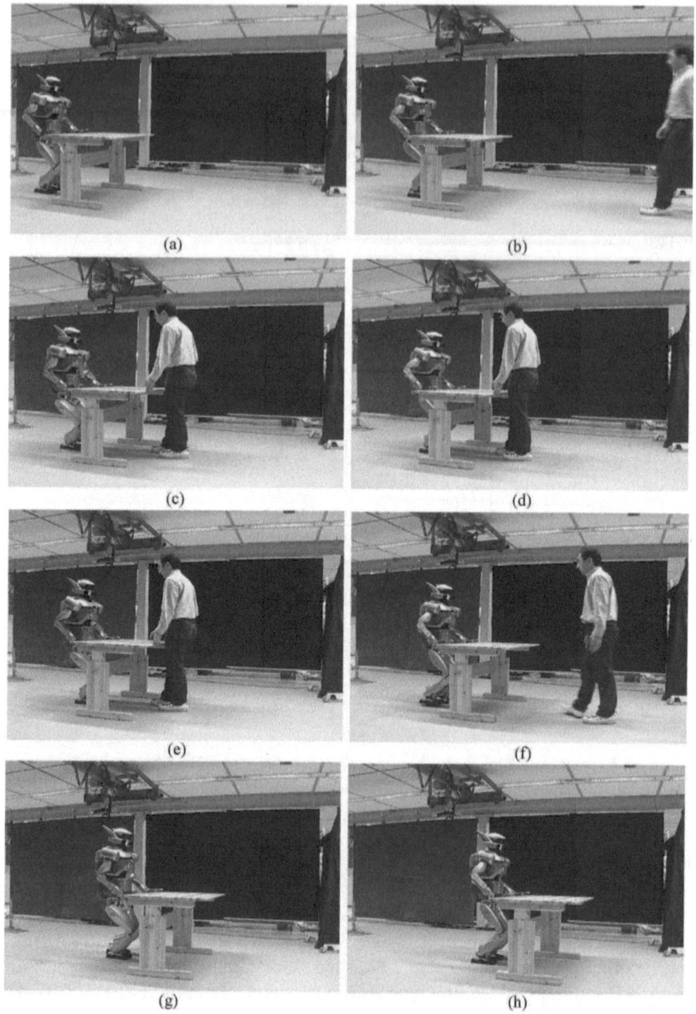

Figure 7.17 Pushing manipulation experiment: (a) $t = 0.0$ s; (b) $t = 10.0$ s; (c) $t = 12.0$ s; (d) $t = 14.0$ s; (e) $t = 18.0$ s; (f) $t = 22.0$ s; (g) $t = 32.0$ s; (h) $t = 42.0$ s;

[17] S.Kajita et al., Resolved Momentum Control: Humanoid Motion Planning based on the Linear and Angular Momentum, In: proceedings of IEEE/RSJ international conference on intelligent robots and systems, pp 1644–1650, 2003.
[18] http://robotics.stanford.edu/ mitul/mpk/

Chapter 8
Autonomous Manipulation of Movable Obstacles

Mike Stilman

Abstract In this chapter we describe recent progress towards autonomous manipulation of environment objects. Many tasks, such as nursing home assistance, construction or search and rescue, require the robot to not only avoid obstacles but also move them out if its way to make space for reaching the goal. We present algorithms that decide which objects should be moved, where to move them and how to move them. Finally, we introduce a complete system that takes into account humanoid balance, joint limits and fullbody constraints to accomplish environment interaction.

8.1 Introduction

Humanoid robots would be much more useful if they could move obstacles out of the way. In this section, we address the challenges of autonomous navigation among movable obstacles (NAMO), present algorithms and evaluate their effectiveness in simulation and present a successful implementation of NAMO on a humanoid robot. Traditional motion planning searches for collision freepaths from a start to a goal. However, realworld domains like search and rescue, construction, home robots and nursing home assistance contain debris, materials clutter, doors and objects that need to be moved by the robot. Theoretically, one can represent all possible interactions between the robot and movable objects as a huge search. We will present two methods that use state space decomposition and heuristic search to simplify the problem and find practical solutions in common environments.

The ability to solve NAMO problems has direct applications in domains that require robot autonomy for safety and assistance. Consider a search and rescue mission to extract the victims of a natural disaster such as Hurricane Katrina. Already robots enter these environments to minimize the exposure of humans to unstable structures, gases and other hazards. However, collapsed rubble and objects displaced by floodwaters cause previously navigable passages to become blocked. In solving NAMO, the robot examines its environment, decides which objects must be moved and clears a way to further progress. Human safety is one of many motivations for

Mike Stilman
Georgia Tech, Atlanta, GA, USA e-mail: mstilman@cc.gatech.edu

Figure 8.1 Planned solution to a *navigation* problem that manipulates an environment object

NAMO. Consider the comfort and self-reliance of the elderly and disabled. In 2006, Nursing Economic reported that the United States shortage of nurses would grow to eight times the current size by the year 2020 [1]. Robots that help patients get around, reach for medicine and food alleviate this need and provide independence as well as personal autonomy. Yet, in contrast to laboratories and factories, human domains have movable objects that are often misplaced. In Figure 8.1, even a chair in front of a table challenges the robot to NAMO.

8.1.1 Planning Challenges

We begin our analysis of the NAMO domain as an instance of geometric motion planning [2]. During planning, we assume that the geometry and kinematics of the environment and the robot are known. We also assume that there is no uncertainty in sensing and the effects of robot actions. These assumptions are softened during implementation through active modeling and replanning. We represent objects and robot links as polyhedrons. The environment objects are classified as either fixed or movable.

Formally, the environment is modeled as a 2D or 3D Euclidian space that contains the following items:

- $O_F = \{F_1, \ldots, F_f\}$ - a set of polyhedral *Fixed Obstacles* that must be avoided.
- $O_M = \{O_1, \ldots, O_m\}$ - a set of *Movable Obstacles* that the robot can manipulate.
- R - a manipulator with n degrees of freedom represented by polyhedral links.

While paths may not be explicitly parameterized by time, we will use the variable t to refer to a chronological ordering on states and operations. At any time t, the world state W^t defines the position and orientation of the robot links and each object. We represent the world state as follows:

$$W^t = (t, r^t, q_1^t, q_2^t, \ldots, q_m^t).$$

Given an initial configuration W^0 of the robot r^0 and each movable obstacle q_i^0, the goal is to achieve a final configuration r^f for the manipulator.

8.1.2 Operators

In order to achieve the goal configuration the robot is permitted to change its own configuration and possibly the configuration of one grasped obstacle at any time step. Between time steps, any change to the robot joints must be continuous. We therefore interpret any "change" as an action that follows a path or trajectory.

We can distinguish between two primitive operators or actions: *Navigate* and *Manipulate*. Each action is parameterized by a path $\tau(r_i, r_j)$ that defines the motion of the robot between two configurations: $\tau : [0,1] \to r$ where $\tau[0] = r_i$ and $\tau[1] = r_j$.

The *Navigate* operator refers to contact-free motion. While the robot may be in sliding contact with an object, its motion must not displace any objects by collision or friction. *Navigate* simply moves the robot joints as specified by τ:

$$Navigate : (W^t, \tau(r^t, r^{t+1})) \to W^{t+1}. \tag{8.1}$$

When the robot motion affects the environment by displacing an object, O_i, we refer to the action as *Manipulate*. The manipulate operator consists of two paths: one for the robot and one for O_i. Since the object is not autonomous, the object path is parameterized by the robot path and the initial contact or grasp $G_i \in G(O_i)$. The set $G(O_i)$ consists of relative transformations between the robot end-effector and the object that constitute contact.

$$Manipulate : (W^t, O_i, G_i, \tau(r^t, r^{t+1})) \to W^{t+1}. \tag{8.2}$$

Distinct G_i lead to different object motions given the same end-effector trajectory. We let $\tau_o = ManipPath(G_i, \tau)$ be the interpretation of the path for the object during manipulation according to the constraints imposed by the contact. *Manipulate* maps a world state, contact and path to a new world state where the robot and O_i have been displaced. The action is valid when neither the robot nor object collide or displace other objects. Validity is also subject to the constraints discussed in Section 8.1.3.

The two action descriptions point to a general formulation for interacting with environment objects. The robot iterates a twostep procedure. First, it moves to a contact state with the *Navigate* operator and then applies a *Manipulate* operator to displace the object. The robot also uses *Navigate* to reach a goal state.

8.1.3 Action Spaces

In Section 8.1.2 we left the parameters for *Manipulate* open to interpretation. This is consistent with our focus on deciding a high-level movement strategy for the robot in Sections 8.2 and 8.3. In this Section we give three classes of parameterizations for manipulating objects. In each case, the class translates the trajectory of the robot into a motion for the object. We point out the relative advantages and disadvantages of the classes with regard to modeling requirements, generality and reliability.

Grasping (Constrained Contact): The simplest method for translating robot motion into object motion can be applied when the object is rigidly grasped by the robot. The *Constrained Contact (CC)* interpetation of our actions relates them directly to *Transit* and *Transfer* operators described by Alami [3]. A grasped object remains at a fixed transform relative to the robot end-effector. To move an object, the robot must first *Navigate* to a grasping configuration and then *Manipulate*.

In addition to requiring collision free paths, a valid *CC Manipulate* operator constrains the initial state of the robot and object. Typically the contact must be a grasp that satisfies form closure. These criteria indicate that a desired motion of the robot will not cause it to release the object [4]. Such grasps may either be specified by the user or computed automatically [5]. It is also possible to constrain grasps with regard to robot force capabilities.

The advantages of *CC* are in interpretability and invariance to modeling error. The disadvantage is in generality. We easily predict the possible displacements for an object by looking at robot and object geometry. Furthermore, an accurate geometric model is typically the only requirement for ensuring reliable execution after grasping. However, some objects, such as large boxes, are difficult or impossible to grasp. Constrained environments may also restrict robot positions to make grasping impossible or require the end-effector to move with respect to the object.

Pushing (Constrained Motion): Manipulation that does not require a grasp with closure is called non-prehensile [4]. *Constrained Motion CM* is a subclass of non-prehensile *Manipulate* operators. Given any contact between the robot and object *CM* restricts the path that the manipulator can follow in order to maintain a fixed transform between the end-effector and the object.

The most studied form of *CM* manipulation relies on static friction during pushing [6, 7]. At sufficiently low velocities static friction prevents the object from slipping with respect to the contact surface. Given the friction coefficient and the friction center of the object we can restrict robot paths to those that apply a force inside the friction cone for the object.[6]

Constrained motion is more general than *CC* manipulation, however it relies on more detailed modeling. In addition to geometry, this method requires knowledge of friction properties. An incorrect assessment would lead to slip causing unplanned online behavior. *CM* is also less interpretable than *CC* since it is difficult to visualize the space of curves generated by local differential constraints.

Manipulation Primitives (MP): The *Manipulate* operator can be unconstrained in both grasp and motion. Morris and Morrow give taxonomies for interactions between a robot end-effector and an object [8, 9]. These include grasping/pushing and add other modes of interaction that allow translational or rotational slip.

Methods that do not guarantee a fixed transform between the robot and the object rely on forward simulation of the object dynamics. Interaction with the robot is simulated to determine the displacement of the object. We experimented with *MP* in [10] by commanding translation and allowing slip in rotation. In some cases a parameterized *MP* yields solutions where grasping and pushing cannot.

Further generality requires even more accurate dynamic modeling of the object for forward simulation. By allowing slip it is desirable that online sensing be used to close the loop and ensure that the object path is closely followed. The motions generated by constraining different degrees of freedom are not unique. Primitives can be restricted to a subset of significant motions in a given environment.

All three classes of object motion interpret the robot trajectory as an object trajectory. We presented them in order of increasing generality and decreasing reliability. Intuitively, less constraint in contact and motion leads to greater demands on model accuracy. In theory one can combine these operators to achieve generality and stability whenever possible. Practically, *Manipulate* operators can be selected based on the geometry of the environment. A smaller set of permitted interactions restricts the possible displacements for the object and reduces the branching factor of search.

8.1.4 Complexity of Search

In contrast to a mobile robot, the branching factor of search for humanoid robots relies on two modes of interaction. In addition to choosing paths for *Navigate*, the robot chooses contacts and paths for *Manipulate* operators. An even greater challenge to branching comes from the size of the state space. The robot must plan the state of the environment including all the positions of movable objects. Wilfong first showed that a simplified variant of this domain is NP-hard, making complete planning for the full domain computationally infeasible [11]. More recent work demonstrated NP-completeness results for trivial problems where square blocks can translated on a planar grid [12].

In order to better understand the source of complexity, consider a simplified grid world that contains a robot and m obstacles as shown in Figure 8.2. Suppose we attempt a complete search over the robot's action space. Let the configuration of the robot base be represented by $\mathscr{C}_R = (x, y)$, with resolution n in each direction. The generic size of the robot $\mathscr{C}$-space $|\mathscr{C}_R| = O(n^2)$. Analogously, for each object $|\mathscr{O}_i| = O(n^2)$. The full space of possible environment configurations is the product of these subspaces, $\mathscr{C}_R \times \mathscr{O}_1 \times \mathscr{O}_2 \times \ldots \times \mathscr{O}_N$, and therefore has $O(n^{2(N+1)})$ world states. The size of the search space is exponential in the number of objects that occupy the robot's environment.

The size of the configuration space relates directly to the complexity of a complete search over the robot's actions. In Figure 8.2, the robot can translate to an adjacent square, grasp an adjacent movable block and move while holding the block. The total number of possible actions at any time is nine: four directions of motion,

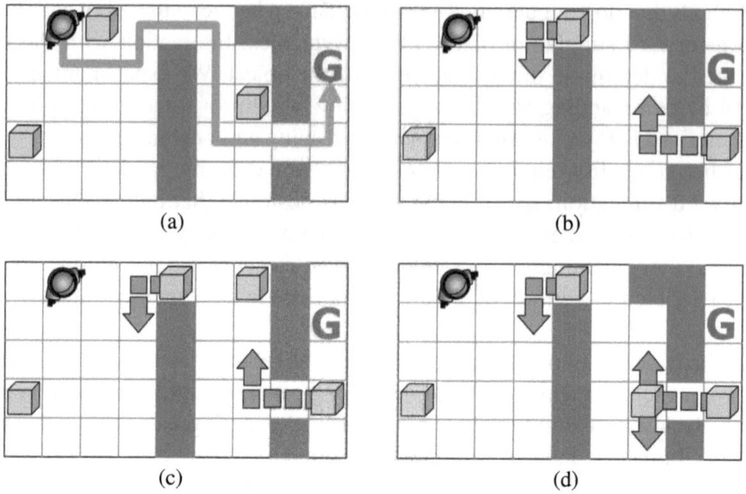

Figure 8.2 Simple NAMO problems on a planar grid: (a) N: 3, Moved: 0, States Searched: 4,840; (b) N: 3, Moved: 2, States Searched: 14,637; (c) N: 4, Moved: 2, States Searched: 48,264; (d) N: 4, Moved: 3, States Searched: 32,258

four directions of grasping and one release. For simplicity, we assign equal cost to all actions and apply breadth-first search (BFS) to find a sequence of actions that gives the robot access to the goal.

In Figure 8.2(a), although no obstacles are moved, the search examines 4840 distinct configurations before reaching the goal. Note that our search remembers and does not revisit previously explored world states. While there are only $9 \times 5 = 45$ possible placements of the robot, every obstacle displacement generates a new world state where every robot position must be reconsidered. A change in the world configuration may open formerly unreachable directions of motion for the robot or for the objects.

To a human, Figure 8.2(c) may look more like Figure 8.2(b) than Figure 8.2(d). A previously static obstacle is replaced by a movable one. The additional movable obstacle is not in the way of reaching the goal. However, breadth-first search takes three times longer to find the solution. These results generalize from simple search to the most recent domain independent action planners. Junghanns [13] applied the *Blackbox* [14] planner to Sokoban problems where a robot pushes blocks on a grid. The AIPS planning competition winner showed a similar rise in planning time when adding only a single block.

The most successful planners for puzzle problems such as the 15-puzzle, Towers of Hanoi and Sokoban benefit significantly from identifying and pre-computing solutions to patterns of states [15, 16, 13]. In highly structured problems, these patterns can be recognized and used as heuristics or macros for search. Such methods are likely to also succeed in grid world NAMO. However, our interest is in NAMO as a practical robot domain. Arbitrary object geometry and higher granularity action sets make pre-computation infeasible for any significant portion of subproblems.

In the context of geometric planning, sampling based methods such as probabilistic roadmaps and rapidly-exploring random trees have been applied to problems with exponentially large search spaces [17, 18]. These methods are most effective in *expansive* spaces where it is easy to sample points that significantly expand the search tree [19]. NAMO problems typically have numerous narrow passages in the state space. In Figure 8.2, among thousands of explored actions only one or two object grasps and translations make progress to the goal. Section 8.2 will discuss the narrow passages that exist in NAMO planning and use them to guide search.

8.2 NAMO Planning

8.2.1 Overview

In Section 8.1 we defined the NAMO domain and noted the complexity of motion planning with movable obstacles. While complete planning for NAMO may be infeasible, this section gives a resolution complete algorithm for an intuitive subclass of problems. To arrive at this algorithm we first give a *configuration space* interpretation of our domain. We observe the inherent structure of environments that consist of disjoint components of free space. This observation leads to the definition of a linear class of problems and an abstract graph algorithm for solving them. Finally, we give a practical variant of this algorithm, prove its validity and give experimental results in domains with nearly 100 movable objects.

8.2.2 Configuration Space

To understand the structure of NAMO and related spatial reasoning tasks, let us first interpret this problem in terms of the configuration space [20]. Let $\mathscr{C}_W$ be the space of all possible W^t. During a *Navigate* operation, the robot can only change its own configuration. Hence we denote a subspace or *slice* of $\mathscr{C}_W$:

$$\mathscr{C}_R(W^t) = (\{r\}, q_1^t, q_2^t, \ldots, q_m^t). \tag{8.3}$$

While $\mathscr{C}_R$ includes all possible configurations for the robot, some collide with static or movable obstacles. The free space of valid robot configurations is parameterized by the locations of movable obstacles. To make this relationship explicit:

$$A(q) = \{x \in \mathrm{R}^k | x \text{ is a point of object } A \text{ in configuration } q\}. \tag{8.4}$$

For any set of obstacle points S in R^k, a configuration space obstacle in $\mathscr{C}_A$ is the set $\mathscr{X}_A(S) = \{q \in \mathscr{C}_A | A(q) \cap S \neq \emptyset\}$. Let q be a configuration of A and p be a configuration of object B. Since two objects cannot occupy the same space in R^n, $\mathscr{X}$ is symmetric:

$$p \in \mathscr{X}_B(A(q)) \Rightarrow B(p) \cap A(q) \neq \emptyset \Rightarrow q \in \mathscr{X}_A(B(p)). \tag{8.5}$$

To simplify notation we define the following: $\mathscr{X}_R^{O_i}(W^t) = \mathscr{X}_R(O_i(q_i^t))$ and $\mathscr{X}_{O_j}^{O_i}(W^t) = \mathscr{X}_{O_j}(O_i(q_i^t))$ represent obstacles due to O_i in $\mathscr{C}_R$ and $\mathscr{C}_{O_j}$, respectively. For any set of obstacle points S in R^k, a configuration space obstacle in $\mathscr{C}_R$ is the set $\mathscr{X}_R(S) = \{r \in \mathscr{C}_R | R(r) \cap S \neq \emptyset\}$. Lastly, $\overline{\mathscr{X}_A(B)}$ is the complement of $\mathscr{X}_A(B)$ in $\mathscr{C}_A$.

The free space of a robot, $\mathscr{C}_A^{free}(W^t)$, is the set of configurations where the object is not in collision with fixed or movable obstacles. Eq. 8.6 defines $\mathscr{C}_R^{free}(W^t)$ and $\mathscr{C}_{O_i}^{free}(W^t)$ as sets of collision free configuration for the robot and movable objects. Figure 8.3(a) shows the free configuration space $\mathscr{C}_R^{free}(W^t)$ for a circular robot.

$$\mathscr{C}_R^{free}(W^t) = \bigcap_k \overline{\mathscr{X}_R(F_k)} \bigcap_i \overline{\mathscr{X}_R^{O_i}(W^t)} \qquad \mathscr{C}_{O_i}^{free}(W^t) = \bigcap_k \overline{\mathscr{X}_{O_i}(F_k)} \bigcap_{O_j \neq O_i} \overline{\mathscr{X}_{O_i}^{O_j}(W^t)} \tag{8.6}$$

We can use the $\mathscr{C}$-space representation to identify valid *Manipulate* and *Navigate* operators. *Navigate* is valid if and only if its path is collision free:

$$\tau(s) \in \mathscr{C}_R^{free}(W^t) \qquad \forall s \in [0,1]. \tag{8.7}$$

Equations 8.8 – 8.12 validate a *Manipulate* operator. In addition to satisfying collision free motion manipulation must end with the object in a statically stable *placement* $\mathscr{C}_{O_i}^{place}(W^t) \subseteq \mathscr{C}_{O_i}^{free}(W^t)$ (Equation 8.11). Equation Eq. 8.12 ensures that the robot does not collide with obstacle O_i. In our 2D examples, we assume gravity is orthogonal to the object plane and hence $\mathscr{C}_{O_i}^{place}(W^t) = \mathscr{C}_{O_i}^{free}(W^t)$.

$$\tau(s) \in \bigcap_k \overline{\mathscr{X}_R(F_k)} \bigcap_{j \neq i} \overline{\mathscr{X}_R^{O_j}(W^t)} \qquad \forall s \in [0,1] \tag{8.8}$$

$$\tau_{O_i}(s) \in \mathscr{C}_{O_i}^{free}(W^t) \qquad \forall s \in [0,1] \tag{8.9}$$

$$\tau_{O_i}(0) = q_i^t \tag{8.10}$$

$$\tau_{O_i}(1) \in \mathscr{C}_{O_i}^{place}(W^t) \tag{8.11}$$

$$R(\tau(s)) \cap O_i(\tau_{O_i}(s)) = \emptyset \qquad \forall s \in [0,1]. \tag{8.12}$$

8.2.3 Goals for Navigation

Having defined our domain in terms of configuration space we can begin to make useful observations about the planning problem. So far, we have looked at *Manipulate* and *Navigate* operators independently. However, the most interesting aspect of NAMO is the interdependence of actions. For instance, in order for the robot to manipulate an object it must be within reach of the object. It is not always possible to make contact with an object without previously moving another one. In this section we define *non-trivial* goals for navigation and use them to constrain subsequent manipulation.

First, consider the two robot configurations depicted in Figure 8.3(a). There exists a collision free path $\tau(r_0, r_1)$ that validates $Navigate(W^0, \tau(r_0, r_1))$. Finding this path is a typical problem for existing motion planners. Equation 8.13 generalizes this relationship to all robot configurations that are accessible from r^t in one step of *Navigate*:

$$\mathscr{C}_R^{acc}(W^t) = \{r_j \in \mathscr{C}_R^{free}(W^t) | \text{ exists } \tau(r^t, r_j) \text{ s.t. } \forall s(\tau[s] \in \mathscr{C}_R^{free}(W^t)\}. \quad (8.13)$$

Furthermore, the configurations in $\mathscr{C}_R^{acc}(W^t)$ can all be reached from one another.

$$\forall r_i, r_j \in \mathscr{C}_R^{acc}(W^t) \text{ exists } \tau(r_i, r_j) \text{ s.t. } \forall s(\tau[s] \in \mathscr{C}_R^{acc}(W^t)). \quad (8.14)$$

The space of accessible configurations is lightly shaded in Figure 8.3(a). We can compute configurations that belong to this set using grid-based wavefront propagation [21].

Accessible configurations are feasible goals for navigation. Likewise, contact configurations are feasible starting states for manipulation. Any class of *Manipulate* operators in Section 8.1.3 restricts the initial robot configuration such that robot motion will result in the displacement of the object. For instance, form closure requires the end-effector to surround part of the object, and pushing demands surface contact.

We defined valid contacts $G(O_i)$ as a set of end-effector positions relative to O_i. Given the object configuration, q_i^t, this set maps to absolute positions for the end-effector. Absolute positions map to robot configurations via inverse kinematics (IK):

$$\mathscr{C}_R^{cont}(O_i, W^t) = IK(G(O_i), W^t). \quad (8.15)$$

In Figure 8.3(b) the table can be grasped at any point on the perimeter. The shaded area represents $\mathscr{C}_R^{cont}(Table_1, W^0)$ as a set of feasible positions for the robot base that make it possible to grasp the object. Figure 8.3(c) illustrates the intersubsection of $\mathscr{C}_R^{acc}$ and $\mathscr{C}_R^{cont}$. These configurations are object contacts that are accessible to the robot in W^t.

$$\mathscr{C}_R^{AC}(O_i, W^t) = \mathscr{C}_R^{acc}(W^t) \bigcap \mathscr{C}_R^{cont}(O_i, W^t). \quad (8.16)$$

The set $\mathscr{C}_R^{AC}(O_i, W^t)$ represents useful goals for the *Navigate* operator in W^t. There are two cases. When $\mathscr{C}_R^{acc}(W^t)$ contains the goal state, NAMO reduces to a path plan

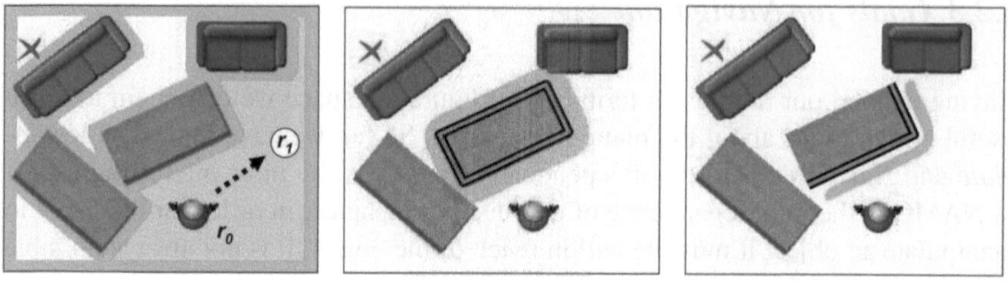

(a) (b) (c)

Figure 8.3 Simulated 2D NAMO $\mathscr{C}_R$-space for a circular robot: (a) $A, B \in \mathscr{C}_R^{acc}(W^0)$; (b) $G(O_i) \rightarrow \mathscr{C}_R^{cont}(O_i, W^t)$; (c) $\mathscr{C}_R^{acc}(W^t) \cap \mathscr{C}_R^{cont}(O_i, W^t)$

to the goal. Otherwise, at least one object must be manipulated prior to reaching the goal. Since *Navigate* only displaces the robot, $\mathscr{C}_R^{acc}(W^t)$ and $\mathscr{C}_R^{cont}(O_i, W^t)$ do not change after the operation for any O_i. By definition $\mathscr{C}_R^{AC}(O_i, W^t)$ is not affected. $Navigate(W^t, \tau(r^t, r^{t+1}))$ must satisfy $r^{t+1} \in \mathscr{C}_R^{AC}(O_i, W^t)$ for some O_i in order to make progress.

8.2.4 Goals for Manipulation

In Section 8.2.3 we saw that the contact states for manipulation constrain useful goals for *Navigate*. We also know that the subsequent *Manipulate* operation can only be applied from one of these states to one of the accessible objects. In this section we look at how the navigation space can be used to select goals for manipulation. These concepts form the basis for our first planner in the NAMO domain.

In general, there is no method for identifying non-trivial manipulation. Unlike *Navigate*, any valid *Manipulate* operator changes the state of the world and the set of accessible configurations. Even if the change seems to decrease immediate accessibility it is possible that manipulation opens space for a future displacement of another object. Some manipulation actions, however, are more clearly useful than others. To identify them, let us return to the sets of configurations defined in Section 8.2.3.

We already observed that $\mathscr{C}_R^{acc} \in \mathscr{C}_R^{free}$ is a subspace of configurations that can be reached from one another by a single *Navigate* action. Suppose that the robot was in a free configuration outside of $\mathscr{C}_R^{acc}$. There would also be a set of configurations accessible to the robot. In fact, as shown in Figure 8.4(b), we can partition the robot free space, $\mathscr{C}_R^{free}$, into disjoint sets of robot configurations that are closed under *Navigate* operators: $\{C_1, C_2, \ldots, C_d\}$. The goal configuration lies in one of these subsets, C_G.

Partitioning the free space results in an abstraction for identifying useful manipulation subgoals. Consider the effect of *Manipulate* in Figure 8.4(c). After the action,

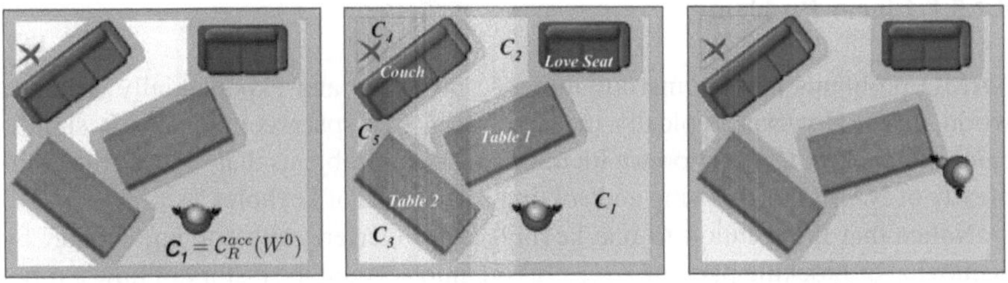

(a) (b) (c)

Figure 8.4 NAMO $\mathscr{C}_R$-space partitioned into components: (a) $C_1 = \mathscr{C}_R^{acc}(W^0)$; (b) $\mathscr{C}_R^{free}$ Components; (c) Keyhole solution

configurations in C_2 become valid goals for *Navigate*. The illustration shows part of the solution to a *keyhole* in this NAMO problem.

Definition 8.1. A *keyhole*, $K(W^1, C_i)$, is a subgoal problem in NAMO that specifies a start state, W^1, and a component of free space, C_i. The goal is to find a sequence of operators that results in W^2 s.t. every free configuration in C_i is accessible to the robot:

$$C_i \cap \mathscr{C}_R^{free}(W^2) \subset \mathscr{C}_R^{acc}(W^2) \qquad \text{and} \qquad C_i \cap \mathscr{C}_R^{free}(W^2) \neq \emptyset. \tag{8.17}$$

To restrict the number of obstacles moved in solving a keyhole we define a *keyhole solution* according to the maximum number of permitted *Manipulate* operators.

Definition 8.2. A *k-solution* to $K(W^1, C_i)$ is a sequence of valid actions including at most *k* *Manipulate* operators that results in W^2 satisfying Equation 8.17.

Manipulating obstacles to solve keyholes is useful because it creates passages in the robot's navigation space which open to entire sets of valid navigation goals. Rather than considering individual actions, we can simplify a NAMO task as follows: *The robot must resolve a sequence of keyholes until* $r^T \in C_G$ *at some future time* T.

8.2.5 Planning as Graph Search

With the introduction of useful subgoals for *Navigate* and *Manipulate* operators we now describe a conceptual planner that applies these goals to solving NAMO. First, we present an intuitive class of *linear* problems for which the planner is resolution complete. Next we describe the local search method used by our planner and give the planning algorithm.

8.2.5.1 Linear Problems

NAMO problems have numerous movable obstacles and exponentially as many world states. However, typically the number of free space components is significantly smaller. In this section we identify a class of problems that reduces the complexity of NAMO to choosing and solving a sequence of keyholes.

Notice that the solution to one keyhole may interfere with solving another by occupying or blocking parts of $\mathscr{C}_{free}$. Taking into account the history of how a robot entered a given free space component returns planning complexity to a search over all obstacle displacements. We therefore introduce a class of problems where keyholes can be solved independently.

Definition 8.3. A NAMO problem, (W_0, r^f), is *linear of degree k* or L_k if and only if:

1. There exists an ordered set of $n \geq 1$ distinct configuration space components, $\{C_1, \ldots, C_n\}$ such that $C_n = C_G$ and $C_1 = \mathscr{C}_R^{acc}(W_0)$.
2. For any $i (0 < i < n)$, any sequence of *k-solutions* to $\{K(W_{j-1}, C_j) | j < i\}$ results in W_{i-1} such that there exists a *k-solution* to $K(W_{i-1}, C_i)$.
3. If $n > 1$, any sequence of *k-solutions* to $\{K(W_{j-1}, C_j) | j < n\}$ results in W_{n-1} s.t. there exists a *k-solution* to $K(W_{n-1}, C_n)$ that results in W_n where $r^f \in \mathscr{C}_R^{free}(W_n)$.

For at least one sequence of free space components this inductive definition ensures that once the robot has reached a configuration in C_i it can always access C_{i+1}. Condition (3) guarantees the existence of a final keyhole solution such that the goal is in $C_G \cap \mathscr{C}_R^{free}(W^n)$. Although this definition allows for arbitrary *k*, in this section we will focus on problems that are *linear of degree 1* or L_1.

Even when only one *Manipulation* operator is permitted, it is often possible to find a keyhole solution that blocks a subsequent keyhole. For instance, in Figure 8.4(c), consider turning the table to block a future displacement of the couch. We propose to broaden the scope of problems that can be represented as L_1 by constraining the action space of the robot. We have considered two restrictions on the placement of objects:

Occupancy Bounds: Any displacement of an object must not occupy additional free space that is not accessible to the robot: $\mathscr{X}_R^{O_i}(W^{t+1}) \subset (\mathscr{X}_R^{O_i}(W^t) \cup \mathscr{C}_R^{acc}(W^t))$. This implies that any solution to (W^{t-1}, C_t) results in W^t s.t. $C_t \cap \mathscr{C}_R^{free}(W^t) = C_t$.

Minimal Intrusion: Paths used for *Manipulate* operators are restricted to being minimal with respect to a chosen criterion. The criteria could include the dimension of inaccessible $\mathscr{C}_R$ occupied by a displaced object, the distance of the displacement or a simulation of expected work to be used in manipulation.

Occupancy bounds create an intuitive classification for problems. If there is a sequence of free space components where each C_i can be accessed using only the space of C_{i-1} then the problem is *linear*. Minimal intrusion is less intuitive since it requires some notion of the extent to which a previous component could be altered and still yield sufficient space for a keyhole solution. However, this method is more powerful since it allows the robot to take advantage of space in both C_i in addition to C_{i-1} for placing objects.

Regardless of restrictions, L_1 includes many realworld problems. Generally, $\mathscr{C}_R^{free}$ is connected and allows navigation. L_1 problems arise when the state is perturbed due to the motion of some object. An L_1 planner should detect this object and restore connectivity.

8.2.5.2 Local Manipulation Search

Solving a linear NAMO problem requires us to determine a sequence of keyholes that connect free space components. First, we introduce a simple search routine for accessing $C_2 \subset \mathscr{C}_R^{free}$ from $r^t \in C_R^{acc}(W^t)$ by moving a single object O_i. The routine returns W^{t+2}, a robot path, τ_M, and the total cost c or NIL when no path is found.

MANIP-SEARCH(W^t, O_i, C_2): we apply Section 8.2.3 to find a discrete or sampled set of contact states for *Manipulate*: $\mathscr{C}_R^{AC}(W^t)$. Starting from all distinct configurations $r^{t+1} = G_i$ we perform a uniform cost (breadth first) search over paths τ_M that validate *Manipulate*$(W^t, O_i, G_i, \tau_M(r^{t+1}, r^{t+2}))$. The object path τ_o is determined by the specific action space. The search terminates when W^{t+2} satisfies Equation 8.17 or every reachable state has been examined. If $r^f \in C_2$ the search must also ensure that $r^f \in C_R^{free}(W^{t+2})$.

Local uniform cost search operates with any cost function and ensures that the returned path will minimize this function over the discrete action set. When restricting the action space to minimize intrusion (8.2.5.1) the cost should reflect the intrusion criteria such as occupied inaccessible space. We have optimized for path length in quasi-static planning and work or energy usage for dynamic manipulation.

Although MANIP-SEARCH is a simple method, efficient recognition of goal satisfaction is non-trivial. We provide one possible solution in [10].

8.2.5.3 Connecting Free Space

Given a routine for accessing a component of free space we turn to the global task of solving linear problems. In this section we introduce the global CONNECTFS algorithm.

First, define a set FC of disjoint free space components $C_i \subset \mathscr{C}_R^{free}$. C_S and C_G refer to the components containing the robot and the goal, respectively. Our algorithm keeps a search tree, FCT, and a queue of unexpanded nodes Q. Each node is a pair (C_i, W^t) and each edge is an obstacle O_l. FCT and Q are initialized with the node (C_S, W^0). At each step we remove a node from Q, $N = (C_i, W^t)$ and expand the tree:

1. Construct the set $C_N \subset FC$ of all free space components, C_j, such that (C_j, W') is not an ancestor of N in FCT for any W'.

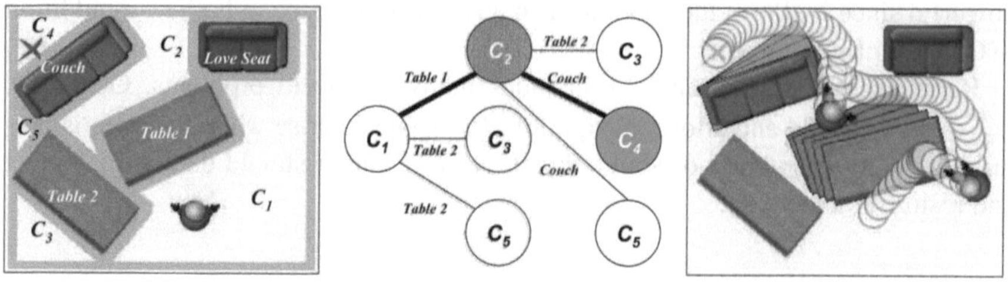

(a) (b) (c)

Figure 8.5 Construction of FCT structures the solution for L_1 problems: (a) L_1 NAMO Problem; (b) *FCT* with solution path; (c) Solution

2. For each $C_j \in C_N$ find the set of *neighboring* obstacles $O_N(C_j)$ such that there exists a path from r^t to some r_j in C_j that collides only with the obstacle:

$$O_N(C_j) = \{O_l | \exists r_j \in C_j, \tau(r^t, r_j) : \forall s(\tau[s] \in \bigcap_k \overline{\mathscr{X}_R(F_k)} \bigcap_{j \neq l} \overline{\mathscr{X}_R^{O_j}(W^t)})\}. \quad (8.18)$$

3. Remove $C_j \in C_N$. For each $O_l \in O_N(C_j)$ let $S(O_l) = (W^{t+2}, \tau, c)$ be the result of calling $Manip-Search(W^t, O_l, C_j)$. If at least one call succeeds choose $S(O_l)$ with least cost and add the node $N' = (C_j, W^{t+2})$, edge $e(N, N') = O_l$ to FCT.
4. Repeat step 3 until $C_N = \emptyset$.

We repeat the expansion of nodes until $N^G = (C_G, W^T)$ is added to FCT for some W^T or every node has been removed from Q (Figure 8.5).

8.2.5.4 Analysis

Our description of CONNECTFS illustrates the potential for using a configuration space decomposition of NAMO problems to select subproblems that can easily be solved by a motion planner. Furthermore, the construction of CONNECTFS allows for a simple proof of its relationship to the L_1 problem class. First, we must show that any *1-solution* to a keyhole must move a *neighboring* object as defined in step 2.

Lemma 8.1. *If there exists a* 1-solution *to $K(W^T, C_j)$ which displaces O_l then there exist $r_j \in C_j$ and $\tau(r^T, r_j)$ such that for all s, $\tau[s] \in \bigcap_k \overline{\mathscr{X}_R(F_k)} \bigcap_{j \neq l} \overline{\mathscr{X}_R^{O_j}(W^T)}$.*

Proof. To simplify notation let $A^t = \bigcap_k \overline{\mathscr{X}_R(F_k)} \bigcap_{j \neq l} \overline{\mathscr{X}_R^{O_j}(W^t)}$. Assume there exists a *1-solution* consisting of a sequence of n operators, each parameterized by a robot path $\tau(r^t, r^{t+1})$ where t refers to any time step such that $(T \leq t \leq T+n)$.

Since only O_l is displaced between T and $T+n$, all other obstacle configurations are unchanged: $q_j^t = q_j^T$ for all $j \neq l$ and t. Hence, $A^t = A^T$ for all t. We now show that every operator path in the *1-solution* satisfies $\tau[s] \in A^T$ for all s.

Manipulate$(W^t, O_l, G_i, \tau_M(r^t, r^{t+1}))$: the operator is valid only if the path satisfies $\tau_M[s] \in A^t$ for all s (Equation 8.8). Hence, the path satisfies $\tau_M[s] \in A^T$ for all s.

Navigate$(W^t, \tau_N(r^t, r^{t+1}))$: the operator is valid only if the path satisfies $\tau_N[s] \in \mathscr{C}_R^{free}(W^t)$ for all s (Equation 8.13). Since $\mathscr{C}_R^{free}(W^t) = \bigcap_k \overline{\mathscr{X}_R(F_k)} \bigcap_j \overline{\mathscr{X}_R^{O_j}(W^t)}$, we have $\mathscr{C}_R^{free}(W^t) \subset A^t$. Therefore, $\tau_N[s] \in A^t$ and consequently $\tau_N[s] \in A^T$ for all s.

Furthermore, Equation 8.17, guarantees the existence of $r_j \in C_j \cap \mathscr{C}_R^{free}(W^{T+n})$. Since $C_j \cap \mathscr{C}_R^{free}(W^{T+n}) \subset \mathscr{C}_R^{acc}(W^{T+n})$, there exists $\tau_j(r^{T+n}, r_j)$ such that $\tau_j[s] \in \mathscr{C}_R^{free}(W^{T+n})$ for all s(Equation 8.13). Therefore $\tau_j[s] \in A^{T+n} = A^T$ for all s.

Finally, we construct a path $\tau_D(r^T, \ldots, r^{T+n}, r_j)$ by composing the operator paths from the *1-solution* with τ_j. Since $\tau[s] \in A^T$ for all s in all subpaths of τ_D, we also have $\tau_D[s] \in A^T$ for all s.

Lemma 8.2. CONNECTFS *is resolution complete for problems in* L_1.

Proof. Let Π be a solvable problem in L_1. We show that CONNECTFS will find a solution. By definition of L_1, there exists an ordered set Ω of n disjoint $\mathscr{C}_R^{free}$ components $\{C_S, \ldots, C_G\}$. We show that CONNECTFS will add (C_G, W^T) to FCT by induction. In the base case, $(C_S, W^0) \in FCT$.

Assume (C_{i-1}, W^t) has been added to FCT such that $C_{i-1} \in \Omega, i < n$ and W^t was produced by a sequence of *1-solutions* to $\{K(W_{j-1}, C_j) | 0 < j < i\}$. By the definition of L_1 there exists a *1-solution* with n operators to $K(W^t, C_i)$ (Equation 8.17). Hence, starting in W^t, there is an operator sequence with one *Manipulate* that displaces some obstacle O_l and results in W_i such that $C_i \cap \mathscr{C}_R^{free}(W_i) \subset \mathscr{C}_R^{acc}(W_i)$.

When (C_{i-1}, W^t) is expanded, Lemma 8.1 guarantees that there exist $r_i \in C_i$ and a path $\tau_D(r^t, r_i)$ in W^t that only passes through O_l. Consequently, in step 2 of CONNECTFS, O_l will be added to $O_N(C_i)$. Step 3 will call W^{t+2} = MANIP-SEARCH(W^t, O_l, C_i). Since MANIP-SEARCH is resolution complete over the action space, it will find a *Manipulate* path that satisfies Equation 8.17. Therefore, (C_i, W^{t+2}) will be added to FCT.

By induction, (C_G, W^T) will be added to FCT. Regarding termination, let d be the number of free space components in **FC**. Since all nodes added to the tree must contain distinct free space components from their ancestors the tree has a maximum depth of d (step 1). Furthermore, each node at depth $i (1 \leq i \leq d)$ has at most $d - i$ children. Either (C_G, W^T) is added or all nodes are expanded to depth d and no further nodes can be added to Q. Hence CONNECTFS terminates in finite time and is resolution complete for problems in L_1.

The construction of FCT in CONNECTFS also allows us to optimize for various criteria. First of all, we can find solutions that minimize the number of objects moved simply by choosing nodes from Q according to their depth in the tree. Nodes of lesser depth in the tree correspond to less displaced objects. We can also associate a cost with each node by summing the cost of the parent node and the cost returned by MANIP-SEARCH. When a solution of cost c is found, we would continue the search until all nodes in Q have cost $c' > c$. The lowest cost solution would be minimal with respect to MANIP-SEARCH costs.

8.2.5.5 Challenges of CONNECTFS

Although CONNECTFS reduces the search space from brute force action-space search, its primary purpose is to convey the utility of the reduced dimensional configuration space structure. Practically, a NAMO domain could have large numbers of objects and free-space components. Constructing the tree would require an algorithm to determine all *neighboring* objects for all components of free-space. Furthermore, we would need to call MANIP-SEARCH to verify every potential access. This seems unreasonable for a navigation planner since we may only need to move one or two objects to reach the goal.

We believe that this observation is critical for developing a real-time system. In particular, the complexity of the problem should depend on the complexity of resolving *keyholes* that lie on a reasonable navigation path, not on the complexity of unrelated components of the world. In other words, since the purpose of NAMO is navigation, a NAMO problem should only be difficult when navigation is difficult.

Section 8.2.6 gives one answer to this challenge. We introduce a heuristic algorithm that implicitly follows the structure of CONNECTFS without graph construction. To do so, we again turn to the navigational substructure of the problem. Previously, we observed that every L_1 plan is a sequence of actions that access entire components of $\mathscr{C}_R^{free}$. Lemma 8.1 also showed that obstacles that lie in the path of *Navigate* should be considered for motion. We will now consider extending such paths through the rest of $\mathscr{C}_R$ until they reach the goal and using them to guide the planner in subgoal selection.

8.2.6 Planner Prototype

With the results from the previous section, we now formulate a simple and effective planner for the NAMO domain: SELECTCONNECT. The planner is a best-first search that generates fast plans in L_1 environments. Its heuristic, RCH, is a navigation planner with relaxed constraints. RCH selects obstacles to consider for motion. Following the algorithm description, we show that its search space is equivalent to that of CONNECTFS (Section 8.2.5.3). With best-first search, optimality is no longer guaranteed. However, the planner is much more efficient and resolution complete for problems in L_1. If memory and efficiency are less of a concern, optimality can be regained with uniform-cost search.

8.2.6.1 Relaxed Constraint Heuristic

The relaxed constraint heuristic (RCH) is a navigation planner that allows collisions with movable obstacles. It selects obstacles for SC to displace. RCH is parameterized by W^t, *AvoidList* of (O_i, C_i) pairs to avoid, and *PrevList* of visited C_j. After generating a path estimate to the goal, it returns the pair (O_1, C_1) of the first obstacle in the path, and the first component of free space. If no valid path exists, RCH returns NIL. Since RCH does not move obstacles, all obstacle positions are given in terms of W^t.

RCH is a modified A^* search from r^t to r^f on a dense regular grid of $\mathscr{C}_R(W^t)$ which plans *through* movable obstacles. RCH keeps a priority queue of grid cells x_i. Each cell is associated with a robot configuration, r_i, the cost of reaching it, $g(x_i)$, and the estimated cost for a solution path that passes through r_i, $f(x_i)$. On each iteration, RCH removes x_i with least $f(x_i)$ from the queue and *expands* it by adding valid adjacent cells x_j with the costs in Equation 8.19. Let $e(r_i, r_j)$ be the transition cost of entering a cell $r_j \in \mathscr{X}_R^{O_j}$ from $r_i \notin \mathscr{X}_R^{O_j}$, where e is estimated from the size or mass of O_j. $h(x_j, x^f)$ estimates goal distance and α relates the estimated cost of moving an object to the cost of navigation:

$$\begin{aligned} g(x_j) &= g(x_i) + (1-\alpha) + \alpha e(r_i, r_j), \\ f(x_j) &= g(x_j) + h(x_j, x^f). \end{aligned} \tag{8.19}$$

As shown in Algorithm 8.1, RCH restricts valid transitions. We use *exclusively contained* to refer to r_i contained in only one obstacle: $r_i \in \textbf{exc}\ \mathscr{X}_R^{O_i} \Rightarrow (O_j \neq O_i$ then $r_i \notin \mathscr{X}_R^{O_j})$.

Definition 8.4. A path $P = \{r_1, \ldots, r_n = r^f\}$ is $Valid(O_i, C_j)$ if and only if $(O_i, C_j) \notin$ *AvoidList* and $O_j \notin$ *PrevList* and the path collides with exactly one obstacle prior to entering C_j. Alternatively, there exists $m < n$ such that: $r_m \in C_j$ and for all r_i where $i < m$ either $r_i \in \mathscr{X}_R^{O_i}$ or $r_i \in \mathscr{C}_R^{free}$.

Since the validity of a transition depends on the history of objects/free-space encountered along a path, we expand the state with a dimension for each obstacle and $\mathscr{C}_R^{free}$ component. The states searched are triples $x_i = (r_i, O_i, C_j)$. O_i is the first obstacle encountered or 0 if one has not been encountered. C_j is likewise 0 or the first free space component.

Lemma 8.3. *If there exists a* $Valid(O_F, C_F)$ *path then RCH will find a solution.*

Proof. Assume there exists a $Valid(O_F, C_F)$ path P for some O_F, C_F then the RCH will find a path. After adding $(r^t, 0, 0)$ to Q, line 14 expands the children of this state adding all $r_i \in \mathscr{C}_R^{acc}$ to Q as $(r_i, 0, 0)$. Line 15 allows the transition from any $(r_i, 0, 0)$ to any adjacent $(r_j, O_A, 0)$ where $r_j \in \mathscr{X}_R^{O_A}$. Lines 10 and 11 expand any state $(r_i, O_A, 0)$ to $(r_j, O_A, 0)$ where $r_j \in \mathscr{C}_R^{free}$ or r_j is exclusively in O_A. Hence every state that can be reached after entering only one obstacle will be added to Q as $(r_i, O_A, 0)$.

Algorithm 8.1 Pseudo-code for RCH.

Algorithm:RCH (W^t, *AvoidList*, *PrevList*, r^f)
1 $Closed \leftarrow \emptyset$
2 $Q \leftarrow$ MAKE-PRIORITY-QUEUE(r^t,0,0)
3 **while** $Q \neq$ **empty do**
4 $x_1 = (r_1, O_F, C_F) =$ REMOVE-FIRST(Q)
5 **if**($x_1 \in Closed$) **continue**
6 **if**($r_1 = r^f$ **and** $O_F \neq 0$ **and** $C_F \neq 0$)**return** (O_F, C_F)
7 *Closed* **append** (x_1)
8 **foreach** $r_2 \in$ ADJACENT(r_1) **do**
9 **if**($C_F \neq 0$) ENQUEUE($Q, (r_2, O_F, C_F)$); **continue**
10 **if**($O_F \neq 0$ **and** $r_2 \in \mathscr{C}_R^{free}$) ENQUEUE($Q, (r_2, O_F, 0)$)
11 **if**($O_F \neq 0$ **and** $r_2 \in exc\mathscr{X}_R^{O_F}$) ENQUEUE($Q, (r_2, O_F, 0)$)
12 **if**($O_F \neq 0$ **and** $r_2 \in C_i s.t. C_i \notin PrevList$ **and** $(O_F, C_i) \notin AvoidList$)
13 ENQUEUE($Q, (r_2, O_F, C_i)$)
14 **if**($O_F = 0$ **and** $r_2 \in \mathscr{C}_R^{free}$)ENQUEUE($Q, (r_2, 0, 0)$)
15 **if**($O_F = 0$ **and** $r_2 \in exc\mathscr{X}_R^{O_i}$)ENQUEUE($Q, (r_2, O_i, 0)$)
end
end
16 **return** NIL

From Definition 8.4 we know that there exists such a state $(r_{m-1}, O_F, 0)$ in P that can be reached after entering only one obstacle. Furthermore, when $(r_{m-1}, O_F, 0)$ is expanded to r_m the conditions on line 12 will be satisfied since $r_m \in C_F$ and $C_F \notin PrevList$ and $(O_F, C_F) \notin AvoidList$. Hence (r_m, O_F, C_F) will be added to Q on line 13. Finally, line 9 will continue to expand this state to all adjacent states as (r_i, O_F, C_F). Since path P exists, there exists some path from r_m to the goal and therefore (r^f, O_F, C_F) will be found. We know the process will terminate since states are not revisited by addition to *Closed* and the sizes of $\{r_i\}$, $\{O_i\}$ and $\{C_i\}$ are finite.

Lemma 8.4. *If RCH finds a solution then there exists a* $Valid(O_F, C_F)$ *path.*

Proof. Suppose RCH has found a solution path P terminating in (r_j, O_F, C_F) on line 6. Lines 10–15 do not permit any transition from a state (r_j, O_F, C_F) to $(r_i, O_F, 0)$ or from $(r_j, O_F, 0)$ to $(r_i, 0, 0)$. Consequently we can separate P into three segments. P_1 consists of states $(r_i, 0, 0)$, P_2 contains $(r_i, O_F, 0)$ and P_3 contains (r_i, O_F, C_F). Any transition from the last state in P_2 to the first in P_3 must have satisfied the condition on line 12, hence $C_F \notin PrevList$ and $(O_F, C_F) \notin AvoidList$. Furthermore, every state in P_2 must be either in $\mathscr{C}_R^{free}$ or exclusively in $\mathscr{X}_R^{O_F}$ as added by lines 13, 10 and 11. Finally, every state in P_1 must be in $\mathscr{C}_R^{free}$ since it was added by line 15. Hence the path P satisfies the second criterion of Definition 8.4.

While this algorithm appears more complex than A^* the branching factor is unchanged. Furthermore, only objects that can be reached without colliding with other objects are taken into account. To increase efficiency, membership in *Closed* on line

6 can be checked using (r_i, O_i) rather than the full state. Since there are no restrictions on transitions from non-zero C_i path existence will not depend on its value.

8.2.6.2 High-level Planner

Algorithm 8.2 gives the pseudo-code for SELECTCONNECT (SC). The planner makes use of RCH and MANIP-SEARCH, as described in Section 8.2.5.2. It is a greedy heuristic search with backtracking. The planner backtracks locally when the object selected by RCH cannot be moved to merge the selected $C_i \subset \mathscr{C}_{free}$. It backtracks globally when all the paths identified by RCH from C_i are unsuccessful.

SC calls FIND-PATH to determine a *Navigate* path from r^t to a contact, r^{t+1}. The existence of $\tau_N(r^t, r^{t+1})$ is guaranteed by the choice of contacts in MANIP-SEARCH.

Algorithm 8.2 Pseudo-code for *SelectConnect.*

```
   Algorithm:SelectConnect(W^t, PrevList, r^f)
1  AvoidList ← ∅
2  if x^f ∈ C_R^acc(W^t) then
3      return FIND-PATH(W^t, x^f))
   end
4  while (O_1, C_1) ← RCH(W^t, AvoidList, PrevList, r^f) ≠ NIL do
5      (W^{t+2}, τ_M, c) ← MANIP-SEARCH(W^t, O_1, C_1)
6      if τ_M ≠ NIL then
7          FuturePlan ← SELECTCONNECT(W^{t+2}, PrevList append C_1, r^f)
8          if FuturePlan ≠ NIL then
9              τ_N ← FIND-PATH(W^t, τ_M[0])
10             return ((τ_N, τ_M) append FuturePlan)
           end
       end
11     AvoidList append(O_1, C_1)
   end
12 return NIL
```

8.2.6.3 Examples and Experimental Results

We have implemented the proposed NAMO planner in a dynamic simulation environment. The intuitive nature of SELECTCONNECT is best illustrated by a sample problem solution generated by the planner. In Figure 8.6(a), we see that $\mathscr{C}_R^{free}$ is disjoint–making this a NAMO problem. Line 4 of SC calls RCH, the heuristic subplanner. RCH finds that the least cost $Valid(O_i, C_j)$ path to the goal lies through $O_i = Couch$. The path is shown in Figure 8.6(b). RCH also determines that the free-space component to be connected contains the goal. Line 6 calls MANIP-SEARCH to find a motion for the couch. Figure 8.6(c) shows the minimum cost manipulation

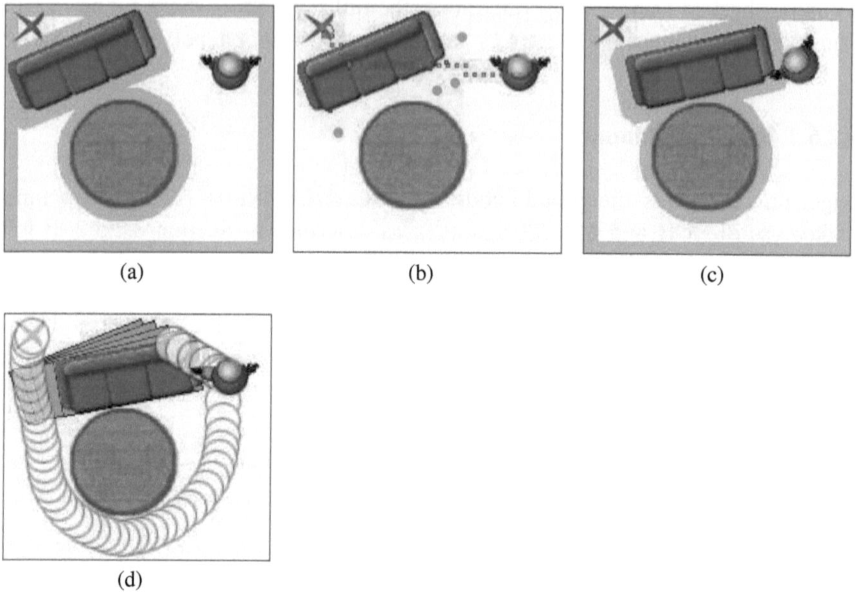

Figure 8.6 Walk-through of an autonomous SELECTCONNECT: (a) Problem; (b) RCH; (c) Keyhole solution; (d) Final plan

path that opens the goal free-space. Finally, SELECTCONNECT is called recursively. Since r^f is accessible, line 3 finds a plan to the goal and completes the procedure (Figure 8.6(d)). The remainder of the pseudo-code iterates this process until the goal is reached and backtracks when a space cannot be connected.

Figure 8.6(d) is particularly interesting because it demonstrates our use of $\mathscr{C}_R^{free}$ connectivity. As opposed to the local planner approach employed in PLR [22], MANIP-SEARCH does not directly attempt to connect two neighboring points in $\mathscr{C}_R$. MANIP-SEARCH searches all actions in the manipulation space to join the configuration space components occupied by the robot and the subgoal. The procedure finds that it is easiest to pull the couch from one side and then go around the table for access. This decision resembles human reasoning and cannot be reached with existing navigation planners.

Figure 8.6(a) also demonstrates a weakness of L_1 planning. Suppose the couch was further constrained by the table such that there was no way to move it. Although the table is obstructing the couch, the table does not explicitly disconnect any free-space and would therefore not be considered for motion.

Figure 8.7 is a more complex example with backtracking. In the lower frame, we changed the initial configuration of the table. The initial call to RCH still plans through the couch, however, MANIP-SEARCH finds that it cannot be moved. The planner backtracks, calling RCH again and selects an alternative route.

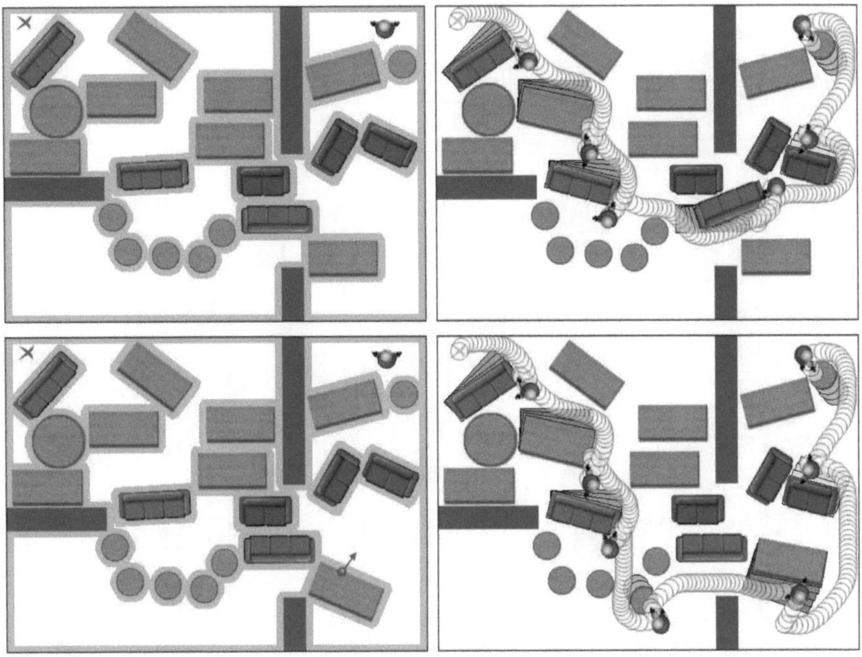

Figure 8.7 The generated plan output by our dynamic simulation NAMO planner is illustrated by the time-lapse sequences on the right

Figures 8.7 and 8.8 show the scalability of our algorithm to problems with more movable objects. While computation time for Figure 8.6(a) is < 1s, the solutions for Figures 8.7 and 8.8 were found in 6.5 and 9s, respectively (on a Pentium 4 3 GHz). Notice that the planning time depends primarily on the number of manipulation plans that need to be generated for a solution. Although the largest example contains 90 movable obstacles, compared with 20 in Figure 8.7, there is no sizable increase in the solution time.

Finally, consider the simple examples for which BFS examined tens of thousands of states in Figure 8.2. The solution to (a) is found instantly by the first heuristic search after examining 15 states. Both (b) and (c) are solved after considering 147 states. This number includes states considered during heuristic search, *Navigate* path search and the verification of connectivity at each step of *Manipulate*. (d) is solved after 190 states.

8.2.6.4 Analysis

SELECTCONNECT has clear advantages over CONNECTFS in terms of both average computation time and ease of implementation. Implemented as best first search, SELECTCONNECT is not globally optimal. Note, however, that for each choice of

Figure 8.8 A larger scale example consisting of 90 movable obstacles. Two separate plans are computed and demonstrated in our dynamic simulation.

obstacle, the planner still selects a motion with least cost. If planning efficiency and space are not the primary concern, uniform cost or A^* variants would restore optimality. We now prove L_1 completeness for SELECTCONNECT.

Lemma 8.5. *Any solution found by* $\mathrm{SC}(W^t, PrevList, r^f)$ *is valid in NAMO.*

Proof. This can be seen from the construction of the algorithm. By induction: in the base case SELECTCONNECT returns a single valid $Navigate(W^t, \tau_N(r^t, r^f))$ on line 3. Such a path exists by definition of $\mathscr{C}_R^{acc}(W^t)$ (8.13).

Assume that the call to $\mathrm{SELECTCONNECT}(W^{t+2}, PrevList, r^f)$ on line 11 returns a valid *FuturePlan*. $\mathrm{SC}(W^t, \ldots)$ pre-pends *FuturePlan* with $Navigate(W^t, \tau_N(r^t, r^{t+1}))$ and $Manipulate(W^{t+1}, O_l, G_i, \tau_M(r^{t+1}, r^{t+2}))$ operators. τ_M is valid due to the completeness of MANIP-SEARCH (8.2.5.2) and τ_N is valid since $r^{t+1} = \tau_M[0] \in \mathscr{C}_R^{AC}(W^t)$ by construction of MANIP-SEARCH and Definition 8.16. Hence, $\mathrm{SC}(W^t, PrevList, r^f)$ also returns a valid plan. By induction every plan returned by SELECTCONNECT is valid.

Lemma 8.6. *Suppose there exists a* 1-solution *to* $K(W^t, C_F)$ *which displaces* O_F *and* $C_F \notin PrevList$*. Then the call to* $\mathrm{SELECTCONNECT}(W^t, PrevList, r^f)$ *will either find a valid solution to* $NAMO(W^t, r^f)$ *or call* $\mathrm{MANIP\text{-}SEARCH}(W^t, O_F, C_F)$.

Proof. Since there exists a *1-solution* to $K(W^t, C_F)$, Lemma 8.1 ensures that there exist $r_F \in C_F$ and $\tau_1(r^t, r_F)$ such that for all s, $\tau_1[s] \in \bigcap_k \overline{\mathscr{X}_R(F_k)} \bigcap_{j \neq F} \overline{\mathscr{X}_R^{O_j}(W^T)}$. Let τ_2 be any path in $\mathscr{C}_R$ from r_F to r^f. Let τ_F be the compound path (τ_1, τ_2).

On the first call to RCH(W^t,*AvoidList*,*PrevList*,r^f) (line 4), *AvoidList* is empty. We are given that $C_F \notin$ *PrevList*. Let $r_m \in C_F$ be the first state in τ_1 that is in C_F. Such a state must exist since $r_F \in C_F$. Since all τ_F states, r_i $(i < m)$, cannot be in any obstacle other than O_F, they are either in $\mathscr{C}_R^{free}$ or exclusively in O_F, satisfying the conditions of Definition 8.4. Hence, τ_F is $Valid(O_F, C_F)$. Since a $Valid(O_F, C_F)$ path exists, RCH will find a path (Lemma 8.3).

The loop on lines 4-12 will terminate only if SC succeeds in solving NAMO on line 8 or RCH fails to find a path on line 4. Line 12 of RCH ensures that on each iteration the pair (O_j, C_j) returned by RCH is distinct from any in *AvoidList*. This pair is added to *AvoidList*. Since there are finite combinations of obstacles and free space components, the loop must terminate. However, τ_R will remain $Valid(O_F, C_F)$ and RCH will find paths until it returns (O_F, C_F). Therefore either a NAMO solution will be found or RCH will return (O_F, C_F). In the latter case, line 6 of SC calls MANIP-SEARCH(W^t, O_F, C_F).

Theorem 8.1. SELECTCONNECT *is resolution complete for problems in* L_1.

Proof. Let NAMO(W^0, r^f) be an L_1 problem. We will show that SELECTCONNECT(W^0, r^f) finds a solution. In the base case, $x^f \in \mathscr{C}_R^{acc}(W^t)$ and line 3 yields the simple *Navigate* plan. In the following, let $\Omega = \{C_1, \ldots, C_n\}$ be an ordered set of disjoint free space components that satisfies Definition 8.3 for the given problem.

Assume SELECTCONNECT(W_{i-1}, r^f) has been called such that W_{i-1} is a world state resulting from a sequence of *1-solutions* to $K(W_{j-1}, C_j) | C_j \in \Omega, j < i)$. By definition of L_1 there exists a *1-solution* to $K(W_{i-1}, C_i)$ that moves some obstacle O_F (Definition 8.3). From Lemma 8.6 we have that SC(W_{i-1}, r^f) will either find a sequence of valid actions that solve NAMO(W_{i-1}, r^f) or call MANIP-SEARCH(W_{i-1}, O_F, C_i) on line 6. Since MANIP-SEARCH is resolution complete it will return a solution (τ_M, W^i, c) where W_i is the next state in the sequence of solutions to $K(W_{j-1}, C_j) | C_j \in \Omega, j \leq i)$. SELECTCONNECT($W_{i-1}$, r^f) will call SELECTCONNECT(W_i, r^f).

By induction, if SELECTCONNECT does not find another solution it will find the solution indicated by Ω. Each recursive call to SC adds a C_i to *PrevList*. When all C_i are added to *PrevList*, there are no $Valid(O_F, C_i)$ paths and RCH will return NIL (Lemma 8.4). Hence the maximum depth of recursion is the number of C_i. Analogously, each loop in lines 4–12 adds a distinct pair (C_i, C_j) to *AvoidList* such that when all pairs are added RCH will return NIL. Hence, the maximum number of calls made from each loop is the number of such pairs and the algorithm will terminate.

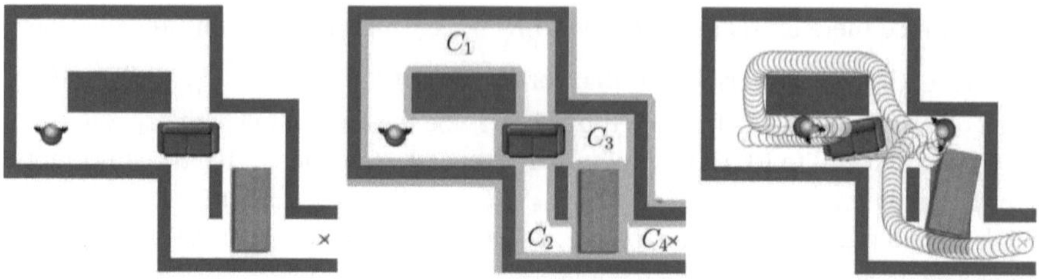

(a) (b) (c)

Figure 8.9 SELECTCONNECT solves the open problem considered difficult by Chen

8.2.7 Summary

This section describes initial progress towards planning for NAMO. First, we gave a configuration space representation for NAMO problems. Our analysis of the relationship between action spaces for *Navigate* and *Manipulate* operators gave us tools for constructing a conceptual planner and a practical solution for problems in the intuitive L_1 problem class. Search complexity was reduced from the number of objects in the scene to the difficulty of the *Navigation* task. The planner solved problems with nearly 100 movable obstacles in seconds.

In addition to high-dimensional problems, SELECTCONNECT is also effective in domains with complex geometry. Previously, Chen [22] presented one difficult puzzle problem shown in Figure 8.9. The PLR planner pushes the love seat into C_3 and cannot recover. Using $\mathscr{C}_R^{free}$ connectivity, SELECTCONNECT considers connecting C_3 as one of the options and successfully solves the example.

Clearly, there are many problems that do not fall into the L_1 class. These problems require us to consider cases where moving one object affects the robot's ability to move another. Further work on this topic is presented in [23]. Sections 8.3 and 8.4 will show how decisions to move objects are applied in autonomous execution.

8.3 Humanoid Manipulation

So far our investigation of NAMO has been largely theoretical. We showed that it is possible to decide the movement strategy for a robot that can manipulate obstacles in a large cluttered environment. In this section we will address the control problem of executing the desired motion on a humanoid robot. The robot used in our experiments is the Kawada HRP-2. This robot has the capacity for navigation and manipulation of large objects. Its anthropomorphic kinematics make it suitable for interacting in human environments. Furthermore, implementation on HRP-2 allowed us to study the interaction between navigation and manipulation from the perspective of multi-objective control.

We are primarily interested in manipulation of large objects such as carts, tables, doors and construction materials. Small objects can be lifted by the robot and modeled as additional robot links. Heavy objects are typically supported against gravity by external sources such as carts, door hinges or construction cranes. Yet, neither wheeled objects nor suspended objects are reliable sources of support for the robot. Large, heavy objects are interesting because they require the robot to handle significant forces while maintaining balance. We present a method that generates trajectories for the robot's torso, hands and feet that result in dynamically stable walking in the presence of known external forces.

In NAMO, the robot interacts with unspecified objects. Consequently the interaction forces are rarely known. While small variations can be removed by impedance control and online trajectory modification, larger correlated errors must be taken into account during trajectory generation. To account for this, we give a method for learning dynamic models of objects and applying them to trajectory generation. By using learned models we show that even 55 kg objects, equal to the robot's mass, can be moved along specified trajectories.

8.3.1 Background

Early results in humanoid manipulation considered balance due to robot dynamics. Inoue [24] changed posture and stance for increased manipulability and Kuffner [25] found collision free motions that also satisfied balance constraints. These methods did not take into account object dynamics. In contrast, Harada [26] extended the ZMP balance criterion for pushing on an object with known dynamics. Harada [27] also proposed an impedance control strategy for pushing objects during the double support phase of walking. We focus on continuous manipulation during all walking phases.

With the introduction of preview control by Kajita [28], Takubo [29] applied this method to adapting step positioning while pushing on an object. Nishiwaki [30, 31] proposed that the external forces from pushing could be handled by rapid trajectory regeneration. Yoshida [32, 33] locally modified the planned path for a light carried object to avoid collisions introduced by applying preview control. Our work extends beyond pushing and modification to realizing a desired trajectory for a heavy object. Furthermore, in contrast to assuming that objects are known or external sources of error, we learn about their response to our force inputs.

Recently, most studies of interaction with unknown objects have been kinematic. Krotkov [34] and Fitzpatrick [35] studied impulsive manipulation to detect the affordances of various objects through different sensing modalities. Stoychev [36] considered learning to use objects for specific behaviors and Christiansen [37] learned to manipulate an object between a set of discrete states. However, for large objects the controller must account for the continuous dynamic effect they have on balance and stability.

While our focus is on learning the dynamic model of an unknown *object*, this paper is closely related to modeling *robot* dynamics. Atkeson [38] summarizes approaches to learning or adapting parameters to achieve precise trajectory following.

Friction modeling has been studied extensively as summarized by Canudas [39] and Olsson [40]. More sophisticated methods for learning the dynamics of tasks in high dimensional spaces are studied by Atkeson, Moore and Schaal [41, 42].

8.3.2 Biped Control with External Forces

Biped locomotion keeps the robot upright by pushing on the ground with the robot's legs. To generate any desired vertical forces the stance foot must be in contact with the ground. Let the center of pressure or ZMP be the point about which the torques resulting from all internal and external forces acting on the robot sum to zero. A necessary condition for maintaining ground contact is that the ZMP be within the area of the stance foot [43, 44]. If the ZMP leaves the foot, the robot tips about a foot edge.

The most common method for maintaining the position of the ZMP is by generating and following trajectories for the robot torso. This method accomplishes two goals. First, it achieves the desired displacement of the torso and satisfies any kinematic constraints. Second, it ensures a desired position for the ZMP throughout the duration of trajectory execution and therefore implies that the vertical forces necessary for supporting the robot can be effected continuously. In our case, trajectories are re-computed at 0.15s intervals.

This Section details the control strategy for walking manipulation that takes into account external forces. The controller consists of three significant elements:

- decoupling the object and robot trajectories;
- trajectory generation satisfying ZMP and object motion;
- online feedback for balance and compliance.

Instantiating these three components lifts control from the 30D robot joint space to a higher level abstract system that realizes a single object trajectory.

8.3.2.1 Decoupled Positioning

At the highest level, we represent the manipulation task as a system of two bodies. The object, o, and robot, r, are attached by horizontal prismatic joints to a grounded stance foot. The stance foot position changes in discrete steps at a constant rate $k = 900$ ms. Section 8.3.2.2 computes independent workspace trajectories for x_r and x_o. To implement this abstraction we describe how workspace trajectories map to joint space.

We start the mapping by defining the trajectories for hands and feet relative to the object. Due to rigid grasp manipulation, the hand positions, p_{lh} and p_{lr} remain at their initial displacements from x_o. For simpler analysis, the stance foot p_{st} is fixed relative to x_o at each impact. The robot swing foot, p_{sw} follows a cubic spline connecting its prior and future stance positions. To achieve a fixed displacement

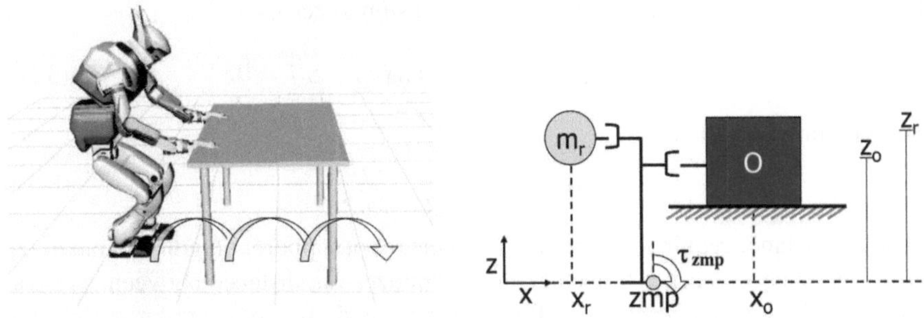

(a) (b)

Figure 8.10 Model of the robot and object used in our work: (a) Geometric model; (b) Computational model

from the object on each step, the object velocity is bounded by the maximum stride length and step rate. We restrict the values of $\dot{x}_o$ in advance.

We also fix the trajectory for the robot torso, p_{torso} relative to x_r. Although the center of mass position, x_r, is a function of all the robot links we assume that x_r remains fixed to p_{torso} after grasp. This assumption is relaxed in Section 8.3.2.2 through iterative controller optimization. Notice that although many of the link positions are highly coupled, the two positions of interest x_r and x_o are not.

Suppose we have workspace trajectories for both x_o and x_r. The former specifies trajectories for hands and feet and the latter defines x_{torso}. Joint values that position the four ungrounded links are found with resolved rate control [45].

$p_{st} \rightarrow x_r$	6 Stance leg	$x_r \rightarrow p_{lh}$	7 L arm
$x_r \rightarrow p_{sw}$	6 Swing leg	$x_r \rightarrow p_{rh}$	7 R arm

These solutions complete the mapping from any valid workspace placement of x_r and x_o to robot joints. We compared analytical inverse kinematics (IK) to a gradient method based on the pseudo-inverse of the robot Jacobian. Analytical IK allowed faster computation, avoided drift and assured that the solutions would satisfy joint limit constraints. The two chest joint values were constants that maximize the workspace. Redundancy in the arms is resolved by fixing elbow rotation about the line connecting the wrist and shoulder.

8.3.2.2 Trajectory Generation

Section 8.3.2.1 gave a mapping from commanded workspace positions of x_r and x_o to joint positions. We now focus on workspace control. Given a commanded trajectory for x_o we compute a trajectory for x_r that satisfies balance constraints.

We relate ZMPt to stance foot position. Let x be the direction of object motion. z_o is the height of the hands and f is the reflected force. Equation 8.20 introduces zmp as the ground point around which the torques due to gravity acting on x_r, reflected force from accelerating x_r and from the object sum to zero.

$$\tau_{zmp} = m_r g(x_r - zmp) - m_r \ddot{x}_r z_r - z_o f = 0. \tag{8.20}$$

Solving for zmp yields:

$$zmp = x_r - \ddot{x}_r \frac{z_r}{g} - \frac{z_o f}{m_r g}. \tag{8.21}$$

Dynamic balance requires zmp to remain in the robot support polygon. To maximize error tolerance we seek a trajectory that minimizes the distance between zmp and the stance foot center $zmp_d = x_{st}$. Recall that x_{st}, and thus zmp_d are known given a trajectory for x_o (subsection 8.3.2.1)

Let $J_0 = \sum_t (zmp^t - zmp^t_d)^2$ be the performance index for balance. Equation 8.22 further defines β and β_d as functions of zmp_d and x_r respectively.

$$\beta_d = zmp_d + \frac{z_o f}{m_r g} \quad \beta = x_r - \ddot{x}_r \frac{z_r}{g}. \tag{8.22}$$

Substitution yields $J_0 = \sum_t (\beta^t - \beta^t_d)^2$. Notice that zmp_d is the trajectory of foot centers and $\{z_o, m_r, g\}$ are constants. Hence, assuming that f is known, the trajectory of future values for β_d is fully determined.

Suppose we interpret β as the observation of a simple linear system in x_r with the input $\dddot{x}_r$. For smoothness, we add squared input change to the performance index.

$$J = \sum_{t=1}^{\infty} Q_e(\beta^t - \beta^t_d)^2 + R(\dddot{x}^t - \dddot{x}^{t-1})^2. \tag{8.23}$$

We can now determine the optimal $\dddot{x}_r$ with preview control [28]. At any time t we know the error $e(t) = \beta^t - \beta^t_d$, state $x(t) = [\, x^t_r \; \dot{x}^t_r \; \ddot{x}^t_r \,]^T$ and N future β^i_d. Stilman [23] gives the procedure for pre-computing the gains G_1, G_2 and G_3 such that the incremental control in Equation 8.24 minimizes J:

$$\Delta \dddot{x}^t_r = -G_1 e(t) - G_2 \Delta x^t_r - \sum_{i=1}^{N} G^i_3 (\beta^{t+i}_d - \beta^{t+i-1}_d). \tag{8.24}$$

The control $\Delta \dddot{x}_r$ is discretely integrated to generate the trajectory $\{\ddot{x}_r, \dot{x}_r$ and $x_r\}$ for x_r. The trajectory for y_r is found by direct application of preview control since the object reflects no forces tangent to x.

Since x_r is assumed to be fixed to the robot torso, the generated joint space trajectory still results in zmp tracking error. We incorporate this error into the reference trajectory and iterate optimization.

8.3.2.3 Online Feedback

Section 8.3.2.2 described the generation of a balanced trajectory for x_r given x_o. To handle online errors we modify these trajectories online prior to realization with robot joints. Online feedback operates at a 1 ms cycle rate.

Accumulated ZMP tracking error can lead to instability over the course of execution. Therefore, a proportional controller modifies the acceleration of x_r to compensate for ZMP errors perceived through the force sensors at the feet. These corrections are discretely integrated to achieve x_r position.

The trajectory for x_o, or the robot hands, is modified by impedance. We use a discrete implementation of the virtual dynamic system in Equation 8.25 to compute the offset for x_o that results from integrating the measured force error F:

$$F = m_i\ddot{x}_o + d_i\dot{x}_o + k_i(x_o - x_o^d). \tag{8.25}$$

Impedance serves two goals. First of all, we ensure that hand positioning errors do not lead to large forces pushing down on the object. Since the robot does not use the object for support, d_i and k_i are set low for the z direction.

Second, we prevent the robot from exceeding torque limits when the trajectory cannot be executed due to un-modeled dynamics. The position gain for the x direction trades a displacement of 10 cm for a 100 N steady state force. This allows for precise trajectory following and soft termination when the trajectory offset exceeds force limits.

8.3.3 Modeling Object Dynamics

Section 8.3.2 described our implementation of whole body manipulation given a known external force. However, when the humanoid interacts with an unspecified object, the reflected forces may not be known in advance. A reactive strategy for handling external forces might assume that the force experienced when acting on the object will remain constant for 0.15 s seconds, or the duration of trajectory execution. In this section we present an alternative method that improves performance by learning the mapping from a manipulation trajectory to the reflected object forces.

8.3.3.1 Motivation for Learning Models

Modeling addresses two challenges: noise in force sensor readings and the dependence of balance control on future information. The former is common to many robot systems. While high frequency forces have no significant impact on balance, low frequency force response must be compensated. The complication is that online filtering introduces a time delay of up to 500 ms for noise free data. Modeling can be used to estimate forces without time delay.

For balancing robots such as humanoids, we not only require estimates of current state but also of future forces. Typically, a balance criterion such as center of pressure location (ZMP) is achieved by commanding a smooth trajectory for the robot COM. [28] demonstrates that accurate positioning of ZMP requires up to 2s of future information about its placement. Since external forces at the hands create torques that affect the ZMP, they should be taken into account 2s earlier, during trajectory generation. Hence, the purpose of modeling is to use known information such as the target object trajectory to accurately predict its low frequency force response in advance. The predicted response is used to generate a smooth trajectory for the robot COM that satisfies the desired ZMP.

8.3.3.2 Modeling Method

Environment objects can exhibit various kinematics and dynamics including compliance, mechanical structure and different forms of friction. For instance, the tables and chairs used in our experiments are on casters. Each caster has two joints for wheel orientation and motion. Depending on the initial orientation of the wheels the object may exhibit different dynamics. Currently, we do not have a perception system that can detect and interpret this level of modeling detail. Consequently we approach this problem from the perspective of finding a simple and effective modeling strategy.

First, we observe that despite the complex kinematic structure of a humanoid robot, the robot is typically modeled as a point mass attached to the stance foot with prismatic joints. Likewise, an object can be modeled as a point mass in Equation 8.26. Given experimental data we could compute the mass and friction for an object and use them to predict force. However, due to uncertainty in caster orientation and the low velocities of manipulation our experiments showed that even this model was unnecessarily complex. We did not find a consistent relationship between acceleration and force. Consequently we chose to base our model solely on viscous friction as shown in Equation 8.27.

$$f^t = m_o \ddot{x}_o^t + c\,\dot{x}_o^t. \tag{8.26}$$

$$f^t = c\,\dot{x}_o^t. \tag{8.27}$$

To find c that satisfies this relationship we applied least squares regression on collected data. We executed a trajectory that displaced the object at distinct velocities, $\dot{x}_o^t$, and measured the force at HRP-2's hands, f^t, at 1 ms intervals. The collected data is represented in Equation 8.28. The term b was used to remove bias, which appeared as a constant force offset allowed by impedance control after grasp.

$$\begin{bmatrix} \dot{x}^1 & \dot{x}^2 & \cdots & \dot{x}^n \\ 1 & 1 & \cdots & 1 \end{bmatrix}^T \begin{bmatrix} c \\ b \end{bmatrix} = \begin{bmatrix} f^1 & f^2 & \cdots & f^n \end{bmatrix}^T. \tag{8.28}$$

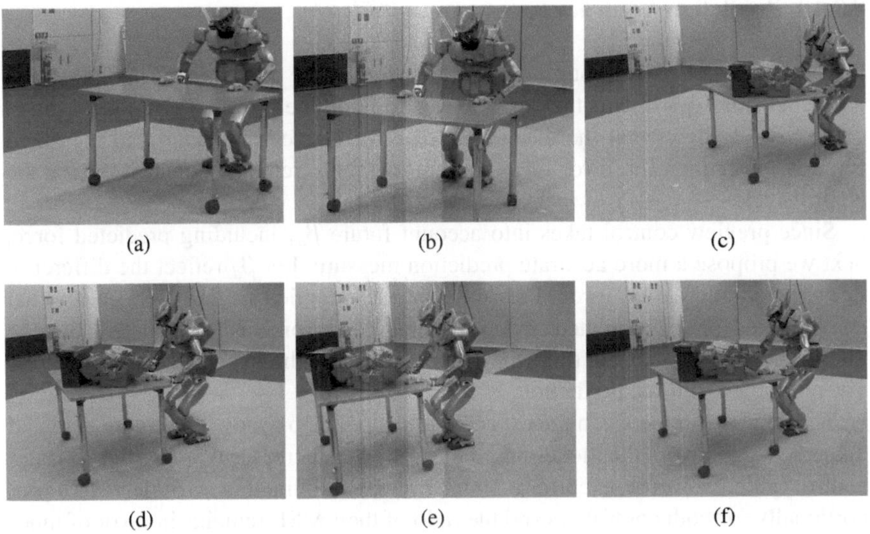

Figure 8.11 HRP-2 pushes then pulls 30 and 55kg loaded tables: (a) t=0s; (b) t=20s; (c) t=0s; (d) t=20s; (e) t=28s; (f) t=38s

The solution to this set of over-constrained equations is found simply by applying the right pseudo-inverse. During data collection we applied a reactive balancing strategy which assumed a constant force response during a 0.15 s trajectory cycle. This approach was sufficiently stable for brief interactions.

8.3.4 Experiments and Results

We conducted experiments on model-based whole body manipulation using a loaded table on casters, as shown in Figure 8.11. The robot grasped the table and followed a smooth trajectory for x_o as generated from a joystick input. Our results were compared with a rigid grasp implementation of a reactive approach to handling external forces presented in [31], which assumed that sensed forces would remain constant. Both methods recomputed the trajectory for x_r every 0.15 s, at which time the reactive strategy updated its estimated force. The reactive method was applied first to gather data and learn an object model from Section 8.3.3.2. Brief experiments of less than 10 s we necessary to collect the data. We applied both methods on a series of experiments that included a change of load such that the total mass ranged from 30 kg to 55 kg.

8.3.4.1 Prediction Accuracy

First, we look at how well our model predicts force. The comparisons in this section use data from experiments that are not used to build the model. The comparison in Table 8.3.4.1 shows that the mean squared error between modeled and measured force is lower than the error of assuming that force remains constant during the control cycle.

Since preview control takes into account future β_d, including predicted force, next we propose a more accurate prediction measure. Let β_d reflect the difference in predicted and actual force. Preview control is applied to find a trajectory that compensates for the simulated error. It generates an erroneous x_r displacement, x_{err}^{PC}, during the 150 ms that the trajectory is active. x_{err}^{PC} is the expected trajectory error given the error in force prediction.

The comparison between the expected trajectory error, shown in Figure 8.13 and Table 8.1, also favors the model-based method. x_{err}^{PC} decreases if we assume a faster control cycle. However, even for a 20 ms cycle, we found that error decreases proportionally for both controllers and the ratio of their MSE remains in favor of modeling.

Table 8.1 Prediction Accuracy

	MSE $F_{err}(N)$		MSE $x_{err}^{PC}(m)$	
	model	react	model	react
$30kg$	4.44	9.19	.427	.674
$55kg$	5.21	12.5	.523	.971

Table 8.2 System Stability

	ZMP SD (m)		Force SD (F)	
	model	react	model	react
$30kg$	.0214	.0312	11.06	15.79
$55kg$	.0231	.0312	12.15	46.25

8.3.4.2 System Stability

The accuracy of prediction has significant effect on the overall stability of the controlled system. Incorrect predictions affect the trajectories for x_r and x_o. First consider the resulting ZMP of the robot. While both controllers exhibit a slight offset in ZMP from grasping the object, the constant error can be removed with integral or adaptive control. A greater concern is the variance in this error. Figure 8.14 shows the increased noise in the control signal for ZMP when using the reactive control. Table 8.2 summarizes this effect.

An even clearer distinction between the two methods is directly reflected in the noise of the perceived force data. Table 8.2 also shows the variance in the noise given the offline filtered signal. The difference in noise is clearly seen in Figure 8.12.

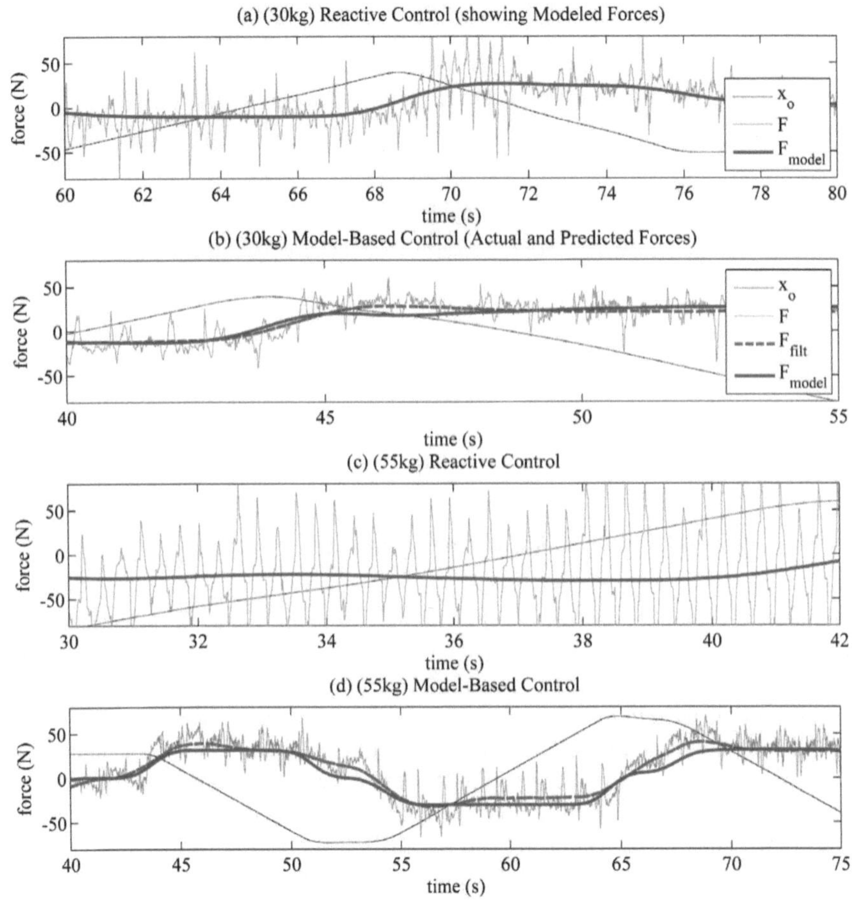

Figure 8.12 Comparison of forces experienced with reactive and model-based control.

8.3.5 Summary

We have shown that it is possible to reliably manipulate unknown, large, heavy objects, such as tables, along specified trajectories with existing humanoid robots. Our experiments demonstrate a significant improvement both in prediction accuracy and system stability when using the learned object model for control. We found that statistical methods such as least squares regression can be used to learn a dynamic model for the unknown object and use it to improve balance during manipulation. One of the most convincing results is the accurate force prediction for a 55 kg object in Figure 8.12(d). Additionally, notice the low noise variance in sensed forces when using the model based controller.

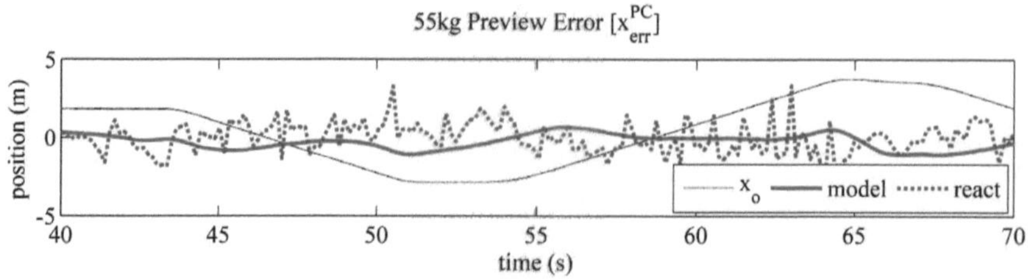

Figure 8.13 Trajectory error introduced by preview control with erroneous prediction

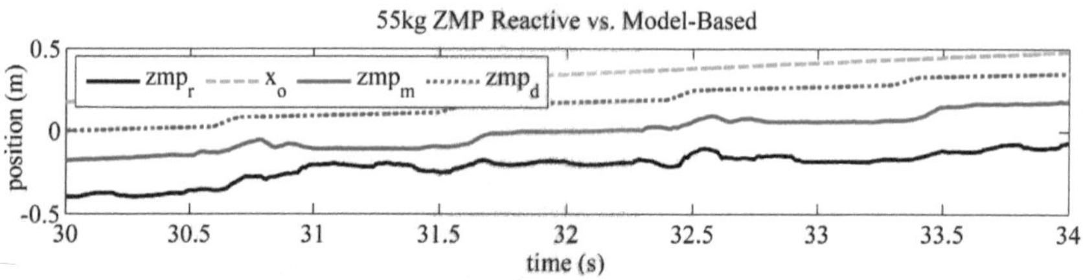

Figure 8.14 Realized ZMP for identical reference trajectories

Future work should consider merging the feed-forward model with feedback information. State estimators such as Kalman filters could be used to maximize the performance of the robot by combining information sources. Furthermore, adaptive control could be used to handle online changes in friction and mass.

While our experiments were restricted to rigid grasp manipulation for objects on casters, similar techniques can be applied to very different scenarios. Two important classes are objects that can be lifted by the robot and objects that are suspended by cranes rather than carts. The robot can statically estimate the former object mass by measuring the load after lifting. The latter cannot be lifted and would require a similar experimental trajectory execution. For suspended objects inertial terms will likely dominate friction. In this case, we propose estimation of mass and inertia.

We have now presented methods for planning object motion and controlling a robot to perform the desired movement. Section 8.4 will detail our complete architecture for NAMO execution.

8.4 System Integration

8.4.1 From Planning to Execution

Having investigated a strategy for NAMO planning and a method for mobile manipulation by humanoid robots we now turn our attention to merging these elements into a complete system for NAMO. This section introduces the architecture for our implementation of NAMO using the humanoid robot HRP-2 shown in Figure 8.15.

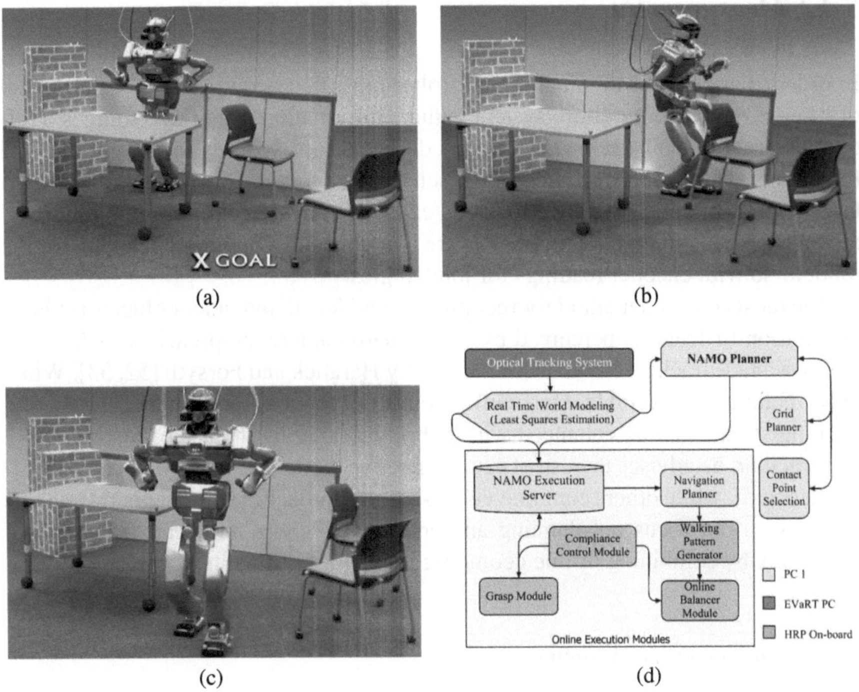

Figure 8.15 Autonomous execution of NAMO with architecture diagram: (a) t=0s; (b) t=45s; (c) t=180s; (d) NAMO Architecture

We present the methods used for measuring the environment, mapping world objects into a planar search space and constructing motion plans for robots. The NAMO planner used in this section is derived from Section 8.3. The details of control for robot walking and whole body manipulation were addressed in Section 8.4.

The architecture in this section is distinct from previous humanoid systems planners which focus on specified tasks such as navigation and manipulation. Sakagami, Chestnutt and Gutmann and plan navigation using stereo vision [46, 47, 48]. Kuffner, Harada, Takubo, Yoshida and Nishiwaki generate dynamically stable trajectories for manipulation given an environment model [26, 49, 31, 29]. Brooks recognizes doors and defines an opening behavior [50]. Existing systems do not allow the robot to perceive the environment and perform arbitrary manipulation. Our domain requires autonomous localization, planning and control for walking, grasping and object manipulation.

Our implementation was performed in a 25 m^2 office setting consisting of tables and chairs. All movable objects are on casters to simplify the task of manipulation. This domain is representative of hospitals, homes and nursing homes where heavy non-wheeled objects are typically stationary. Furthermore, our instrumentation approach to measurement is feasible in these environments.

8.4.2 *Measurement*

In order to apply NAMO planning the robot must first acquire a geometric model of its environment. Our entire system gathers information from three sources: real-time external optical tracking, joint encoders and four six-axis force sensors. The force sensors at the feet and hands are discussed in Section 8.6 with regard to closed loop control. In this Section we focus on external optical tracking for the robot and movable objects. Individual robot links are positioned by combining tracking for the robot torso with encoder readings for joint angles.

The most common method for recognizing and localizing indoor objects is visual registration of features perceived by an onboard camera. Approaches such as the Lucas-Kanade tracker [51] are summarized by Haralick and Forsyth [52, 53]. While these methods are portable, Dorfmuller points out that the speed and accuracy of a tracking system can be enhanced with hybrid tracking by the use of markers [54]. In particular, he advises the use of retro-reflective markers. Some systems use LED markers [55], while others combine vision-based approaches with magnetic trackers [56]. Given our focus on planning and control, we chose an accurate method of perception by combining offline geometric modeling with online localization.

8.4.2.1 Object Mesh Modeling

The robot world model consists of the robot and two types of objects: movable and static. Static objects cannot be repositioned and must always be avoided. Limited interaction with static objects prompted us to use approximate bounding box models to represent their geometry. Movable objects require manipulation during which the robot must come close to the object and execute a precise grasp. These objects were represented internally by 3D triangular mesh models.

Our experimental environment contained two types of movable objects (chairs and tables). To construct accurate models of these objects we used the Minolta Vivid laser scanner. The resulting meshes shown in Figure 8.16(b) were edited for holes and processed to minimize the overall polygon count while ensuring that at least one vertex exists in every 0.125 m^3 voxel of the mesh. This simplified object detection in any given region of 3D space to vertex inclusion.

8.4.2.2 Recognition and Localization

Precise object localization and model fitting was achieved using the EVa Real-Time Software (EVaRT) with the Eagle Motion Analysis optical tracking system. Each model was assigned a unique arrangement of retro-reflective markers. Under rigid body assumptions, any 6D configuration of the object corresponds to a unique set of configurations for the markers. We define a *template* as the set of x,y and z coordinates of each marker in a reference configuration.

(a) (b) (c)

Figure 8.16 Off-line modeling and online localization of a chair: (a) Chair and markers; (b) Mesh model; (c) Real-time overlay

---

Given positions for unoccluded template, $\{a_1, ..., a_n\}$, and localized markers $\{b_1, ..., b_n\}$:

1. Find the centroids (c_a and c_b) of the template markers points and the observed marker locations. Estimate the translational offset $\hat{t} = c_b - c_a$. Removing this offset, $b'_i = b_i - \hat{t}$ places the markers at a common origin.
2. Next we define a linear system for the orientation of the object,

$$\begin{bmatrix} a_1^T & & 0 \\ 0 & a_1^T & \\ & & a_1^T \\ & \cdots & \\ a_n^T & & 0 \\ 0 & a_n^T & \\ & & a_n^T \end{bmatrix} \begin{bmatrix} \hat{r}_1 \\ \hat{r}_2 \\ \hat{r}_3 \end{bmatrix} = \begin{bmatrix} b'_1 \\ b'_2 \\ \vdots \\ b'_n \end{bmatrix} \qquad \text{such that} \qquad b_i = \begin{bmatrix} - & \hat{r}_1^T & - & \\ - & \hat{r}_2^T & - & \hat{t} \\ - & \hat{r}_3^T & - & \\ 0 & 0 & 0 & 1 \end{bmatrix} a_i$$

expresses an estimate of the object transform. We solve this system for $\hat{\mathscr{R}}$ online using LQ decomposition.

Figure 8.17 Object localization procedure.

EVaRT continuously tracks the locations of all the markers to a height of 2 m. Distances between markers are calculated to 0.3 tracking approximately 60 markers the acquisition rate is 60 Hz. Matching the distances between markers, EVaRT partitioned them among objects and provided our system with sets of marker locations and identities for each object. The detected markers are rigidly transformed from the template and permit a linear relationship in the form of a transformation matrix.

Since some markers can be occluded from camera view, we add redundant markers to the objects and perform pose estimation using only the visible set. Estimation is a two-step procedure given in Figure 8.17 At this time, we do not enforce rigidity constraints. Even for a system with only four markers, the accuracy of marker tracking yields negligible shear and scaling in the estimated transformation. The transform is optimal in minimizing the summed squared 3D marker error.

We refer to the collection of transformed meshes for the movable objects, static objects and the robot as the *world model*. Poses for individual robot links are found by combining joint encoder readings with the tracked position and orientation of the robot torso. The entire model is continuously updated at 30 hz. Further details and applications of our approach to mixed reality experimentation can be found in [57].

8.4.3 Planning

The world model, constructed in Section 8.4.2.2, combined with kinematic and geometric models of the robot is sufficient to implement NAMO planning. However, the search space for such a planner would have 38 degrees of freedom for robot motion alone. The size of this space requires additional considerations which are taken into account in Section 8. Presently, we observe that for many navigation tasks the 2D subspace consisting of the walking surface is sufficiently expressive. In fact, larger objects such as tables and chairs that inhibit the robot's path must be pushed rather than lifted [26, 49]. Hence, their configurations are also restricted to a planar manifold. Reducing the search space to two dimensions makes it possible to apply our NAMO implementations from Sections 8.3 and 8.4 directly to real environments. To do so, we map the world model to a planar configuration space. We also introduce a definition for contact and an abstract action space for the robot.

8.4.3.1 Configuration Space

Mapping the world model into a configuration space that coincides with our previous NAMO implementations requires us to project all objects onto the ground plane. Each object is associated with the planar convex hull of the projected mesh vertices. Figure 8.18 shows the model of a chair projected onto the ground plane. While our algorithms are general for any spatial representation, the projected space is computationally advantageous since it considers only three degrees of freedom for the robot and objects. Currently, the space does not allow interpenetration between objects. Alternative implementations could use multiple planes to allow penetration at distinct heights.

The robot is represented by a vertical cylinder centered at the robot torso. A 0.3 m radius safely encloses torso motion. The cylinder projects to a circle on the ground plane. We pre-compute the navigational configuration space, $\mathscr{C}_R$ of the robot. Each $\mathscr{C}_R$ obstacle (O_i) is a Minkowski sum of the robot bounds with the corresponding planar object. For navigation, the robot can be treated as a point that moves through the $\mathscr{C}$-space. Lighter shaded ground regions around projected obstacles in Figure 8.18 are $\mathscr{C}_R$ obstacles. The robot may walk on the lighter regions but its centroid must not enter them.

To further decrease computation costs, we used the Bentley–Faust–Preparata (BFP) approximate convex hull algorithm to compute $\mathscr{C}_R$ obstacles [58]. The resulting obstacles have significantly fewer vertices and edges while subject to only

(a) (b) (c)

Figure 8.18 Mapping objects into the planning domain: (a) Mesh model; (b) Projection; (c) Contact selection

1% error. Decreasing the number of edges reduces the cost of testing for robot collision. The error is acceptable since we can adjust the radius of robot safety bounds.

8.4.3.2 Contact Selection

As part of NAMO planning, the robot must select locations for contacting movable objects. Our implementation uses rigid grasp manipulation and automatically selects points for grasps. In our work, we have found three essential criteria for this selection:

1. *Proximity to object perimeter* – ensure that the point is in the robot's workspace when standing next to the object.
2. *Restricted quantity* – limit the number of possible interactions to a discrete set of grasp points to increase the efficiency of planning.
3. *Uniform dispersion* – provide contact points for any proximal robot configuration.

We introduce one solution for locating such points by interpreting the convex hull from the previous section as a representation of the object perimeter. Starting at an arbitrary vertex of this hull, our algorithm places reference points at equidistant intervals along the edges. The interval is a parameter that we set to 0.2 m.

For each reference point, we find the closest vertex by Euclidian distance in the full 3D object mesh along the horizonal model axes. The selected vertices are restricted to lie in the vertical range [0.5 m, 1.0 m]. When interpreted as contact points, we have found that these vertices satisfy the desired criteria for objects such as tables and chairs. Selected points can be seen in Figure 8.18(c). The robot successfully performed power grasps of our objects at the computed locations.

More complex objects may be of width outside the operating range or may have geometry that restricts the locations and orientations of valid grasps. These cases can be handled by applying a grasp planner such as GraspIt to determine valid contacts. [5] The proposed criteria can still be optimized by using proximity to the contour points as a heuristic for selecting grasps.

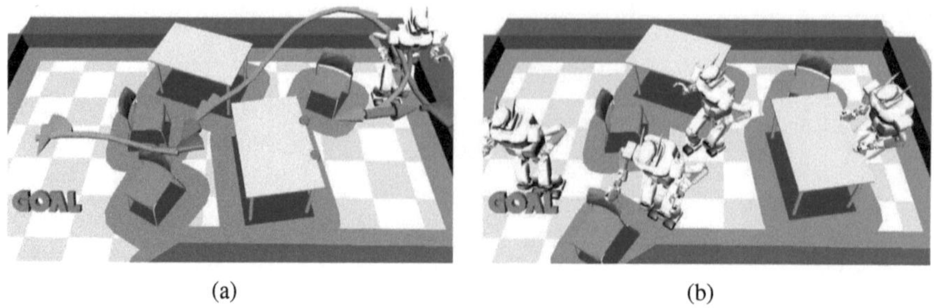

Figure 8.19 Simulated example shows NAMO plan and traces of execution: (a) Initial state and NAMO plan; (b) Traces of execution

8.4.3.3 Action Spaces

So far we have given definitions for the geometry of the configuration space and for allowable contacts. The NAMO planner also requires us to define the action space of the robot that will be searched in selecting *Navigate* and *Manipulate* operators.

Humanoid walking has the property of being inherently discrete since a stable motion should not terminate in mid-stride. Each footstep is a displacement of the robot. One option for planning is to associate the robot base with the stance foot and directly plan the locations of footsteps [59, 47]. This creates discrete jumps in the location of the robot base at each change of supporting foot. In our approach, we define an abstract *base* and plan its motion along continuous paths. The simplest mapping of the base trajectory to footsteps places feet at fixed transforms with respect to the base on every step cycle. We found that a horizontal offset of 9 cm from the base along a line orthogonal to the base trajectory yields repeatable, safe and predictable robot motion.

The *Navigate* space consists of base displacements. Since the foot trajectory generated from base motion must be realizable by the controller, we restricted base motions to ones that translate into feasible footstep gaits. From any base configuration, the robot is permitted 41 discrete actions which are tested for collisions with the $\mathscr{C}$-space obstacles:

1. One 0.1 m translation backward.
2. Twenty rotations in place in the range of 30°.
3. Twenty 0.2 m translations forward with a rotation in the range of 30°.

During search, the planner keeps track of visited states. A* finds least cost paths and terminates when all states have been visited.

Having grasped an object *Manipulate* searches the same space as *Navigate*. During planning we assume that the object remains at a fixed transform with respect to the robot base and therefore the stance foot by construction. Our planner distinguishes large objects, such as tables, from smaller objects such as chairs. Smaller objects have less inertia and can be manipulated safely with one arm. Tables, however, have a significant impact on the robot dynamics and are grasped with two hands. While both grasps are restricted to the computed contact points, there are fewer contact configurations for the robot base during a two-handed grasp.

Action definitions complete the description of a NAMO problem and allow us to apply the planner to solve problems such as the one in Figure 8.19. In this case we applied SELECTCONNECT. The figure shows light arrows to indicate planned navigation and dark arrows for manipulation. The robot joint configurations shown in Figure 8.19(b) interpret planned actions as described in Section 8.3.

8.4.4 Uncertainty

The NAMO planner presented in Section 8.2 assumes that the world conforms to the actions determined by the planner. While precise modeling and localization of objects serves to decrease error, significant uncertainty remains during execution. Planning with uncertainty or in partially known environments is a complex problem. Erdmann corners the system into a desired state [60]. Stentz efficiently replans while expanding knowledge of the environment [61]. Kaelbling finds optimally successful plans [62]. One or more of these approaches can be adapted to NAMO planning. In this thesis we present only the necessary solution to uncertainty that makes execution possible.

We consider the complete NAMO implementation as a hierarchy of high level action planning, lower level path planning and online joint control. At the lowest end of the spectrum, insignificant positioning errors can be handled by joint-level servo controllers. At the highest, a movable obstacle that cannot be moved may require us to recompute the entire NAMO plan. We propose a minimalist strategy to error recovery. Each level of the hierarchy is implemented with a strategy to reduce uncertainty. When this strategy fails, execution is terminated. Further development should consider reporting the error and considering alternatives at higher levels of the architecture. Presently, we describe three of the strategies applied in our work.

8.4.4.1 Impedance Control

At the lowest level of the hierarchy uncertainty in our estimate of object mass and robot positioning is handled with impedance control as described in Section 8.3.2.3. This controller handles small positioning errors that occur due to robot vibration and environment interaction, ensuring that the robot does not damage itself or the obstacle. Impedance limits the forces that the robot can exert in order to achieve the precise positioning demanded by the planner.

8.4.4.2 Replanning Walking Paths

Since the lowest level of execution does not guarantee precise placement of objects, it is possible that the navigation paths computed by the NAMO planner will not be possible after the displacement of some object. Furthermore, since robot trajectories are executed open loop with regard to localization sensors, the objects and the robot may not reach their desired placements.

In order to compensate for positioning errors, the path plans for subsequent *Navigate* and *Manipulate* actions are re-computed at the termination of each NAMO action. At this time the planner updates the world model from the optical tracker and finds suitable paths for the robot. Notice that we do not change the abstract action plan with regard to object choices and orderings. We simply adapt the actions that the plan will take to ensure their success. For *Navigate* paths we iterate state estimation, planning and execution to bring the robot closer to the desired goal. This iteration acts as a low rate feedback loop from the tracker to the walking control and significantly improves the robot's positioning at the time of grasp.

8.4.4.3 Guarded Grasping

Although we have described the execution of *Navigate* and *Manipulate* actions as two essential components of NAMO, bridging these two actions is also an interesting problem. Having navigated to a grasp location, the robot is required to execute a power grasp of the object at a given contact point. Positioning errors in grasping can be more severe than manipulation since the object is not yet fixed to the robot and its location is uncertain.

Having walked up to an object, the robot must grasp it and then walk with it. HRP-2 reacquires the world model and determines the workspace position for its hand. It preshapes the hand to lie close to this position and then performs a guarded move to compensate for any perception error. This process is not only functionally successful but also approximates human grasping behavior as described in [63].

The initial workspace hand position for grasping is obtained by selecting a point $.05m$ above the object and 0.05 m closer to the robot (in the direction of robot motion). The robot moves its hands to this position via cubic spline interpolation from its estimated state. Subsequently the robot moves its hand downward and forward to close the 0.05 m gaps. This guarded move is executed using impedance control to prevent hard collisions and is terminated when the force sensors reach a desired threshold. We ensure that the robot's palm is in contact with the top surface of the object and the thumb is in contact with the outer edge. The remaining fingers are closed in a power grasp until the finger strain gauges exceed the desired threshold.

Table 8.3 NAMO implementation: run times for planning and execution

	NAMO	Navigation	Execution
Figure 8.15 (real)	0.06 s	0.09 s	63.0 s
Figure 8.19 (simulated)	0.13 s	0.93 s	153.0 s

8.4.5 Results

Our complete NAMO system was successfully applied to a number of simulated examples such as the one presented in Figure 8.19 as well as the real robot control problem in Figure 8.15(a). The simulated evaluations involved all planning aspects of our work, including dynamically stable walking pattern generation for *Navigate* and *Manipulate* actions. In our laboratory experiments, the robot perceived its environment and autonomously constructed a plan for reaching the goal. After walking to approach an object, HRP-2 successfully grasped it and walked to create the planned displacement. Subsequently it was able to reach the desired goal.

Table 8.3 details the high-level NAMO planning time, replanning time for navigation and the total execution time for the two presented examples. Notice that the NAMO planning time is three orders of magnitude smaller than the actual time for executing the computed motion. This makes it feasible to view the NAMO system as a real-time generalization of modern path planning techniques to worlds where strictly collision free paths are not available.

References

[1] P. Buerhaus, K. Donelan, B. Ulrich, L. Norman, and B. Dittus. State of the Registered Nurse Workforce in the United States. Nurs Econ, 24(1):6–12, 2006.

[2] J.C. Latombe. Robot Motion Planning. Springer, 1991.

[3] R. Alami, J.P. Laumond, and T. Sim'eon. Two manipulation planning algorithms. In: workshop on the algorithmic foundations of robotics, 1994.

[4] M.T. Mason. Mechanics of Robotic Manipulation. MIT Press, 2001.

[5] Andrew T. Miller. GraspIt: A Versatile Simulator for Robotic Grasping. PhD thesis, Dept. of Computer Science, Columbia University, 2001.

[6] K. M. Lynch and M. T. Mason. Stable pushing: Mechanics, controllability, and planning. Int J of Robotics Research, 15(6):533–556, 1996.

[7] Michael Erdmann. An exploration of nonprehensile two-palm manipulation. Int J of Robotics Research, 17(5), 1998.

[8] G. Morris and L. Haynes. Robotic assembly by constraints. In: proceedings of IEEE international conference on robotics and automation., 4, 1987.

[9] JD Morrow and PK Khosla. Manipulation task primitives for composing robot skills. In: proceedings of IEEE international conference on robotics and automation, 4, 1997.

[10] M. Stilman and J.J. Kuffner. Navigation among movable obstacles: Real-time reasoning in complex environments. In: proceedings of IEEE international conference on humanoid robotics (Humanoids'04) http://www.golems.org/NAMO, 2004.

[11] G. Wilfong. Motion panning in the presence of movable obstacles. In: proceedings of ACM symposium on computational geometry, pp 279–288, 1988.

[12] E. Demaine and et. al. Pushpush and push-1 are np-complete. Technical Report 064, Smith, 1999.
[13] A. Junghanns and J. Schaeffer. Sokoban: Enhancing general single-agent search methods using domain knowledge. Artificial Intelligence, 129(1):219–251, 2001.
[14] H. Kautz and B. Selman. BLACKBOX: A new approach to the application of theorem proving to problem solving. In: AIPS98 workshop on planning as combinatorial search, pp 58–60, 1998.
[15] J.C. Culberson and J. Schaeffer. Searching with pattern databases. Lecture Notes in Computer Science, 981:101–112, 2001.
[16] R.E. Korf. Finding optimal solutions to Rubiks Cube using pattern databases. In: proceedings of the fourteenth national conference on artificial intelligence (AAAI-97), pp 700–705, 1997.
[17] L. Kavraki, P. Svestka, J. C. Latombe, and M. H. Overmars. Probabilistic roadmaps for path planning high-dimensional configuration spaces. IEEE Trans Robotics Automat, 12(4), 1996.
[18] S.M. LaValle and J.J. Kuffner. Rapidly exploring random trees: Progress and prospects. In: workshop on the algorithmic foundations of robotics, 2000.
[19] D. Hsu, J.C. Latombe, and R. Motwani. Path planning in expansive configuration spaces. In: proceedings of IEEE international conference on robotics and automation, 3, 1997.
[20] T. Lozano-Perez. Spatial planning: a configuration space approach. IEEE Trans Comput, pp 108–120, 1983.
[21] S.M. LaValle. Planning Algorithms. Cambridge University Press, 2006.
[22] P.C.Chen and Y.K.Hwang. Pracitcal path planning among movable obstacles. In: proceedings of IEEE international conference on robotics and automation, pp 444–449, 1991.
[23] M. Stilman. PhD Thesis: Navigation Among Movable Obstacles. Technical report, Technical Report CMU-RI-TR-07-37, Robotics Institute, Carnegie Mellon University, Pittsburgh, PA, October 2007.
[24] K. Inoue, H. Yoshida, T. Arai, and Y. Mae. Mobile manipulation of humanoids: Real-time control based on manipulability and stabilty. In: proceedings of IEEE international conference robotics and automation (ICRA), pp 2217–2222, 2000.
[25] J. Kuffner. Dynamically-stable motion planning for humanoid robots. Autonomous Robots, 12(1), 2002.
[26] K. Harada, S. Kajita, K. Kaneko, and H. Hirukawa. Pushing manipulation by humanoid considering two-kinds of zmps. In: IEEE international conference on robotics and automation, pp 1627–1632, 2003.
[27] K. Harada, S. Kajita, F. Kanehiro, K.Fujiwara, K. Kaneko, K.Yokoi, and H. Hirukawa. Real-time planning of humanoid robot's gait for force controlled manipulation. In: IEEE international conference on robotics and automation, pp 616–622, 2004.
[28] Shuuji Kajita and et. al. Biped walking pattern generation by using preview control of zero-moment point. In: IEEE international conference on robotics and automation, pp 1620–1626, 2003.
[29] T. Takubo, K. Inoue, and T. Arai. Pushing an object considering the hand reflect forces by humanoid robot in dynamic walking. In: IEEE international conference on robotics and automation, pp 1718–1723, 2005.
[30] K. Nishiwaki, W-K. Yoon, and S. Kagami. Motion control system that realizes physical interaction between robot's hands and environment during walk. In: IEEE international conference on Humanoid Robotics, 2006.
[31] K. Nishiwaki and S. Kagami. High frequency walking pattern generation based on preview control of zmp. In: IEEE international conference on robotics and automation (ICRA'06), 2006.
[32] E. Yoshida, I. Belousov, Claudia Esteves, and J-P. Laumond. Humanoid motion planning for dynamic tasks. In: IEEE international conference on Humanoid Robotics (Humanoids'05), 2005.

[33] E. Yoshida, C. Esteves, T. Sakaguchi, J-P. Laumond, and K. Yokoi. Smooth collision avoidance: Practical issues in dynamic humanoid motion. In: proceedings of IEEE/RSJ international conference on intelligent robots and systems, 2006.

[34] E. Krotkov. Robotic perception of material. IJCAI, pp 88–95, 1995.

[35] P. Fitzpatrick, G. Metta, L. Natale, S. Rao, and G. Sandini. Learning about objects through interaction - initial steps towards artificial cognition. In: proceedings of IEEE international conference on robotics and automation, pp 3140–3145, 2005.

[36] A. Stoytchev. Behavior-grounded representation of tool affordances. In: proceedings of IEEE international conference on robotics and automation, pp 3060–3065, 2005.

[37] A. Christiansen, T. M. Mitchell, and M. T. Mason. Learning reliable manipulation strategies without initial physical models. In: proceedings of IEEE international conference on robotics and automation, 1990.

[38] C.H. An, C.G. Atkeson, and J.M. Hollerbach. Model-Based Control of a Robot Manipulator. MIT Press, 1988.

[39] C. Canudas de Wit, P. No, A. Aubin, and B. Brogliato. Adaptive friction compensation in robot manipulators: low velocities.

[40] H. Olsson, KJ Astrom, CC. de Wit, M. Gafvert, and P. Lischinsky. Friction models and friction compensation.

[41] Stefan Schaal Chris Atkeson, Andrew Moore. Locally weighted learning. AI Review, 11:11–73, April 1997.

[42] Andrew Moore, C. G. Atkeson, and S. A. Schaal. Locally weighted learning for control. AI Review, 11:75–113, 1997.

[43] M. Vukobratovic, B. Borovac, D. Surla, and D. Stokic. Biped Locomotion: Dynamics, Stability, Control and Application. Springer, 1990.

[44] M. Vukobratovic and B. Borovac. Zero-moment point-thirty five years of its life. Int J of Humanoid Robotics, 1(1):157–173, 2004.

[45] D.E. Whitney. Resolved motion rate control of manipulators and human prostheses. IEEE Trans on Man Machine Systems, 10:47–53, 1969.

[46] Y. Sakagami, R. Watanabe, C. Aoyama, S. Matsunaga, N. Higaki, K. Fujimura, H.R.D.C. Ltd, and J. Saitama. The intelligent ASIMO: system overview and integration. In: proceedings of IEEE/RSJ international conference on intelligent robots and system, 3, 2002.

[47] J. Chestnutt, J. Kuffner, K. Nishiwaki, and S. Kagami. Planning biped navigation strategies in complex environments. In: 2003 international conference on humanoid robots, 2003.

[48] J. Gutmann, M. Fukuchi, and M. Fujita. Real-time path planning for humanoid robot navigation. In: international joint conference on artificial intelligence, 2005.

[49] E. Yoshida, P. Blazevic, and V. Hugel. Pivoting manipulation of a large object. In: IEEE international conference on robotics and automation, pp 1052–1057, 2005.

[50] R. Brooks, L. Aryananda, A. Edsinger, P. Fitzpatrick, C. Kemp, U.M. OReilly, E. Torres-Jara, P. Varshavskaya, and J. Weber. Sensing and Manipulating Built-for-Human Environments. Int J Humanoid Robotics, 1(1):1–28, 2004.

[51] B.D. Lucas and T. Kanade. An iterative image registration technique with an application to stereo vision. In: proceedings of DARPA Image Understanding workshop, 121:130, 1981.

[52] R.M. Haralick and L.G. Shapiro. Computer and Robot Vision. Addison-Wesley Longman Publishing Co., Inc. Boston, MA, USA, 1992.

[53] D.A. Forsyth and J. Ponce. Computer Vision: A Modern Approach. Prentice Hall, 2003.

[54] K. Dorfmuller. Robust tracking for augmented reality using retroreflective markers. Computers and Graphics, 23(6):795–800, 1999.

[55] Y. Argotti, L. Davis, V. Outters, and J. Rolland. Dynamic superimposition of synthetic objects on rigid and simple-deformable real objects. Computers and Graphics, 26(6):919, 2002.

[56] A. State, G. Hirota, D.T. Chen, W.F. Garrett, and M.A. Livingston. Superior augmented reality registration by integrating landmark tracking and magnetic tracking. In: proceedings of SIGGRAPH'96, page 429, 1996.

[57] M. Stilman, P. Michel, J. Chestnutt, K. Nishiwaki, S. Kagami, and J. Kuffner. Augmented reality for robot development and experimentation. Technical Report CMU-RI-TR-05-55, Robotics Institute, Carnegie Mellon University, November 2005.
[58] J. Bentley, G.M. Faust, and F. Preparata. Approximation algorithms for convex hulls. Comm. of the ACM, 25(1):64–68, 1982.
[59] JJ Kuffner Jr, K. Nishiwaki, S. Kagami, M. Inaba, and H. Inoue. Footstep planning among obstacles for biped robots. In: proceedings of international conference on intelligent robots and systems, page 500, 2001.
[60] M.A. Erdmann. On Motion Planning with Uncertainty. 1984.
[61] A. Stentz. Optimal and efficient path planning for partially-knownenvironments. In: proceedings of 1994 IEEE international conference on robotics and automation, pp 3310–3317, 1994.
[62] L.P. Kaelbling, M.L. Littman, and A.R. Cassandra. Planning and acting in partially observable stochastic domains. Artificial Intelligence, 101(1):99–134, 1998.
[63] M. Jeannerod, M.A. Arbib, G. Rizzolatti, and H. Sakata. Grasping objects: the cortical mechanisms of visuomotor transformation. Trends Neurosci., 18:314–320, 1995.

Chapter 9
Multi-modal Motion Planning for Precision Pushing on a Humanoid Robot

Kris Hauser and Victor Ng-Thow-Hing

Abstract This chapter presents a motion planner that enables a humanoid robot to push an object to a desired location on a cluttered table. To reduce reliance on visual feedback, we restrict the robot to use a class of stable pushes that move the object predictably, such that a plan can be executed precisely with infrequent sensing. The motion of the robot–object system lies in a space with a multi-modal structure, where the motion switches between walking, reaching, and pushing modes. Each mode imposes mode-specific constraints (e.g., dynamic constraints, kinematic limits, obstacle avoidance) such that motion is restricted to a lower dimensional subspace. The multi-modal planner must choose a discrete sequence of mode switches to reach the goal, while performing continuous motion planning to move between them. To address the problem of selecting modes, we present the Random-MMP algorithm, which randomly samples mode transitions to distribute a sparse number of modes across configuration space. The resulting planner solves problems that require several carefully chosen pushes in minutes. Results are presented in simulation and on the Honda ASIMO robot.

9.1 Introduction

We address the problem of enabling the Honda ASIMO robot to push an object to a desired location on a table. Pushing is a useful manipulation capability for a humanoid robot. It may move objects that are too large or heavy to be grasped. Any part of the body, not just the hands, may be used for pushing; this greatly increases the workspace in which manipulation can be applied. Modeling of pushing tasks re-

Kris Hauser
Computer Science School of Informatics and Computing, Indiana University at Bloomington, e-mail: hauserk@indiana.edu

Victor Ng-Thow-Hing
Honda Research Institute e-mail: vng@honda-ri.com

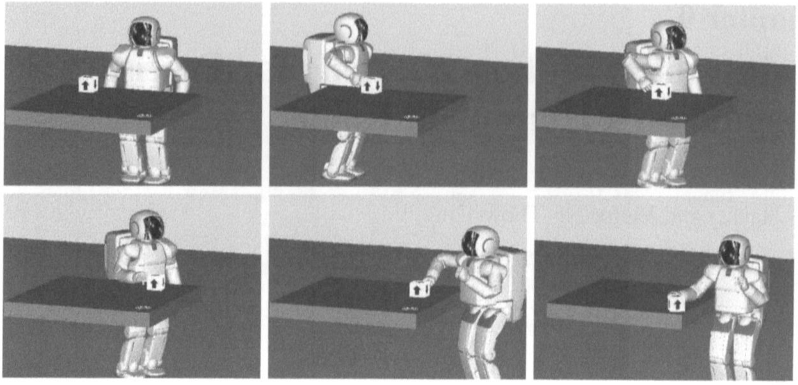

Figure 9.1 To cross a table, an object must be pushed along a table's edges

quires sophisticated modeling of multi-point contact mechanics, and as such can be viewed as a first step toward general full-body object manipulation, such as carrying trays, bulky objects, and manipulation with tools.

But pushing is not as simple as walking to the object and moving the arm; advance planning is crucial. Even simple tasks, like reorienting the object in place, may require a large number of pushes. Between pushes, the robot may need to switch hands or walk to a new location, choosing carefully among alternatives. Furthermore, many tasks cannot be solved by greedily pushing the object toward its target. For example, an object cannot be pushed directly across a large table (Figure 9.1). Once the object is out of reach it cannot be pushed further, and since pushing is nonprehensile, the object cannot be recovered.

Complex geometric constraints are imposed by collision avoidance and joint limit constraints. Integration with the current technology in humanoid robot platforms also poses additional challenges. ASIMO must stand while pushing, because its hand cannot be positioned accurately while it walks. Its reach is limited, so it must repeatedly walk and push to move objects any reasonable distance (say, across a table). ASIMO's cameras cannot be tilted to view nearby objects at table height, so visual feedback is not possible during pushing – it must essentially push blindly. Thus, we restrict the robot to use a class of stable pushing actions which rotate the object predictably.

These constraints introduce a particular *multi-modal* structure to the problem, where the system moves between *walking*, *reaching*, and *pushing* modes. Each mode imposes mode-specific motion constraints that must be satisfied. A motion planner for this problem must therefore produce a discrete sequence of modes, as well a continuous motion through them. This multi-modal planning problem occurs in several areas of robotics. For example, in manipulation planning, motion alternates between transfer and transit (object grasped/not grasped) modes [2, 39, 44]. In legged locomotion, each set of environment contacts defines a mode [6, 24]. Modes also occur in reconfigurable robots [11] and as sets of subassemblies in assembly planning [52].

This chapter presents the application of a recently developed multi-modal planning algorithm, Random-MMP, to the task of ASIMO pushing an object on a table. Originally presented in [27], Random-MMP builds a tree of configurations sampled across the configuration space. Each configuration lies at the *transition* between two modes, and pairs of configurations are connected by single-mode paths. It grows the tree by sampling mode transitions at random, according to a strategy designed to distribute modes sparsely across configuration space.

Random-MMP has two attractive properties. First, it is general enough to handle general multi-modal systems. For example, multi-handed pushing, grasping, and multiple objects could be incorporated in the same framework by adding new modes. Second, it satisfies an asymptotic completeness guarantee that states that the probability it finds a path approaches 1 as more time is spent planning.

Even though the theoretical reliability of Random-MMP is guaranteed by these completeness properties, its practical performance depends greatly on the choice of sampling distribution. Without any prior information, uniform sampling may be a reasonable choice, but we can do much better with problem-specific domain knowledge. Because the reachable workspace of ASIMO's hands is fairly limited, the planner must position the robot carefully to execute a single long-distance push. Thus, the planner faces a challenge in choosing where to stop walking and start pushing. To help choose such locations, we precompute tables of push *utility*, the expected distance the object can be pushed. During planning, we use these tables to sample mode switches that yield higher utility pushes. This *utility-centered expansion* approach greatly improves planning speed and quality, to the point where motions can be planned and executed on ASIMO at interactive rates.

We further improve motion quality with a fast postprocessing technique that smooths arm motions while respecting collision constraints. It repeatedly attempts to replace segments of the unprocessed path with "shortcuts" of time-optimal collision-free trajectories.

The overall planner solves easy problems in less than a minute, and difficult problems with obstacles in minutes on a PC. Results are demonstrated in simulation and in experiments on the real robot.

9.2 Background

9.2.1 Pushing

Pushing has received a great deal of interest for industrial part orientation applications. An object pushed on a plane with one contact point moves with uncertainty, due to indeterminacy in the distribution of support forces [41, 36]. The use of point pushing to control orientation has been achieved using visual and tactile feedback [35]. Two or more contact points enables open-loop control of the object's position and orientation [33, 1]. Multi-contact pushing has been used in sensorless manipulation, where passive mechanisms reduce uncertainty in an object's position and orientation. Sensorless manipulation has been achieved with pushing mechanisms such as fences [55], tilting trays [16], and parallel-jaw grippers [19].

To plan the motion of objects pushed with multiple contacts, Akella and Mason presented a complete algorithm for posing an object in the plane without obstacles [1]. Lynch and Mason presented a planner that avoids collision with obstacles [34]. In our work, we also consider the problem of finding collision-free paths of the manipulator under kinematic limits.

Our planner assumes that the object is light enough such that the balance of the robot is not affected much. To extend our work to heavier objects, the walking controllers developed by Takubo et al. [48] and Harada et al. [22] could be used.

9.2.2 Multi-modal Planning

Pushing is an example of a hybrid system that travels between discrete modes, where each mode has its own feasibility constraints. Other such systems include grasping manipulation, dexterous manipulation, legged locomotion, and reconfigurable robots. For such systems, a multi-modal planner must choose a *discrete* sequence of modes, as well as a *continuous* motion to traverse them.

Hybrid system modeling, verification, controllability, and optimality have been well studied [5, 8, 9, 20, 37]. From a planning perspective, motion planners have been developed for part manipulation using multiple grasp and regrasp operations [2, 17, 29, 44], dexterous manipulation [13, 53, 54], legged locomotion [4, 7, 14, 24, 25, 30, 42], navigation among movable obstacles [39, 47, 49], assembly planning [52], and reconfigurable robots [11, 12]. Recent work has addressed multi-modal planning for general hybrid systems specified with STRIPS-like task languages [10].

Multi-modal planners take a variety of approaches to addressing the discrete and continuous choices that must be made. If the connected components of each mode can be enumerated (for example, in low-dimensional or geometrically tractable problems), the problem reduces entirely to discrete search [2, 4, 52]. Some problems (e.g., pp. 270–274 of [32], [53]) can be addressed by treating mode switching as an action available to the robot, which allows the use of sample-based planners. This approach does not generalize to hybrid systems where mode switches are only applicable in sets of measure zero, and therefore will never be sampled.

Some researchers have taken a decoupled approach. The first stage produces a sequence of modes, and the second stage plans single-mode paths along that sequence [11, 12]. A common approach in manipulation is to plan an object trajectory first, then motions to make or break contact with the object [13, 29, 54]. For this approach to work, however, requires that the modes produced by the first stage contain a path to the goal. In general, this cannot be guaranteed.

A search-based approach first appeared in manipulation planning as a "manipulation graph" [2]. It maintains a search tree to explore a *mode graph*. To expand the tree, an existing mode is selected, and the search expands to neighboring modes that are connected by single-mode motions. In high-dimensional problems, sample-based techniques must be used for single-mode planning [7, 24, 25, 38].

Since mode graph search is highly general, we use it in this work. However, it does not address the fact that some modes have a continuous set of neighbors, which need to be discretized during planning. Any fixed discretization (e.g., [24, 25, 38, 39]) can potentially fail to contain a feasible path, even if a path exists. Some researchers have addressed adaptive discretization of continuous modes using sampling techniques. This technique has been applied to manipulation with grasps and regrasps [2, 44], navigation among movable obstacles [49], and our previous work [27].

9.2.3 Complexity and Completeness

Multi-modal planning may require solving hard problems in both the discrete and continuous domains. Motion planning in high-dimensional continuous spaces is hard; the worst case time of any complete algorithm to plan a collision-free motion of an articulated robot is exponential in the number of degrees of freedom [43]. But multi-modal problems may be hard even if each continuous space has low dimensionality; for example, navigation among moving obstacles in 2D is NP-hard when the number of obstacles is not fixed [51]. And yet, multi-modal planners have been developed for specific problems with some success.

For continuous motion planning in high-dimensional modes, probabilistic roadmap (PRM) methods are typically used. These methods build a network of randomly sampled feasible configurations connected by simple line segments. They are probabilistically complete, which means that the probability that a feasible path is found, if one exists, approaches 1 as more time is spent planning. Recent work has proven probabilistically complete a multi-modal planner that uses PRMs for single-mode planning [26], under the assumption that the set of modes is pre-discretized.

van Der Berg et al. has proven probabilistic completeness of a multi-modal planner that uses uniform sampling for mode discretization [49]. However, they assume that a complete planner is used for single-mode planning. In Section 9.5, we argue that probabilistic completeness also applies to our planner. Furthermore, we propose sampling distributions that seek to explore the space of modes more rapidly than a uniform sampling.

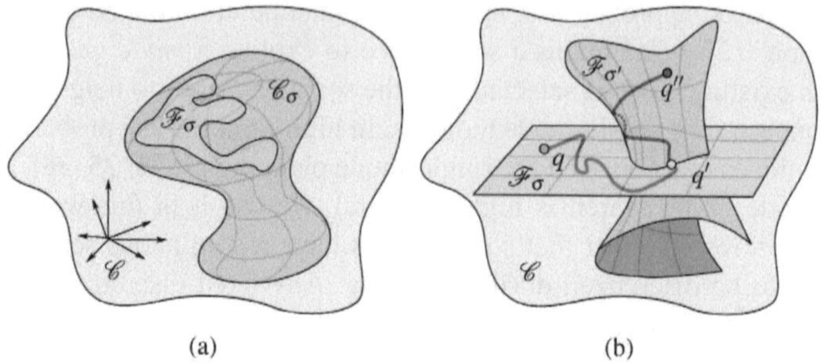

Figure 9.2 Abstract depiction of mode's manifold, feasible space, and transitions. (a) At a mode σ, motion is constrained to a subset $\mathcal{F}_\sigma$ of a submanifold $\mathcal{C}_\sigma$ with lower dimension than $\mathcal{C}$. (b) To move from configuration q at stance σ to q'' at an adjacent stance σ', the motion must pass through a transition configuration q'

9.3 Problem Definition

We plan for ASIMO to push an object across a horizontal table. We move one arm at a time for convenience. We assume the object moves quasi-statically (slides without toppling and comes to rest immediately), and can be pushed without affecting the robot's balance. The planner is given a perfect geometric model of the robot, the object, and all obstacles. Other physical parameters are specified, e.g., the object's mass and the hand–object friction coefficient. Given a desired translation and/or rotation for the object, it computes a path for the robot to follow, and an expected path for the object.

The motion of the robot is segmented into walking, reaching, and pushing modes. The multi-modal planning problem asks for a path that achieves a goal, which may require multiple mode switches. In each mode, the continuous path must satisfy mode-specific geometric and motion constraints.

9.3.1 Configuration Space

A configuration q combines a robot configuration q_{robot} and an object configuration q_{obj}. ASIMO's walking subsystem allows fully controllable motion in the plane, so leg joint angles can be ignored. Thus, q_{robot} consists of a planar transformation $(x_{robot}, y_{robot}, \theta_{robot})$, five joint angles for each arm, and a degree of freedom for each hand ranging from open to closed. Since the object slides on the table, q_{obj} is a planar transformation $(x_{obj}, y_{obj}, \theta_{obj})$. In all, the configuration space $\mathcal{C}$ is 18 D.

The robot is not permitted to collide with itself or obstacles, and may only touch the object with its hands. It must obey kinematic limits. The object may not collide with obstacles or fall off the table. We also require that the object be in front of the robot while pushing to avoid some unnatural motions (e.g. behind-the-back pushes).

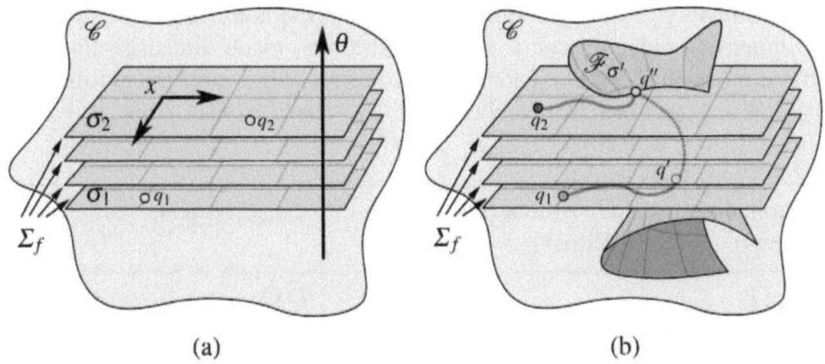

Figure 9.3 Continuous family of modes. (a) A continuous family of modes Σ_f. Each value of the co-parameter θ defines a different mode $\sigma \in \Sigma_f$ (four are shown). Each mode is depicted as a linear subspace parameterized by x (but, in general, modes may be nonlinear). A mode in Σ_f can be identified by a representative configuration, e.g., q_1 identifies σ_1 and q_2 identifies σ_2. (b) Moving within Σ_f, from q_1 to q_2, requires transitioning to a different mode σ' at q', and back again at q''

9.3.2 Modes

The motion of the ASIMO–object system is divided into modes, in which motion is allowed only in a particular subspace of $\mathscr{C}$. The motion constraints are defined as follows:

- *Walking*. Only the base of the robot $(x_{robot}, y_{robot}, \theta_{robot})$ moves. The arms must be raised to a "home configuration" that avoids colliding with the table while walking.
- *Reach*. Only a single arm and its hand may move. The arm must avoid collision with the table or object.
- *Push*. The hand is in contact with the object. The object moves in response to the arm motion according to push dynamics, and reciprocally, the dynamics impose constraints on arm motions (Section 9.4.4.1). Additionally, the object must lie in the robot's field of view.

In all modes, the robot must avoid collision with itself and the table, and may only contact the surface of the object in a push mode.

Using the formal terminology of [23], the system moves between an infinite set of modes, Σ. Each mode $\sigma \in \Sigma$ defines a *feasible space* $\mathscr{F}_\sigma$, the set of configurations that satisfy certain mode-specific constraints. We partition Σ into $F = 5$ disjoint *families*, $\Sigma_1 \ldots \Sigma_F$, which contain the *walking*, *reach left*, *reach right*, *push right*, and *push left* modes.

An important semantic note is that "mode" refers to the particular subspace and motion constraints. A "mode family" is a set of related modes, defined precisely as follows. Every mode $\sigma \in \Sigma_f$ is defined uniquely by a *coparameter* θ in a manifold Θ_f (Figure 9.3(a)). We say $\dim(\Theta_f)$ is the *codimension* of σ. For example, the coparameters of a walk mode σ are the configuration of the object q_{obj}. Each distinct object configuration defines a unique walk mode. The coparameters of a reach

Table 9.1 Dimensionality of mode families. Rows report the non-fixed configuration parameters (Parameters), dimension-reducing constraints (Constraints), mode dimensionality (Dims), coparameters, and codimension (Codims). Parameter subsets are abbreviated are as follows: O = object, R_b = robot base, R_a = robot arm, R_h = robot hand, C_h = hand contact point, C_o = object contact point

Type	Parameters (dims)	Constraints (dims)	Dims	Coparameters	Codims
Walk	R_b (3)		3	O (3)	3
Reach	R_a (5), R_h(1)		6	R_b(3) ,O (3)	6
Push	O (3), R_a (5)	contact (5)	3	R_b(3), C_h (2), C_o(2)	7

Table 9.2 Dimensionality of transitions. Parameter subsets are abbreviated are as follows: O = object, R_b = robot base, R_a = robot arm, R_h = robot hand, C_h = hand contact point, C_o = object contact point

Type	Parameters (dims)	Constraints (dims)	Dims	Coparameters	Codims
Walk→Reach	R_b(3)		3	O (3)	3
Reach→Walk		R_a = home	0	R_b(3), O(3)	6
Reach→Reach		R_a = home	0	R_b(3), O(3)	6
Reach→Push	R_a (5), C_h (2), C_o (2)	contact (5)	4	R_b (3), O (3)	6
Push→Reach	O (3), R_a (5)	contact (5)	3	R_b (3), C_h (2), C_o (2)	7

mode σ consist of both q_{obj} and the robot's body location $(x_{robot}, y_{robot}, \theta_{robot})$. The codimension of σ is therefore 6 (see Table 9.1).

Furthermore, no two modes of a family overlap. This means to switch between two modes of the same family, the system much switch modes to another family (Figure 9.3(b)). Also, any configuration q in $\sigma \in \Sigma_f$ identifies a single coparameter θ (Figure 9.3(a)). So, a mode $\sigma = (f, q)$ can be represented by an integer f to describe the family, and a *representative* configuration q from which the coparameter is uniquely derived. This model is sufficiently general for other hybrid systems.

9.3.3 Transitions

We say modes σ and σ' are adjacent if the system is permitted to transition between them. "Permitted" simply means that a transition is not forbidden, not necessarily that a feasible motion connects σ and σ'. To feasibly switch from σ to σ', some configuration q along the way must satisfy the constraints of both modes. Thus, the intersection of $\mathscr{F}_\sigma \cap \mathscr{F}_{\sigma'}$ must be nonempty (Figure 9.2(b)). We call $q' \in \mathscr{F}_\sigma \cap \mathscr{F}_{\sigma'}$ a transition configuration.

The following mode transitions are permitted (Figure 9.4). Walk-to-reach and push-to-reach are allowed from any feasible configuration. Reach-to-walk and

Figure 9.4 Diagram of permitted mode transitions

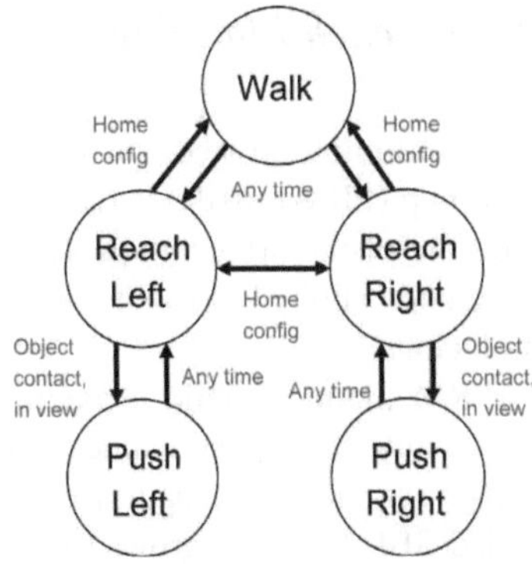

reach-to-reach are allowed if the arms are returned to the home configuration. Reach-to-push is allowed when the hand of the moving arm contacts the object.

Note that transition regions $\mathscr{F}_\sigma \cap \mathscr{F}_{\sigma'}$ may be of lower dimension than $\mathscr{F}_\sigma$, $\mathscr{F}_{\sigma'}$, or both (Table 9.2). This indicates that a transition configuration cannot be found by chance, and must be found explicitly [23]. In particular, reach-to-push transitions require solving inverse kinematics problems. We will discuss this further in Section 9.5.2.

9.4 Single-mode Motion Planning

Within a single mode, motions can be planned quickly with standard techniques. We use the following mode-specific planners. On average, each plan takes a small but not negligible amount of time (typically between 10 and 100 ms).

9.4.1 Collision Checking

To check collisions between the robot and environment geometry, we use a bounding volume hierarchy technique, as in [21, 46].

9.4.2 Walk Planning

ASIMO moves as a mobile robot in the 3D space $(x_{robot}, y_{robot}, \theta_{robot})$. A variety of path planning methods may be used, such as A* footstep search [14].

9.4.3 Reach Planning

The configuration space of reach modes is 6D, requiring the use of probabilistic roadmap (PRM) methods (see Chapter 7 of [15]). PRMs (and numerous variants) are state-of-the-art techniques for motion planning in high-dimensional spaces, and have been applied to robotic manipulators, free flying rigid objects, and biological molecules.

PRM planners build a roadmap of randomly sampled feasible configurations, called milestones, and connect them with straight-line feasible paths. The roadmap is then searched for a path from the start configuration to the goal. We use a variant called SBL [45] that grows two trees of feasible milestones bidirectionally, rooted at the start and the goal configurations. It achieves major speed gains by delaying feasibility tests for straight-line paths until it believes it has found a sequence of milestones connecting the start to the goal.

PRM planners plan quickly when the space has favorable visibility properties [28]. But PRMs cannot distinguish between a hard problem and one with no feasible path, so a time limit must be imposed, after which the planner returns with failure. In our experiments, PRMs plan quickly in reach modes, so this limit does not need to be very high (we use 0.5 s).

9.4.4 Push Planning

We use a forward-simulation planner as follows. Let p_{hand} be a contact on the hand touching a point p_{obj} on the object, with normals n_{hand} and n_{obj}. First, sample a stable center of rotation (COR) c (see Section 9.4.4.1). Rotating the object by distance D about this COR defines an object trajectory, which is checked for collisions. Then, we ensure that the object trajectory can be executed with an arm trajectory that positions p_{hand} at p_{obj} and orients n_{hand} to $-n_{obj}$ along the entire trajectory. This reuires solving an inverse kinematics (IK) problem (see Section 9.4.4.2). Finally, collisions are checked along the object and arm trajectory. We discretize the object trajectory in tiny increments (5 mm in our implementation), discretize the arm trajectory with the same increments, and then run the collision checker at each configuration. If feasible, the process is repeated to push the object further.

9.4.4.1 Stable Push Dynamics

A sliding object pushed with a point contact does not move deterministically, because the friction forces at the support surface are indeterminate [36, 41]. The current ASIMO hardware does not have the visual or tactile feedback needed for stable point-push control. Thus, we restrict the robot to use multi-contact pushes that rotate the object predictably under open-loop control. These *stable pushes* must be applied with at least two simultaneous collinear contacts, such as flat areas on the robot's hand. Lynch and Mason [33] studied the conditions necessary for stable pushing; these will be summarized here.

Given known center of friction, surface friction, and contact points, one can calculate the set of rigid transformations of the hand that yield stable pushes. In other words, if we choose an appropriate motion of the hand, the object will move as though it were rigidly fixed to the hand.

We project all coordinates onto the plane (Figure 9.5(b)). We assume the region of contact is a line of width w. Fix a frame of reference centered at the midpoint of this line, with the x axis parallel to the line, and the y axis pointing towards the object. We assume the following quantities are known:

- the coefficient of friction μ between the hand and object, or a conservative lower bound (Figure 9.5(c)).
- the object's center of friction $c_f = [x_f, y_f]^T$.
- the object's base of support $\mathscr{B} \subset \mathbb{R}^2$.

We assume the object moves quasistatically, i.e., the object's velocity is low enough such that dynamic effects are negligible, and it stops sliding immediately after contact forces are removed. Here we represent rigid transformations in the plane as an angle of rotation θ, and a point c, called the center of rotation (COR). Pure translations are represented by a COR at infinity.

A COR $c = [x, y]^T$ is in STABLE if it satisfies the following constraints:

- For vectors $u = [1, \mu]^T$ and $v = [1, -\mu]^T$, either (a) $c^T u \geq \max_{p \in \mathscr{B}} p^T u$ and $c^T v \leq \min_{p \in \mathscr{B}} p^T v$, or (b) $c^T u \leq \min_{p \in \mathscr{B}} p^T u$ and $c^T v \geq \max_{p \in \mathscr{B}} p^T v$ hold. Condition (a) must hold for CCW rotations, while (b) must hold for CW rotations. (Figure 9.5(d))
- Let $g = [x_f + w/2, y_f]^T$ and $h = [x_f - w/2, y_f]^T$, respectively, be vectors starting at the left and right ends of the contact and ending at c_f. Let s and t be the midpoints of g and h. Let r be the maximum distance from c_f to any point on the boundary of $\mathscr{B}$ Then, either (a) $c^T g \leq s^T g$ and $(c - c_f)^T h \geq r^2/(h^T h)$ or (b) $c^T h \leq t^T h$ and $(c - c_f)^T g \geq r^2/(g^T g)$ hold. (Figure 9.5(e))
- The prior two conditions guarantee that, when rotating the hand about c, at least one solution to the dynamic equations, maintains the line contact. To rule out the possibility of breaking off contact to only one endpoint, we also enforce the condition $|x_f| \leq w/2$.

For c in STABLE, rotating the hand about c (in the CW direction if x is positive, and CCW if x is negative) yields a stable push. The object's configuration space trajectory consists of a helix with axis at c.

9.4.4.2 Inverse Kinematics

Given specified contact points and the object configuration q_{obj}, the inverse kinematics subroutine solves for the five arm joint angles q_{arm} to maintain contact between the hand and the object. If there is no solution, the subroutine returns failure.

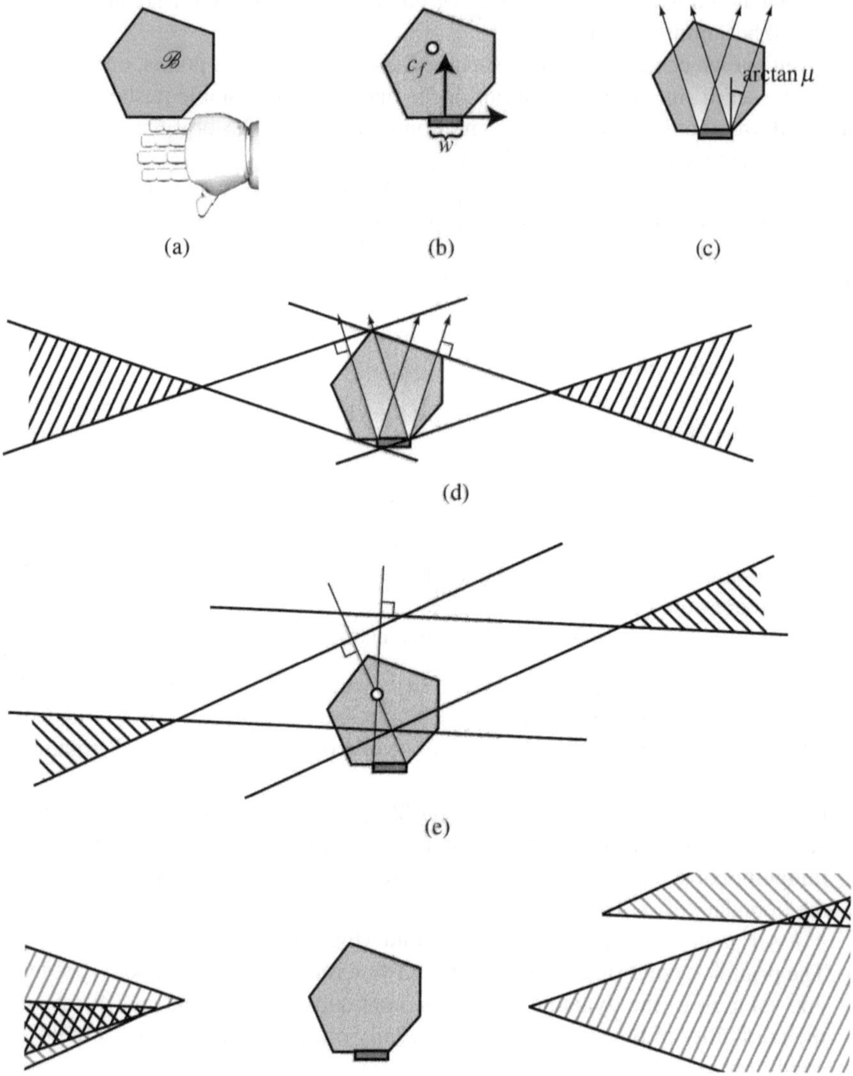

Figure 9.5 Illustrating the conditions necessary for a stable push (adapted from [33]). (a) Line contact with an object. (b) Width of line, and center of friction. (c) Contact friction cones. (d) Centers of rotation that can be achieved by forces within the friction cone. (e) Centers of rotation that can be achieved by forces along the contact. (f) Intersecting the constraints give the set of stable rotations

Denote the contact point and normal in the hand-local frame as p^L_{hand} and n^L_{hand}. Let R_{hand} and t_{hand} denote, respectively, the hand frame's rotation and translation as determined by forward kinematics, such that the world-space coordinates of the hand contact and normal are $p_{hand} = R_{hand}(q_{arm})p^L_{hand} + t_{hand}(q_{arm})$ and $n_{hand} = R_{hand}(q_{arm})n^L_{hand}$. Denote the contact point and normal on the object as p_{obj} and n_{obj}, respectively. Then, the inverse kinematic subroutine must solve for q_{arm} that simultaneously satisfy the six nonlinear equations:

$$R_{hand}(q_{arm})p^L_{hand} + t_{hand}(q_{arm}) = p_{obj} \tag{9.1}$$

$$R_{hand}(q_{arm})n^L_{hand} = -n_{obj}. \tag{9.2}$$

Or more compactly, $C(q_{arm}) = 0$. At first glance this seems like an overdetermined system (5 parameters, 6 constraints). However, the normal-matching constraint is degenerate, and the Jacobian of C is therefore at most rank 5.

Algorithm 9.1 Newton–Raphson inverse kinematics subroutine. The initial configuration is denoted q_0

Algorithm: Newton–Raphson–IK(q_0)

Let $C(q)$ denote the loop closure equations (9.1) and (9.2).
for $k = 0,\dots,n$ **do**
  if $||C(q_k)|| < \varepsilon$ **then**
    return q_k
  end
  $d_k = -\nabla C(q_k)^\dagger C(q_k)$
  Find $\alpha_k > 0$ such that $||C(q_k + \alpha_k d_k)|| < ||C(q_k)||$.
  if no such α_k can be found, **return** "failure"
  $q_{k+1} = q_k + \alpha_k d_k$
end
return "failure"

We solve this system of equations using a Newton–Raphson numerical solver [40], which iteratively moves a start configuration q_0 onto the solution set. At each step, it computes a Jacobian $\nabla C(q)$ and its pseudoinverse $\nabla C(q)^\dagger$ to choose a search direction. It terminates when the residual reaches some tolerance ε, or the number of iterations exceeds a limit n. Pseudocode is given in Algorithm 9.1.

9.5 Multi-modal Planning with Random-MMP

Random-MMP maintains a tree of configurations T and extends it with a single-mode transition. But each extension picks a node from T at random with probability Φ, and expands to a single adjacent mode sampled at random with probability Ψ. But how to select Φ and Ψ? Some intuition can be gained by examining how tree-growing PRMs try to sample in low-density areas. Two early planners have been

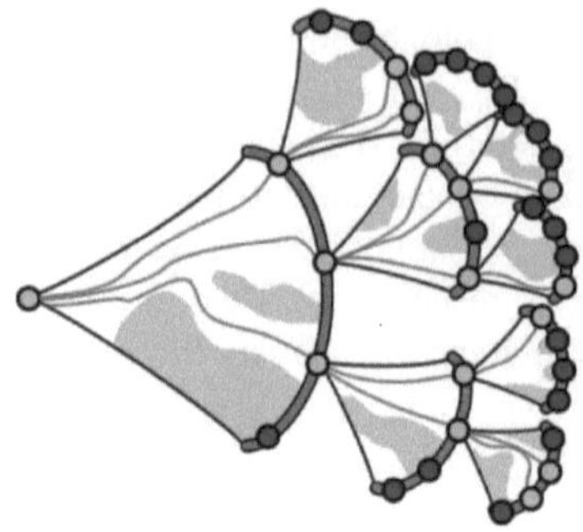

Figure 9.6 Growing a multi-modal planning tree. Each "fan" depicts a mode's configuration space, with transition regions at the distal end. Light nodes are feasible configurations, connected with feasible single-mode paths. Dark nodes are infeasible transition attempts

empirically shown to have good performance over a wide range of problems, and have inspired dozens of variations existing in the literature. The EST planner expands from a node with probability inversely proportional to the number of nearby nodes [28]. The RRT planner tries to distribute configurations uniformly across the space by picking a random point in space, and expanding the closest node toward that point [31]. We use a simple RRT-like implementation, expanding the tree as follows:

RANDOM-MMP

1. Sample a random configuration q_{rand}.
2. Pick a node (q, σ) that minimizes a distance metric $d(q, q_{rand})$. We define d to be the distance of the object configurations, ignoring the robot entirely. Like RRT, step 2 implicitly defines Φ proportional to the size of the Voronoi cell of q.
3. The tree is expanded from (q, σ) to new mode(s) using an expansion strategy *Expand*, which implicitly defines Ψ.

9.5.1 Effects of the Expansion Strategy

A variant of Random-MMP has been shown to be probabilistically complete, when using an expansion strategy that employs a complete single-mode planner [49]. Specifically, the expansion strategy randomly samples a mode m' adjacent to (q, m), and if a feasible single-mode path connects (q, m) to a point q' in m', the tree is expanded to include (q', m'). This result can be extended to state that Random-MMP is probabilistically complete, as long as the single-mode planner always succeeds with nonzero probability. Thus, when PRMs are used for single mode planning, the time cutoff must be set sufficiently high.

Even though Random-MMP is theoretically complete for various expansion strategies, its practical performance is heavily affected by the choice of strategy. We will compare three variants of the expansion strategy. The latter two use precomputed *utility tables*, which will be described below.

- *Blind.* Picks a mode adjacent to σ at random, and plans a single-mode path to it.
- *Utility-based importance sampling.* Same as the blind strategy, but samples contacts for reach-to-push transitions according to expected utility.
- *Utility-centered.* Uses the utility tables to select a high-utility push that moves q_{obj} toward a target configuration q_{des}, and a sequence of modes to achieve that push.

Our experiments in Section 9.5.5 show that Utility-centered exploration is an order of magnitude faster than the Blind strategy. It also has an additional parameter q_{des}, which can be chosen to help heuristics guide the distribution of configurations in T, for example, toward the goal or sparsely sampled regions in configuration space. Our planner implementation achieves large speedups with these heuristics.

9.5.2 Blind Expansion

Given a configuration q at mode σ, blind expansion samples a transition configuration q' at an adjacent mode σ', and if q' is feasible, plans a single-mode path y to connect q and q' as usual. We first choose an adjacent mode family, then sample q' to achieve that type of transition. We sample q' as follows:

- *Walk-to-reach.* Sample a body position between a minimum and maximum distance from the object and a body orientation such that the object lies within the robot's field of view.
- *Reach-to-walk.* Move the arm to the home configuration.
- *Reach-to-reach.* Use any configuration.
- *Reach-to-push.* Let S_{hand} and S_{obj} be the surfaces of the hand and the object. We predefine a number of contact points $C_{cand} \subset S_{hand}$ that are candidates for stable pushes. Sample a point p_{hand} from C_{cand} and a point p_{obj} from S_{obj}, with normals n_{hand} and n_{obj}. Starting from a random arm configuration, use numerical IK to simultaneously place p_{hand} at p_{obj} and orient n_{hand} to $-n_{obj}$.
- *Push-to-reach.* A push-to-reach transition can be taken at any time. So we choose a transition implicitly by planning a single-mode path y, and set q' to the endpoint of y.

After q' is picked, a single-mode path is planned using the techniques of Section 9.4.

9.5.3 Utility computation

In reach-to-push sampling, only a small portion of S_{obj} is reachable from a given point on the hand. For each p in C_{cand}, we precompute one table that identifies the reachable region R on S_{obj}, and another table that estimates the utility of points in R.

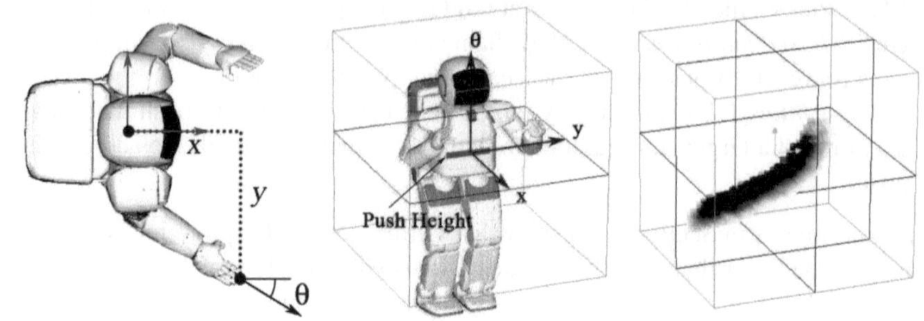

(a) Workspace coordinates of the right fingers. (b) Reference frame F. (c) Utility table for the right fingers.

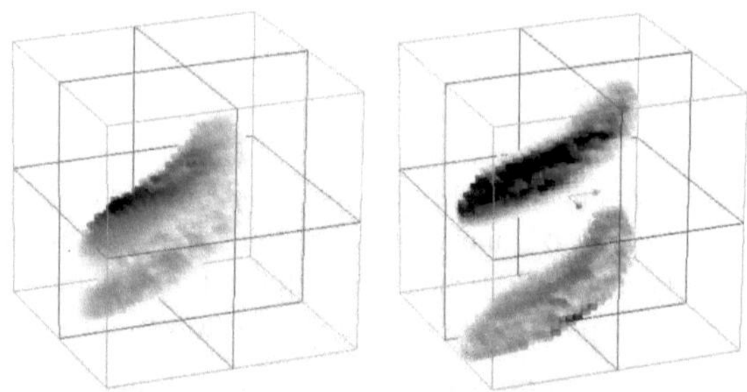

(d) Utility table for the right knuckles. (e) Utility table for the right palm.

Figure 9.7 Workspace coordinates and utility tables. Utility tables (d–f) are drawn in frame F (c). Reachable cells are drawn with darkness increasing with utility

When pushing, the normal n at p must be horizontal in world space. We fix a height of pushing h, constraining the vertical coordinate of p. This define a 3D workspace $\mathscr{W}$ of points (x, y, θ), where (x, y) are the horizontal coordinates of p and θ is the orientation of n, relative to the robot's frame (Figure 9.7(a)). We precompute two grids, `Reachable` and `Utility`, over $\mathscr{W}$ as follows.

`Reachable` stores 1 if the contact is reachable and 0 otherwise (Figure 9.7). We initialize `Reachable` to 0, and then sample the 5D space of the arm joints in a grid. Starting at each sampled configuration, we run IK to bring the height of p to h and reorient n to be horizontal. If successful, we check if the arm avoids collision with the body and the point p is in the robot's field of view. If so, we mark `Reachable`$[(x, y, \theta)]$ with 1, where (x, y, θ) are the workspace coordinates of p and $[\cdot]$ denotes grid indexing.

`Utility` stores the expected distance the contact can be pushed in the absence of obstacles, calculated by Monte Carlo integration through `Reachable` (Figure 9.7). In $\mathscr{W}$, a push traces out a helix that rotates around some COR. We assume a prior probability distribution P over stable CORs for a reasonable range of physical parameters of the object. Starting from $w_0 = (x, y, \theta)$ we generate a path $w_0, w1, \ldots, w_K$. For all k, w_{k+1} is computed by rotating w_k a short distance along some COR sampled from P. The sequence terminates when `Reachable`$[w_{K+1}]$ becomes 0. After generating N paths, we record the average length in `Utility`$[(x, y, \theta)]$.

Given robot and object positions, contacts along S_{obj} at height h form a set of curves B in $\mathscr{W}$. Intersecting B with the marked cells of `Reachable` gives the set of reachable object edges R.

When choosing a reach-to-push transition, utility-based importance sampling picks contacts from R with probability proportional to `Utility`. If R is empty, it does not attempt to make the transition.

9.5.4 Utility-centered Expansion

Utility-centered expansion explicitly chooses a body position to execute a high-utility push. Given a start configuration q at stance σ, and a target configuration q_{des}, this strategy (1) chooses a robot's body and arm configuration and a high-utility push for a reach-to-push transition q_{push}, (2) plans a sequence of three modes backwards from q_{push} to q (which requires no search), and (3) plans a push path forward from q_{push}.

We elaborate on step 1. Let q_{obj} be the object configuration in the desired configuration q_{des}. Choose a point p_{obj} on S_{obj} (at height h and normal n_{obj}) and a stable push such that the object will be pushed toward q_{obj}. Next, sample a contact p_{hand} from C_{cand} and a workspace coordinate $(x_{hand}, y_{hand}, \theta_{hand})$ with probability proportional to `Utility`. Then, compute the body coordinates that transform (x_{hand}, y_{hand}) to p_{obj} and rotate θ_{hand} to θ_{obj}, where θ_{obj} is the orientation of $-n_{obj}$. Repeat until the body position is collision free. Fixing the body, sample the arm configuration of q_{push}, using IK to position p_{hand} to p_{obj} and orient n_{hand} to $-n_{obj}$.

9.5.5 Experimental Comparison of Expansion Strategies

Table 9.3 Fraction of feasible walk-to-reach transitions, reach-to-push transitions, acceptable pushes, and walk-reach-push cycles

Strategy	Walk-to-reach	Reach-to-push	Pushes > 1cm	Overall
Blind	0.63	0.20	0.29	0.037
Utility importance	0.59	0.33	0.79	0.16
Utility-centered	1	0.47	0.66	0.31

Table 9.4 Expansion strategy timing experiments. Bold indicates best in column

Strategy	Modes / push	Time / push (s)	Average push (cm)	Push rate (m/s)	Tgt. seek rate (m/s)
Blind	10.0	0.956	6.7	0.070	0.0302
Utility importance	5.99	**0.325**	8.2	0.254	0.111
Utility-centered	**5.08**	0.404	**13.3**	**0.329**	**0.257**

First, we measure the fraction of feasible mode transitions attempted by each strategy. We ran the three strategies starting from a configuration similar to the first frame of Figure 9.11. Each strategy is initialized with a walk mode, and terminates after a single push has been executed (requiring three transitions, from walk to reach to push). We reject any "insignificant" pushes, defined as moving the object less than 1 cm. The results, averaged over 1000 runs, are shown in Table 9.3. The blind strategy produces less than half as many significant pushes than the other strategies. However, reach-to-push transitions are still a bottleneck for utility-based importance sampling.

Next, Table 9.4 measures the expansion time and rate, paying particular attention to the distance the object is pushed per unit of planning time. The first column measures the number of modes explored before terminating, and the second measures the time elapsed. The third and fourth columns measure the distance of the terminal push, and the distance divided by computation time. The final column measures the distance (if positive) the object was pushed toward a target position q_{des}. Here, q_{des} was picked at random.

These results verify that picking contacts blindly is rarely successful, and the utility table precomputation is valuable for selecting contacts. Moreover, utility-centered expansion almost doubles the average distance per push. These observations, coupled with its useful ability to guide planning toward a desired configuration, makes utility-centered expansion the preferred expansion strategy.

9.6 Postprocessing and System Integration

9.6.1 Visual Sensing

We assume the geometry of the object and table are known, but their robot-relative pose is unknown. To estimate pose, we sense the scene using the robot's stereo cameras. We use an open-source augmented reality system (ARToolKitPlus [50]) to estimate the pose of fiducial markers of known size in the scene. Markers are fixed on the top of the object, and the center and corners of the table. The redundant markers on the table help estimate its position even when part of the table is out of the cameras' field of view, and when some markers are occluded. The pose estimates can also be averaged to improve accuracy.

Experiments show that at a distance of 1.5 m, errors of the marker pose estimation are bounded by approximately 2 cm in translation and 4° in orientation. To help avoid colliding with the object and table under this uncertainty, we slightly grow the geometry of the object and table used for collision detection before planning.

9.6.2 Execution of Walking Trajectories

In practice, the footsteps produced by the walk planner are not achieved exactly by ASIMO's low-level footstep controller, because the controller compensates for unmodeled perturbations (e.g., variations in the ground surface, joint friction, residual motion in the arms, etc.). Thus, we employ a high-level walk controller to monitor the execution of the path, and make local corrections to ensure the robot does not stray off course.

The robot's odometry measures the leg joint angles at foot touch-down to estimate the actual footstep taken. The walk controller uses this estimate to determine a new footstep command to steer the robot back toward the planned path. In practice, the odometry's estimate of the footstep has errors on the order of a few millimeters, causing the estimated robot pose to drift. Though drift did not significantly affect performance up to a few pushes (approximately 5 or 6), periodic visual localization is necessary for longer-term executions.

9.6.3 Smooth Execution of Reach Trajectories

The motions produced by the reach planner are jerky, because they are planned by PRM methods which randomly sample the configuration space. To execute such a motion smoothly, the robot arm would need to stop at every milestone of the path, which slows execution significantly. Thus, we employ a postprocessing step to smooth the PRM output. Numerical optimization can smooth the path [3], but is computationally expensive when including collision constraints.

Instead, we use a shortcutting procedure that repeatedly samples two configurations a and b on the path, and attempts to connect them with a feasible shortcut. But standard straight-line shortcuts [18] would cause the arm to stop at a and b, worsening the problem. Instead, we use smooth trajectories that are time-optimal with bounded joint velocity and acceleration. These trajectories are computed very quickly, which makes the shortcutting procedure fast.

We start by converting the kinematic path (with no velocity information) into the time-optimal trajectory $u(t)$ (with velocity information) that stops at every milestone. Then, as usual, we select random configurations $a = u(t_a)$ and $b = u(t_b)$. We also compute their velocities $c = u'(t_a)$ and $d = u'(t_b)$. We then compute the time-optimal trajectory $v(t)$ between states (a,c) and (b,d) as described below. If $v(t)$ is

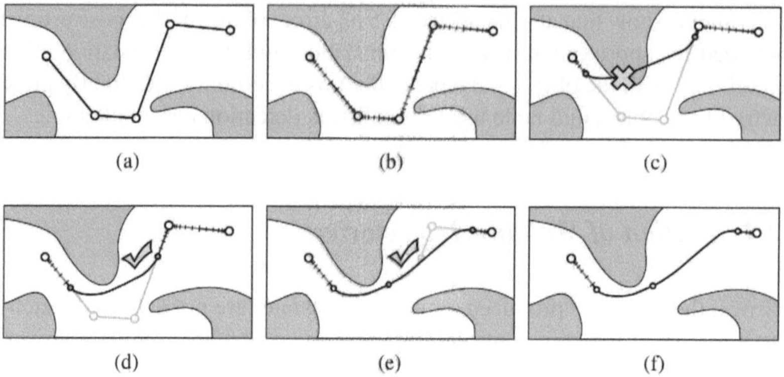

Figure 9.8 Illustrating the smoothing postprocessing step. (a) A jerky path produced by a PRM planner; (b) converted into a trajectory that stops at each milestone. (c) Attempting a random shortcut. The feasibility check fails. (d),(e) Two more shortcuts, which pass the feasibility check. (f) The final trajectory executed by the robot

feasible, we replace the section of $u(t)$ between a and b with $v(t)$. The process is illustrated in Figure 9.8.

9.6.3.1 Time-optimal Joint Trajectories

The time-optimal joint trajectory under bounded velocity e and acceleration f consists of parabolic arcs and straight lines. First, consider the unidimensional case of constructing a trajectory that starts at x_1 with velocity v_1 and ends at x_2 with velocity v_2, under maximum velocity v_{max} and acceleration a_{max}. Let $f(x_1,x_2,v_1,v_2,v_{max},a_{max})$ compute the time of the time-optimal trajectory. Let $g(x_1,x_2,v_1,v_2,v_{max},T,t)$ compute the state at time t of the acceleration-optimal trajectory, given a final time T.

Then letting superscript k denote joint indexing, we first compute the optimal time:

$$T = \max_k f(a^k,b^k,c^k,d^k,e^k,f^k). \tag{9.3}$$

Then we compute the acceleration-optimal joint trajectories

$$v(t)^k = g(a^k,b^k,c^k,d^k,e^k,T,t). \tag{9.4}$$

9.6.3.2 Univariate Time-optimal Trajectories

We show here how to compute $f(x_1,x_2,v_1,v_2,v_{max},a_{max})$, the execution time of the time-optimal, univariate, velocity- and acceleration-bounded trajectory. We compute it by enumerating all bang–bang controls that connect the initial and final states, and picking the one with the lowest execution time. We denote four motion primitives: the parabolas P^+ and P^- accelerating at a_{max} and $-a_{max}$, respectively, and the straight lines S^+ and S^- traveling at v_{max} and $-v_{max}$, respectively. There are four possible combinations of motion primitives that may contain an optimal trajectory: P^+P^-, P^-P^+, $P^+S^+P^-$, and $P^-S^-P^+$. We examine each class for a valid execution time T, and find the class with the minimal execution time.

For class P^+P^-, we must find the switch time t_P when the trajectory stops accelerating, and starts decelerating. A valid t_P is a solution of the equation

$$a_{max}t^2 + 2v_1 t + (v_1^2 - v_2^2)/(2a_{max}) + x_1 - x_2 = 0 \tag{9.5}$$

that also satisfies $0 \le t \le (v_2 - v_1)/a_{max}$. If no solution exists, the class is declared invalid. If a solution exists, the total time is $T = 2t_P + (v_1 - v_2)/a_{max}$. We must also check that the maximum speed of the trajectory $v_1 + t_P a_{max}$ does not exceed v_{max}. The P^-P^+ solution is given by negating a_{max} in the above equations.

For class $P^+S^+P^-$, the compute the time t_S in the straight-line portion:

$$t_S = (v_2^2 + v_1^2 - 2v_{max}^2)/(2v_{max}a_{max}) + (x_2 - x_1)/v_{max}. \tag{9.6}$$

If this value is negative, the class is invalid. Otherwise, the execution time is given by

$$T = (2v_{max} - v_1 - v_2)/a_{max} + t_S. \tag{9.7}$$

The $P^-S^-P^+$ solution is given by negating a_{max} and v_{max} in the above equations.

9.6.3.3 Acceleration-optimal Trajectories

Now we wish to compute the acceleration-optimal trajectory given a fixed end time T. First, we show how to compute the minimal acceleration $a = h(x_1, x_2, v_1, v_2, v_{max}, T)$. Then the optimal trajectory $g(x_1, x_2, v_1, v_2, v_{max}, T, t)$ is defined in a straightforward manner from a. Again, we use the same combinations of motion primitives: P^+P^-, P^-P^+, $P^+S^+P^-$, and $P^-S^-P^+$, and find the class that has minimum acceleration.

For classes P^+P^- and P^-P^+, we compute solutions a to the equation

$$T^2a^2 + \sigma(2T(v_1 + v_2) + 4(x_1 - x_2))a - (v_2 - v_1)^2 = 0 \tag{9.8}$$

whose switch time $t_S = 1/2(T + (v_2 - v_1)/a)$ satisfies the condition $0 \le t_S \le T$. Here, σ is $+1$ for class P^+P^- and -1 for class P^-P^+. We must also check that the maximum speed of the trajectory $|v_1 + at_S|$ does not exceed v_{max}.

For class $P^+S^+P^-$, we have

$$a = \frac{v_{max}^2 - v_{max}(v_1 + v_2) + 0.5(v_1^2 + v_2^2)}{Tv_{max} - (x_2 - x_1)} \tag{9.9}$$

We then check that the value t_P (as computed above) satisfies the condition $0 \le t_P \le T$. For class $P^-S^+P^+$, we negate v_{max} in the above equation.

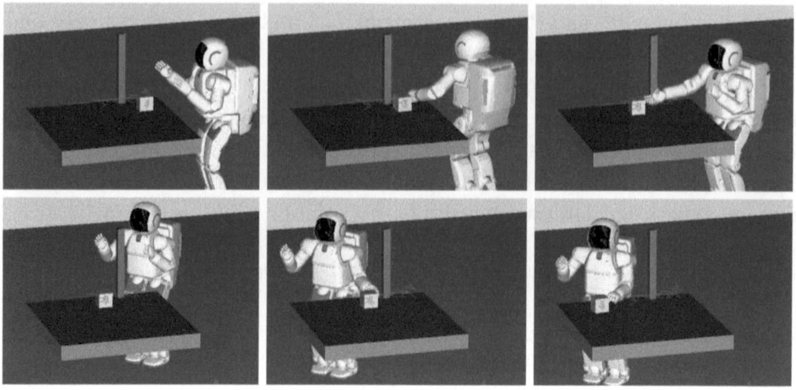

Figure 9.9 Pushing a block while avoiding a vertical obstacle

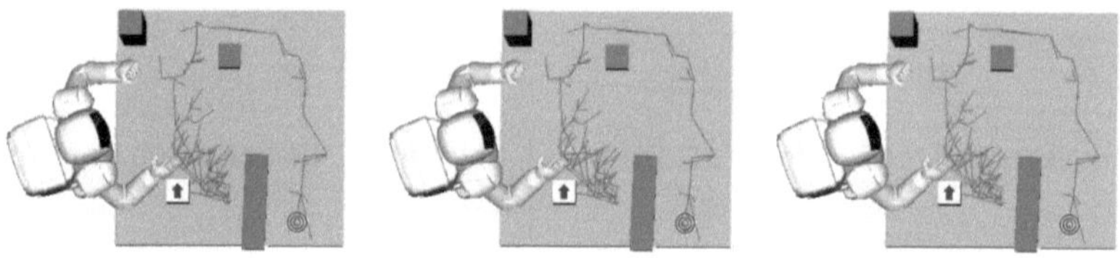

Figure 9.10 Replanning in a changing environment

9.7 Experiments

9.7.1 Simulation Experiments

We tested the planner on several example problems. We specified the goal of moving the object within 5 cm of a target position, with arbitrary orientation. Figure 9.9 shows the solution to a moderately difficult problem, where the robot needs to carefully choose where and how it pushes the object to avoid colliding with the vertical obstacle. The relatively long-distance push chosen in frame 3 enables ASIMO to continue pushing the object in frame 5. This trajectory was found in about four minutes on a 2 GHz PC.

We address uncertainty and errors with replanning. Figure 9.10 shows an example in a changing environment. A moved obstacle invalidates the initial path during execution, forcing the robot to a more difficult alternative. The planner produced a new path in three minutes. For smaller errors encountered during execution, a faster technique might attempt to reuse a large portion of the previously planned path. We hope to implement such a technique in future work.

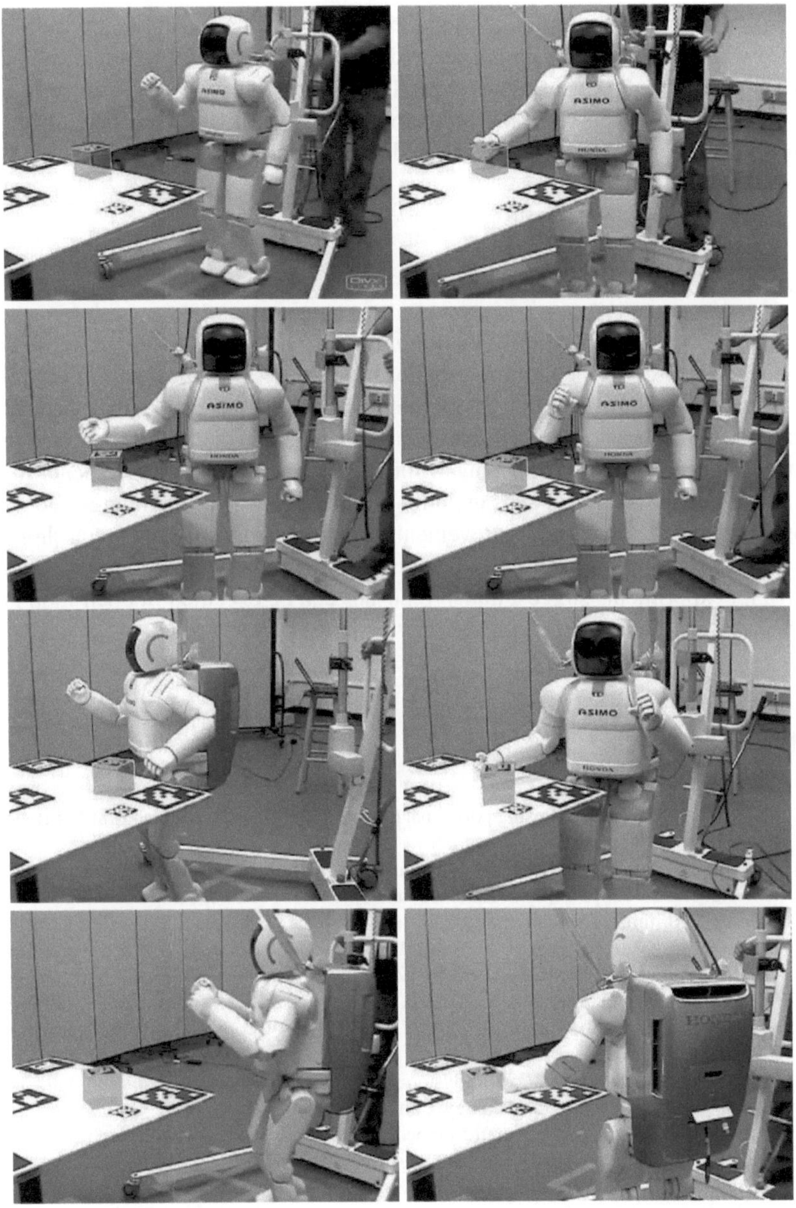

Figure 9.11 Pushing a block with the ASIMO robot. The block is shaded for clarity

9.7.2 Experiments on ASIMO

Figure 9.11 shows a sequence of pushes executed on the physical ASIMO, which was planned in about one minute on an off-board PC. Environment sensing is performed once at the beginning of execution. No additional visual feedback is performed while the planned path is executed. Despite errors in initial sensing and the robot's odometry, ASIMO executes several pushes successfully. In our experiments, 4–6 pushes typically can be executed before the object deviates too much from the planned path (approximately 2–3 cm), and visual resensing is needed.

9.8 Conclusion

This chapter presented a multi-modal motion planner that enables ASIMO to push an object to a precise location on a table. It presented a framework for modeling and planning in multi-modal systems that is extremely general, which should facilitate the future integration of additional manipulation modes, such as grasping, multi-handed pushing, and pushing-while-walking. Various enhancements to the basic technique were presented, including improvements to planning speed using utility-guided sampling, and improvements to motion smoothness using a shortcutting technique. The plans produced by the planner can be executed with limited resensing; experiments suggest that 4–6 pushes can be executed in open-loop fashion before the object significantly deviates from the planned path.

References

[1] S. Akella and M. Mason. Posing polygonal objects in the plane by pushing. Int J of Robotics Res, 17(1):70–88, 1998.

[2] R. Alami, J.-P. Laumond, and T. Siméon. Two manipulation planning algorithms. In K. Goldberg, D. Halperin, J.-C. Latombe, and R. Wilson, editors, In: proceedings of workshop on algorithmic foundation of robotics, pp 109–125. A K Peters, Wellesley, MA, 1995.

[3] J. Bobrow, B. Martin, G. Sohl, E. Wang, F. Park, and J. Kim. Optimal robot motions for physical criteria. J Robotic Systems, 18(12):785–795, 2001.

[4] J.-D. Boissonnat, O. Devillers, L. Donati, and F. Preparata. Motion planning of legged robots: The spider robot problem. Int J Computational Geometry and Applications, 5(1-2):3–20, 1995.

[5] M. S. Branicky. Studies in Hybrid Systems: Modeling, Analysis, and Control. PhD thesis, Massachusetts Institute of Technology, Cambridge, MA, 1995.

[6] T. Bretl. Motion planning of multi-limbed robots subject to equilibrium constraints: The free-climbing robot problem. Int J Robotics Res, 25(4):317–342, 2006.

[7] T. Bretl, J.-C. Latombe, and S. Rock. Toward autonomous free-climbing robots. In: proceedings of international symposium on robotics research, Siena, Italy, 2003.

[8] F. Bullo and M. Žefran. On modeling and locomotion of hybrid mechanical systems with impacts. In: proceedings of IEEE Conf. on Decision and Control, pp 2633–2638, Tampa, FL, 1998.

[9] F. Bullo and M. Žefran. Modeling and controllability for a class of hybrid mechanical systems. IEEE Trans Robot Automat, 18(4):563–573, 2002.

[10] S. Cambon, R. Alami, and F. Gravot. A hybrid approach to intricate motion, manipulation and task planning. Int J of Robotics Res, 28(104), 2009.

[11] A. Casal. Reconfiguration Planning for Modular Self-Reconfigurable Robots. PhD thesis, Stanford University, Stanford, CA, 2001.
[12] A. Casal and M. Yim. Self-reconfiguration planning for a class of modular robots. In: SPIE II, pp 246–257, 1999.
[13] M. Cherif and K. Gupta. Planning for in-hand dextrous manipulation. In P. Agarwal, L. Kavraki, and M. Mason, editors, Robotics: The Algorithmic Perspective, pp 103–117. AK Peters, Ltd., 1998.
[14] J. Chestnutt, J. Kuffner, K. Nishiwaki, and S. Kagami. Planning biped navigation strategies in complex environments. In: proceedings of IEEE-RAS international conference on humanoid robots, Munich, Germany, 2003.
[15] H. Choset, K. Lynch, S. Hutchinson, G. Kantor, W. Burgard, L. Kavraki, and S. Thrun. Principles of Robot Motion: Theory, Algorithms, and Implementations. MIT Press, 2005.
[16] M. Erdmann and M. Mason. An exploration of sensorless manipulation. IEEE Int J of Robot Automat, pp 367–279, 1988.
[17] P. Ferbach and J. Barraquand. A method of progressive constraints for manipulation planning. IEEE Trans Robot Automat, 13(4):473–485, 1997.
[18] R. Geraerts and M. H. Overmars. Creating high-quality paths for motion planning. Int J of Robotics Res, 26(8):845–863, 2007.
[19] K. Goldberg. Orienting polygonal parts without sensors. Algorithmica, 10:201–225, 1993.
[20] B. Goodwine and J. Burdick. Motion planning for kinematic stratified systems with application to quasi-static legged locomotion and finger gaiting. In: workshop on the algorithmic foundations of robotics, Hanover, NH, 2000.
[21] S. Gottschalk, M. Lin, and D. Manocha. OBB-tree: A hierarchical structure for rapid interference detection. In: ACM SIGGRAPH, pp 171–180, 1996.
[22] K. Harada, S. Kajita, K. Kaneko, and H. Hirukawa. Pushing manipulation by humanoid considering two-kinds of zmps. In: proceedings of IEEE international conference on robotics and automation, pp 1627–1632, September 2003.
[23] K. Hauser. Motion Planning for Legged and Humanoid Robots. PhD thesis, Stanford University, 2008.
[24] K. Hauser, T. Bretl, and J.-C. Latombe. Non-gaited humanoid locomotion planning. In: proceedings of IEEE-RAS international conference on humanoid robots, Tsukuba, Japan, 2005.
[25] K. Hauser, T. Bretl, J.-C. Latombe, and B. Wilcox. Motion planning for a six-legged lunar robot. In: workshop on the algorithmic foundations of robotics, New York, NY, 2006.
[26] K. Hauser and J.-C. Latombe. Multi-modal motion planning in non-expansive spaces. In: workshop on the algorithmic foundations of robotics, 2008.
[27] K. Hauser, V. Ng-Thow-Hing, and H. G.-B. nos. Multi-modal planning for a humanoid manipulation task. In: international symposium on robotics research, Hiroshima, Japan, 2007.
[28] D. Hsu, J.-C. Latombe, and R. Motwani. Path planning in expansive configuration spaces. Int J of Computational Geometry and Applications, 9(4-5):495–512, 1999.
[29] Y. Koga and J.-C. Latombe. On multi-arm manipulation planning. In: proceedings of IEEE international conference on robotics and automation, pp 945–952, San Diego, CA, 1994.
[30] J. J. Kuffner, Jr., S. Kagami, M. Inaba, and H. Inoue. Dynamically-stable motion planning for humanoid robots. In: proceedings of first international conferenceon humanoid robotics, Boston, MA, 2000.
[31] S. LaValle and J. Kuffner. Randomized kinodynamic planning. In: proceedings of IEEE international conference on robotics and automation, pp 473–479, 1999.
[32] S. M. LaValle. Planning Algorithms. Cambridge University Press, 2006.
[33] K. Lynch and M. Mason. Controllability of pushing. In: proceedings of IEEE international conference on robotics and automation, 1995.
[34] K. Lynch and M. Mason. Stable pushing: Mechanics, controllability, and planning. Int J Robotics Res, 15(6):533–556, Dec 1996.

[35] K. M. Lynch, H. Maekawa, and K. Tanie. Manipulation and active sensing by pushing using tactile feedback. In: proceedings of IEEE/RSJ international conference on intelligent robots and systems, pp 416–421, 1992.
[36] M. Mason. Mechanics and planning of manipulator pushing operations. Int J Robotics Res, 5(3), 1986.
[37] I. Mitchell and C. Tomlin. Level set methods for computation in hybrid systems. In: Hybrid Systems: Computation and Control, LNCS 1790, Mar 2000.
[38] C. L. Nielsen and L. E. Kavraki. A two level fuzzy prm for manipulation planning. In: proceedings of IEEE/RSJ international conference on intelligent robots and systems, pp 1716–1721, Takamatsu, Japan, 2000.
[39] D. Nieuwenhuisen, A. F. van der Stappen, and M. H. Overmars. An effective framework for path planning amidst movable obstacles. In: proceedings of workshop on algorithmic foundations of robotics, New York, NY, 2006.
[40] J. M. Ortega and W. C. Rheinboldt. Iterative Solution of Nonlinear Equations in Several Variables. Classics in Applied Mathematics. SIAM, 2000.
[41] M. Peshkin and A. Sanderson. The motion of a pushed, sliding workpiece. Int J Robot Automat, 4(6):569–598, Dec 1988.
[42] J. Pettré, J.-P. Laumond, and T. Siméon. A 2-stages locomotion planner for digital actors. In: Eurographics/SIGGRAPH Symp. Comp. Anim., 2003.
[43] J. H. Reif. Complexity of the mover's problem and generalizations. In: 20th Annual IEEE Symposium on Foundations of Computer Science, pp 421–427, San Juan, Puerto Rico, 1979.
[44] A. Sahbani, J. Cortés, and T. Siméon. A probabilistic algorithm for manipulation planning under continuous grasps and placements. In: proceedings of IEEE/RSJ international conference on intelligent robots and systems, pp 1560–1565, Lausanne, Switzerland, 2002.
[45] G. Sánchez and J.-C. Latombe. On delaying collision checking in PRM planning: Application to multi-robot coordination. Int J Robotics Res, 21(1), pp 5–26, 2002.
[46] F. Schwarzer, M. Saha, and J.-C. Latombe. Exact collision checking of robot paths. In: proceedings of workshop on algorithmic foundations of robotics, Nice, France, Dec 2002.
[47] M. Stilman. Navigation Among Movable Obstacles. PhD thesis, Carnegie Mellon University, 2007.
[48] T. Takubo, K. Inoue, and T. Arai. Pushing an object considering the hand reflect forces by humanoid robot in dynamic walking. In: proceedings of IEEE international conference on robotics and automation, pp 1706–1711, 2005.
[49] J. van den Berg, M. Stilman, J. Kuffner, M. Lin, and D. Manocha. Path planning among movable obstacles: a probabilistically complete approach. In: proceedings of workshop on algorithmic foundations of robotics, 2008.
[50] D. Wagner and D. Schmalstieg. Artoolkitplus for pose tracking on mobile devices. In: Computer Vision Winter Workshop 2007, Lambrecht, Austria, February 2007.
[51] G. Wilfong. Motion planning in the presence of movable obstacles. In: proceedings of fourth annual symposium on computational geometry, pp 279–288, Urbana-Champaign, IL, 1988.
[52] R. Wilson and J. Latombe. Geometric reasoning about mechanical assembly. Artificial Intelligence, 71(2):371–396, 1995.
[53] J. Xu, T. J. Koo, and Z. Li. Finger gaits planning for multifingered manipulation. In: proceedings of IEEE/RSJ international conference on intelligent robots and systems, San Diego, CA, 2005.
[54] M. Yashima and H. Yamaguchi. Dynamic motion planning whole arm grasp systems based on switching contact modes. In: proceedings of IEEE international conference on robotics and automation, Washington, D.C., 2002.
[55] N. Zumel and M. Erdmann. Nonprehensile manipulation for orienting parts in the plane. In: proceedings of IEEE international conference on robotics and automation, pp 2433–2439, April 1997.

Chapter 10
A Motion Planning Framework for Skill Coordination and Learning

Marcelo Kallmann and Xiaoxi Jiang

Abstract Coordinated whole-body motions are key for achieving the full potential of humanoids performing tasks in human environments and in cooperation with humans. We present a multi-skill motion planning and learning framework which is able to address several complex whole-body coordinations and as well detection and imitation of motion constraints. The framework is composed of three main parts: first, a minimal set of basic motion skills is defined in order to achieve basic stepping, balance and reaching capabilities. A multi-skill motion planner is then developed for coordinating motion skills in order to solve generic mobile manipulation tasks. Finally, learning strategies are presented for improving skill execution and for learning constraints from imitation. The framework is able to coordinate basic skills in order to solve complex whole-body humanoid tasks and also integrates learning mechanisms for achieving humanlike performances in realistic environments.

10.1 Introduction

Despite several successes in the motion planning domain, achieving whole-body coordinated humanoid motions for solving manipulation tasks remains a challenge. The problem is particularly difficult because most typical humanoid tasks require coordination of different types of motions or skills and therefore cannot be solved by a single planning strategy. One additional challenge is to achieve interactive performances: planning motions from scratch is often computationally intensive and therefore learning has to be addressed in an integrated fashion.

Marcelo Kallmann
University of California, Merced; 5200 N. Lake Road, Merced CA 95343
e-mail: `mkallmann@ucmerced.edu`

Xiaoxi Jiang
University of California, Merced; 5200 N. Lake Road, Merced CA 95343
e-mail: `janexip@gmail.com`

Consider, for example, the coordination of stepping and grasping. While stepping is mainly concerned with balance maintenance for mobility, reaching and grasping skills control the motion of the arms and hands assuming the humanoid to be in a well balanced standing position. Figure 10.1 illustrates a typical scenario where complex body positioning is required in order to support common book relocations in a shelf. It is possible to note that complex torso motions are both used for providing balance and for enlarging the reachable space of the arms. When body motion is not enough, stepping is also employed.

Figure 10.1 Even in simple manipulation tasks complex legs-arm coordinations can be observed

We address the whole-body motion generation problem for reaching tasks as a planning problem. We target object reaching tasks such as the ones illustrated in Figure 10.1. The goal is to plan the coordination of stepping, balance and reaching motion skills in order to reach given target locations with end-effectors.

Figure 10.2 (a) illustrates a situation where the fully articulated humanoid character is not able to grasp the book using its arm because the book is not within reachable range. This problem is even more critical in humanoid robots as they usually have less degrees of freedom (DOFs) and with less range of motion, therefore requiring more complex coordinations. Depending on the motion skills available, humanoids can solve given tasks in different ways, for instance, by performing some steps until the object becomes reachable, by adjusting the body posture (without stepping) until the target is reachable (for example by bending the knees and the torso while keeping in balance), etc. In the proposed framework, a motion planner is employed to search for a solution and to evaluate the best solution whenever several possibilities exist.

It is important to notice that the problem of skill coordination appears frequently in common tasks such as for relocating objects, opening doors, interacting with machines and appliances, etc. Most importantly this class of problems addresses a broad range of real-life tasks and represents a large proportion of people's daily activities. Deciding the sequence of skills to employ for such tasks is not trivial due the large number of possible coordinations and the different capabilities and constraints of each available skill.

Solving the coordination of motion skills is critical for several reasons: (1) to achieve optimal movements exploring the full potential of the humanoid structure; (2) to enable complex mobile manipulations in cluttered environments; and (3) to achieve human-like motions for humanoid assistants interacting and collaborating with people, e.g., for instance: for assistance with physical therapy [49], for performance of collaborative work [6], and for assistance in space exploration [16].

(a) (b)

Figure 10.2 (a) Example of a common grasping task which cannot be solved with a simple arm reaching motion even by a fully articulated humanoid model. (b) In most humanoid robots a balanced torso motion will be needed to enlarge the reachable workspace of the arms. Whole-body coordination is in particular important for less articulated arms, as for example, in Fujitsu's HOAP-3 humanoid model shown which has only 5 degrees of freedom in each arm. In such cases body positioning has a critical role in successfully reaching target locations

We are in particular interested in developing techniques capable of solving the motion generation problem similarly to how humans do, i.e., achieving humanlike performances. Therefore learning is a critical component which has to be addressed. Each time a task is planned and solved it should be analyzed and stored to improve the performance in subsequent similar tasks. In this way motions which have been learned should be quickly executed in subsequent tasks which are similar to the learned ones, similarly to how humans perform tasks. Although achieving generic and robust learning techniques remains a challenge, the proposed framework integrates learning in different ways: to improve skill execution, to learn constraints from imitation and to learn higher-level motion skills.

10.1.1 Related Work

Traditional motion planning approaches [43, 44, 46] are fundamentally based on systematic search in configuration spaces. Among the several techniques, sampling-based methods such as probabilistic roadmaps (PRMs) [35] and rapidly-exploring random trees (RRTs) [45, 41] have become extremely popular for planning in continuous configuration spaces. These and other methods have been applied to humanoid structures, however, as whole-body motion planning for humanoids is inherently a multi-modal problem, most of the approaches have been developed for particular modes or skills, for instance, footstep planning for precise locomotion around obstacles [40, 11, 12], arm reaching motions for manipulation [39, 30, 4, 17, 31, 15], etc.

When planning is limited to a discrete selection among possible actions or predefined motion patterns, discrete planners based on A* and its several variations [47, 38, 37] are well suited to finding the best sequence of actions to be taken. No single planning approach is best in all cases. For instance, while discrete planning is well suited to sequencing stepping motions, arm reaching is better addressed by searching in the continuous configuration space of the arm.

The issue of producing whole-body motions from combination of skills has also been addressed without an explicit motion planner, for example, including balance maintenance, energy minimization and friction [59], and also using dynamic controllers designed for animating characters in physics-based simulations [28, 19].

10.1.1.1 Multi-modal Planning

Multi-modal planning has recently emerged for solving humanoid problems and typically it addresses both discrete and continuous search strategies in an integrated fashion. Multi-modal planning has been in particular developed to achieve locomotion and climbing in difficult terrains [8, 26, 25, 34] but also to sequentially coordinate walking and pushing [27]. Reactive approaches have also been employed for coordinating specific problems such as walking while carrying a large object [18].

We have also addressed the whole-body motion planning problem in previous work. A whole-body reaching skill was developed including coupled spine rotations and knee bending [33], and two main aspects of the coordination problem have also been proposed: (1) the sequencing of movement primitives for climbing around obstacles [34]; and (2) the synchronization of concurrent primitives for simultaneous locomotion and object manipulation [60]. The sequencing problem was solved by a discrete search over multi-modal connections identified by a RRT on the parametric space of each primitive controller. The concurrency problem was solved by including the locomotion time parameterization as one additional (monotone) variable in the planning search space.

The multi-skill framework described here is an evolution of these previous works and aims at achieving a generic framework able to coordinate generic motion skills and as well to incorporate learning.

10.1.1.2 Learning for Motion Planning

Learning has been included in motion planning in a variety of different ways, for example, to select the best planning technique to be applied in a given situation [51], to concentrate sampling toward difficult regions of the configuration space [9], etc. These methods are mainly tailored to improve planning in difficult problems, and not to achieve humanlike performances in humanoid tasks. In contrast, learning reactive behaviors has been extensively addressed in the literature. For instance, reinforcement learning has recently been applied for learning motion transitions

[50] and for mapping state-goal spaces to control walking with collision avoidance [64].

While no work has specifically focused on the development of a learning-capable motion planner framework based on motion skills, the concepts of motion modeling and motion knowledge are known strategies [48, 42], and the idea of learning complex motions from primitive skills [1] is supported by evidence obtained in several cognitive studies [63, 20].

Learning from demonstrations has also been addressed by imitation frameworks [7, 58] and its importance is supported by research on mirror neurons: neurons that enable humans and other primates to imitate actions [14, 21]. Imitation techniques are therefore especially relevant for controlling and programming tasks for humanoid agents [62, 54, 53, 56, 5], and represent a natural approach to human–computer interaction by analogy to the way humans naturally interact with each other. As a consequence, research on humanoid imitation from human motion is currently very popular [62, 13, 23]. An imitation framework based on motion capture is explored here as a way to acquire motion constraints to be associated with objects.

10.1.2 Framework Overview

The presented framework has been mainly designed based on graphical simulations of virtual humanoids and only taking into account kinematic and static balance tests. We believe that such an environment is suitable for planning as long as the motions have low energy, which is mostly the case in the usual reaching and relocation tasks. As an example, we have also successfully applied results of our motion planning framework to control the HOAP-3 humanoid robot (as demonstrated in Figure 10.12).

Besides applications to humanoid robotics, we also believe that motion planning will soon provide important autonomy capabilities for interactive virtual humans, which can impact a number of applications in virtual reality, education, training and also entertainment. A recent tutorial on motion planning for virtual humans well illustrates the potential of motion planning to virtual characters [55].

Figure 10.3 depicts the architecture of the proposed framework. Each box in the figure represents a main module of the architecture. These modules are described in the following paragraphs.

- Skill base: motion skills are organized in a skill base where each skill follows a unified representation and parameterization interface. Skills are designed to generically encapsulate different approaches for motion synthesis and to provide a unified control interface to the motion planner.
- Multi-skill planner: this module implements the planner which will search for coordinations between the available motion skills in order to achieve solutions for given tasks. The output solution will be described as a sequence of instantiated skills needed to accomplish the task at hand.

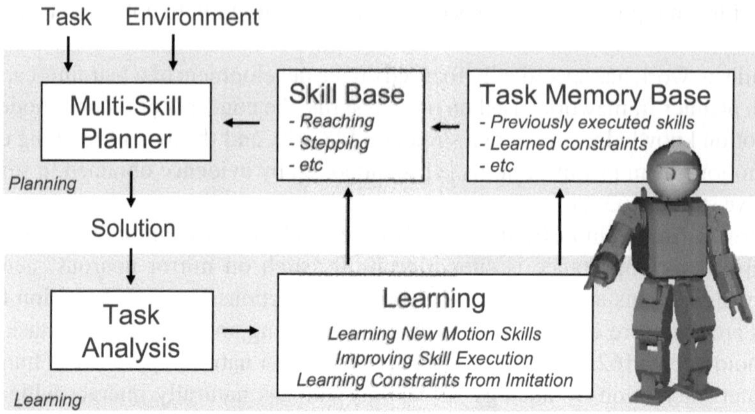

Figure 10.3 Framework main modules

- Task analysis: once a solution is found for a given task, the solution is analyzed and classified with respect to the environment and task at hand, and with respect to previous solutions. The analysis will inform the learning module.
- Learning: this module includes several learning mechanisms supported by the framework. Learning may improve the knowledge of skills in the skill base, may form new skills by interpolation of similar solutions, and may also add learned features to the task memory base which will be used when planning similar future tasks.
- Task memory base: this module stores all relevant features learned from previously planned solutions. Two main aspects are addressed here: (1) learning attractor points to improve the planning of arm reaching motions in new but similar environments; and (2) to learn motion constraints automatically from motion demonstrations. Constraints are typically associated to objects and will inform if an object is supposed to move in a constrained way.

The reminder of this chapter is organized as follows: motion skills are described in the next section and the multi-skill planner is described in Section 10.3. The task analysis, learning and the task memory base are presented in Section 10.4. Section 10.5 concludes this chapter.

10.2 Motion Skills

Motion skills are expected to produce specialized motions efficiently according to their own parameterization, and one of the purposes of employing motion skills is their specialized ability to control the humanoid in a particular mode.

The discrete set M is used to represent all possible humanoid modes. A mode is represented by specifying the state of each end-effector. Three letters are used:

f for free (or unconstrained), s for when the end-effector is being used to support the humanoid, and m for when it is being used for object manipulation. Therefore, assuming that four end-effectors are considered (two feet and two hands), a mode can be represented as a four-letter string, such as: "*ssff*" for a standing rest pose, or "*ssmf*" for a standing pose with one hand grasping an object.

Let $\mathscr{C}$ be the d-dimensional configuration space of the humanoid being controlled and $\mathscr{C}_{free}$ the subspace representing the valid configurations. Configurations in $\mathscr{C}_{free}$ are collision-free, in balance and respect articulation constraints.

Consider now that set $\mathscr{X}$ denotes the state space of the planning problem, which is defined as the Cartesian product $\mathscr{X} = \mathscr{C} \times \mathbb{R}^+$, where $\mathbb{R}^+$ is used to represent time. Therefore $x = (q,t) \in \mathscr{X}$ denotes a configuration q at time t. A function $m(x) \in M$ is also defined for retrieving the mode of the humanoid at state x.

The set $\mathscr{S}$ is defined to represent the skill base of the humanoid and is the finite set of all available motion skills. $\mathscr{S}(x) \subset \mathscr{S}$ represents the subset of skills which can be instantiated at x, i.e., which are applicable to take control over the humanoid in state x. Each skill is essentially a controller specialized to operate in a particular set of modes and is responsible for checking the feasibility of instantiation, for example: a foot placement skill will only be instantiated if there is an unconstrained foot to be controlled, etc.

A skill $\sigma^x \in \mathscr{S}(x)$ is a function of the type $\sigma^x : P_\sigma \times [0,1] \rightarrow \mathscr{X}$, where P_σ is its parametric control space, and $[0,1]$ is the normalized time parameterization of the produced motion, such that $\forall p \in P_\sigma$, $\sigma^x(p,0) = x$. Therefore, once a skill is instantiated it can be evaluated in $[0,1]$ in order to obtain the states traversed by the produced motion.

A skill is said to be explorative if it allows re-instantiation at intermediate poses. Only explorative skills will allow the use of a sampling-based planning strategy to systematically expand a search tree in their parametric space. Skills which are not explorative will be only allowed to be concatenated and their parameterizations will only be traversed in a forward monotone fashion, allowing the encapsulation of algorithms based on forward simulations.

Finally, given an initial state x_i and a final set of goal states X_g, the goal of the multi-skill planner is to produce a sequence of n skills together with their application parameters, such that, after the application of all the skills, the humanoid will have moved from x_i to a state $x_g \in X_g$ only traversing states with configurations in $\mathscr{C}_{free}$.

In this work the specific case of coordinating body motions for object reaching is considered to demonstrate the framework. In this case, the goal of the humanoid is to reach with the hand a pre-defined hand target suitable for grasping the given object. Therefore the final set of goal states X_g represents all body postures with a hand precisely reaching the hand target.

Each skill defines its own parameterization. For instance a balance skill could be parameterized with a single real value to dynamically adjust balance in one main direction, following the approach of *kinematic synergies* [24]. As the main focus here is to address whole-body kinematic reaching problems, skills allowing the precise placement of joints are developed and they are parameterized by the target locations to be reached.

Three basic skills were developed and constitute the minimal set of skills for solving a variety of problems: a reaching skill, a stepping skill, and a body balance skill. These skills were implemented using an analytical inverse kinematics (IK) formulation [32] for arms and legs (see Figure 10.4). The analytical formulation provides a fast closed form solution for the control of end-effectors and is key for allowing the motion planner to quickly evaluate the motion generated by skills during the planning. The main limitation of the analytical formulation is that it cannot be applied for generic linkages and therefore the results presented here focus on controlling anthropomorphic limbs.

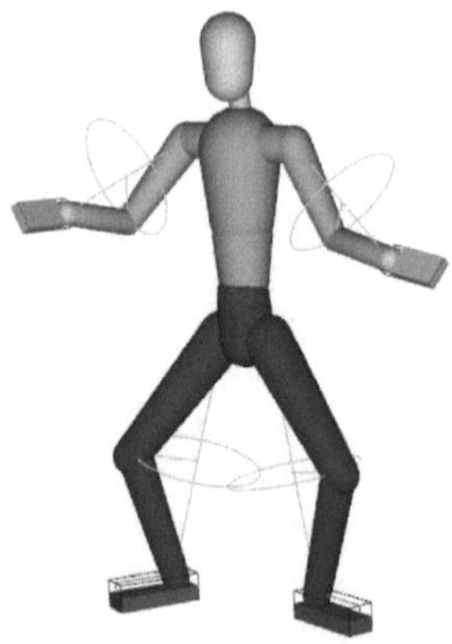

Figure 10.4 The figure illustrates the analytical IK applied to 7-DOF arms and 7-DOF legs and the orbit circles containing the possible positions for the elbows and knees [32]

10.2.1 Reaching Skill

The IK-based reaching skill σ_{reach} is parameterized by the target location to be reached. Given a parameter p, σ_{reach} produces a motion from the current humanoid posture to a final pose with the hand exactly at the target position and orientation encoded in p.

The reaching skill first uses the analytical IK to determine the final arm posture reaching location p. The determination of the final posture includes a fast search algorithm for determining the most suitable swivel angle along the orbit circle of the arm [32]. Once the final posture is determined, the motion between the current posture instantiated by the skill and the final posture is produced by interpolating, from the initial posture to the final posture, the following parameters: (1) the hand position, (2) the hand orientation and (3) the orbit angle. For each intermediate interpolated position and orbit angle, IK is again employed to compute the intermediate arm pose. Interpolation of the orientation is performed with quaternion spherical interpolation.

As a result, a realistic reaching motion with the hand describing a rectilinear trajectory in workspace is obtained (see Figure 10.5). The time parameterization is mapped to a spline curve in order to obtain a bell-shaped velocity profile. These characteristics encode observations of how realistic arm motions are performed by humans [57]. Anatomically-plausible joint parameterizations based on the swing-twist decomposition [22] and range limits based on spherical ellipses [3] are also included in order to represent human-like articulation limits.

This same skill has also been implemented for the HOAP-3 humanoid, the main difference is that the arms and legs of the robot have 5 and 6 DOFs respectively instead of 7 DOFs. The leg IK can therefore be solved exactly but without choice of the orbit angle. The arm IK, however, cannot be solved for arbitrary 6-DOFs targets and our current implementation assumes a pre-defined constrained target orientation to be solved.

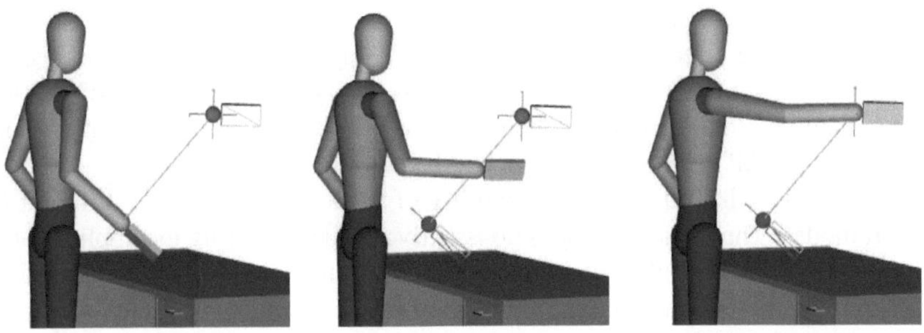

Figure 10.5 The motion obtained by the reaching skill produces a straight-line trajectory of the end-effector in workspace

10.2.2 Stepping Skill

The stepping skill σ_{step} controls one leg of the humanoid to perform the motion of one step towards another foot placement on the floor. Given a parameter p, σ_{step} produces a motion from the current humanoid posture to a final pose where the foot in control will exactly reach the target position and orientation encoded in p, where p is constrained to be a position where the foot aligns with the floor in a suitable manner to support the humanoid with that foot placement.

The motion generation follows the same interpolation strategies used in the reaching skill, however, following a trajectory with a bell-like shape on the vertical plane, in order to achieve the one-leg step motion. Also, the stepping skill can be instantiated only when the leg being controlled is free to move. Therefore a step can only occur when the humanoid is in single support mode. Balance tests are computed by checking if the projection of the center of mass on the floor lies inside the support polygon. Only static balance is considered in the scope of this work.

10.2.3 Balance Skill

The balance skill σ_{bal} allows switching between support modes by adjusting the position of the body while maintaining feet placements. Its parameterization encodes the new target position and orientation for the root joint of the skeleton. Given a target location, the skill will then produce a motion that makes the root joint from the current location to the new desired location, while maintaining the same feet placements with IK.

The motion is computed as follows: first the new configurations of the legs maintaining the feet placements with respect to the target location of the root joint are computed with IK. The motion is then defined by interpolation between the initial configurations and final configurations, which are interpolated in workspace in order to ensure that the feet placements are maintained.

One interesting effect obtained for the 7-DOF legs solution with the orbit angle search mechanism of the analytical IK [32] is the automatic opening of the legs (to avoid ankle joint limits) when the root joint is lowered. This, however, does not apply to the HOAP-3 legs, which have one less DOF.

The balance skill provides the capability of transitioning between different leg support modes. The support mode is constantly monitored during the application of motion skills as it will influence the mode of the humanoid, and therefore also the set of applicable motion skills, i.e., the skills which can be instantiated at a given humanoid configuration.

Figure 10.6 shows two example motions obtained with σ_{bal}. The (a) sequence changes the humanoid mode from both-feet support "ssff" to single feet support "fsff". The (b) sequence changes the mode from "sfff" to "fsff". The balance motion skill can therefore be used to free one leg from supporting the humanoid so that a stepping skill can be applied to the free leg in order to place the (now free) foot in a new placement.

10.2.4 Other Skills and Extensions

The skills presented are enough to coordinate generic stepping and reaching motions. These skills also provide enough flexibility to plan whole-body coordinations for reaching tasks.

Note that if distant targets are given to the motion planner a long sequence of stepping and balance skills will be computed in order for the humanoid to achieve walking. The sequence produced would, however, not be specialized for walking. To achieve distant targets, a specialized walking skill could be included in the framework to provide optimized gait patterns for fast locomotion. Such walking would be specifically implemented for the considered humanoid platform and would typically guarantee dynamic balance during the walking. The skills presented would then be invoked only when precise movement in constrained locations is needed.

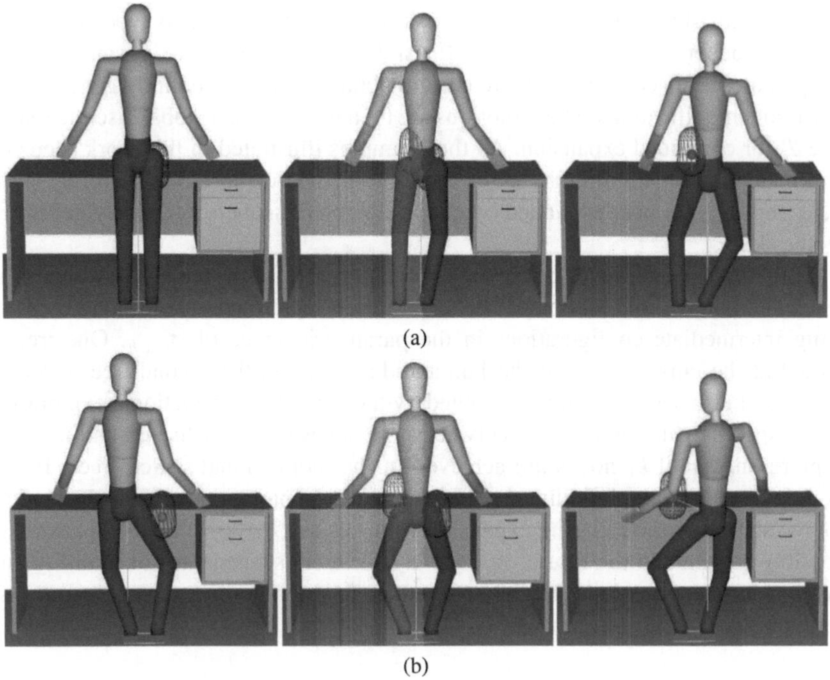

Figure 10.6 Example of motions obtained with the balance skill. (a) From standing pose to single foot support; (b) transitioning the support mode from one foot to the other. The vertical line shows the projection of the center of mass to the floor. The intersection with the support polygon (also shown) reveals the support mode of the humanoid

Note also that the IK-based computations performed in the basic skills are specific to anthropomorphic limbs and cannot easily be extended to generic humanoid platforms. Generic Jacobian-based IK approaches [36, 2, 10] could be employed but would impose more expensive computations as they are iterative methods which require inversion of a Jacobian at every iteration. An alternative solution would be to employ fast *cyclic coordinate descent* iterations [65] to determine target poses and then generate the motions with interpolation in joint angle space.

10.3 Multi-skill Planning

The main procedure of the multi-skill planner repeatedly invokes a skill expansion routine that will select and expand skills. The procedure stops when a solution motion is found, or if too much time has elapsed without success, in which case failure is reported. The pseudo-code for the main skill expansion routine is presented in Algorithm 10.1.

The multi-skill planner integrates in a single framework two types of search strategies depending on the skill type. If a skill σ is not explorative, k_2 samples from the parametric space of the skill are used to obtain k_2 motions entirely generated by σ without modifications. These motions are then included in a global discrete search tree T for continued expansion. As the examples illustrated in this work focus on arm manipulation, all mobility skills are set to be non-explorative. Therefore skills σ_{step} and σ_{bal} are only expanded by concatenation of sequences entirely generated by the skills.

Skill σ_{reach} is, however, set to be explorative and whenever a goal target is in reachable range, a bidirectional *RRT* exploration search tree is expanded by sampling intermediate configurations in the parametric space of σ_{reach}. One tree is rooted at the current state of the humanoid's arm, and the second tree is rooted at the goal arm state, which is computed by the skill. The bidirectional exploration determines if a motion in $\mathscr{C}_{free}$ between both states can be achieved and is set to explore only until k_1 nodes are achieved in the bidirectional search trees. If this limit is reached without finding a solution, the tree exploration is suspended and its source state re-inserted in the expansion front queue Q, allowing the expansion to possibly continue in the future. In this way, different sequencings of mobility and manipulation explorations are able to compete in a search for the best compromise solution.

10.3.1 Algorithm Details

The Expand_Skill procedure basically selects skills and performs the search strategy suitable for each skill. It maintains a tree T storing all states traversed so far and a priority queue Q containing the states in the current expansion front. At each call, the procedure starts by removing the lowest cost (higher priority) state from Q and selecting all skills which are applicable (line 2), i.e., which can be instantiated at the current state x. For each applicable skill, the corresponding expansion method is selected and applied.

The priority queue Q stores values according to a cost metric f_{cost} which can include different characteristics for guiding the search. The cost function used mainly encodes the following terms:

$$f_{cost}(x, p_g, n) = d_c(x_i, x) + w_g d_g(x, p_g) + w_e n.$$

Term $d_c(x_i, x)$ encodes the usual cost-to-come and is computed as the sum of the costs in all the edges in the T branch from the root node to the current node x. Edge costs represent the length of the motions represented in each edge. Term $d_g(x, p_g)$ encodes the cost-to-go heuristic, and is set as the distance between the goal point and the mid-point between the shoulder joints of the humanoid at state x. This cost is weighted by w_g and causes states closer to the goal to be expanded first. Finally, term weighted by w_e penalizes states which have already been expanded (by n expansions) in a bidirectional manipulation search. This term does not affect states reached by mobility skills which will always have $n = 0$.

Algorithm 10.1 Skill expansion of the multi-skill planner.

```
Expand_Skill ( Q, T, p_g, q_g )
x ← Q.remove_lowest_cost ()
S(x) ← applicable_skills ( S )
for ( each σ^x in S(x) ) do
    if ( σ^x type is manipulation and p_g in range ) then
        x_g ← σ^x(p, 1)
        if ( x_g successfully generated by σ^x ) then
            grow_bidirectional_search ( x, x_g, k_1, σ^x )
            if ( connection found ) then
                return SOLUTION_FOUND
            else
                attach the expanded bidirectional trees to x
                n ← total number of nodes in the trees
                Q.insert ( x, f_cost(x, p_g, n) )
            end if
        end if
    else
        // σ^x is of mobility type
        for ( k_2 times ) do
            p ← sample ( P_σ )
            if ( motion generated by σ^x(p) is valid ) then
                x' ← σ^x(p, 1)
                T.append_child ( x, x', σ^x, p )
                Q.insert ( x', f_cost(x', p_g, 0) )
            end if
        end for
    end if
end for
return NOT_YET_FOUND
```

Note also that the planner will request many sample motions from the available skills in order to progress with the search for a solution. Figure 10.7 illustrates some of the motion samples obtained from each of the skills.

In summary, the overall algorithm is able to integrate discrete skill expansion of mobility skills with configuration space exploration of manipulation skills in a single framework. The approach is able to naturally solve the trade-off between performing difficult manipulations or re-adjusting the body placement with mobility skills.

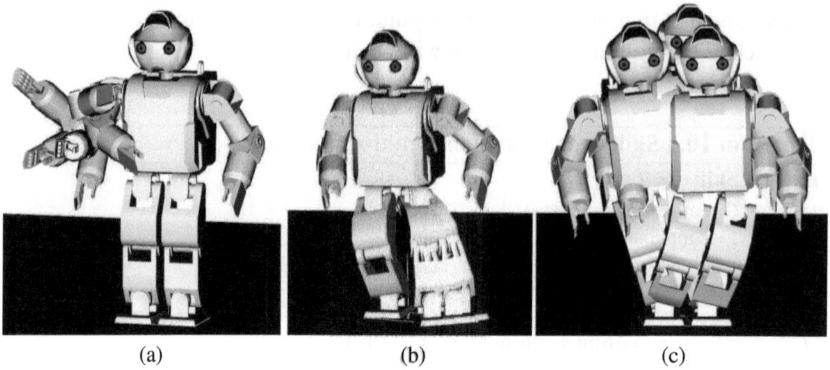

Figure 10.7 Example of end postures from sampled motions. Motions are sampled from each of the considered skills: (a) reaching, (b) stepping, and (c) balance. The HOAP-3 versions of the skills are shown in this example

10.3.2 Results and Discussion

Figure 10.8 shows the coordination obtained by the algorithm for solving a given reaching problem. The first image (a) shows that the goal is not in the reachable range of the arm. The next two images (b,c) show the initial and final postures produced by the balance skill from the initial standing posture to a (non-explorative) sampled body posture favoring approximation to the hand target. The reaching skill is then recruited and instantiated at this posture and a bidirectional exploration tree is expanded (image (d)) to connect the current posture to a posture reaching the target. The solution arm motion found by the reaching skill successfully avoids collisions with the table and reaches the goal, as depicted in the bottom sequence (d–f).

Further examples are illustrated in Figures 10.9 – 10.12. In all examples, the solutions obtained represent valid collision-free motions.

Note that the RRT exploration trees are implemented only using the generic interface for controlling motion skills. Therefore every expansion toward a sampled landmark implies a re-instantiation and application of the skill such that each expansion motion is produced by the skill itself.

In its current version the algorithm is limited to sequencing motion skills. A possible extension is to allow a skill to be activated in parallel with the previous skill, for instance to allow reaching to start while stepping or balance is still being executed. These concurrent executions can also be computed as a post-optimization phase in order to reduce the time required to plan the motions. In most of the examples presented, the computation time taken by the planner was in the range 10 to 40 s. Solutions of longer duration, as the one shown in Figure 10.11, require several minutes of computation.

In the case of applying planned motions to the HOAP-3 humanoid platform (Figure 10.12) we also rely on a reactive controller running in the robot with the purpose to correct the main torso orientation in order to maximize balance stability in

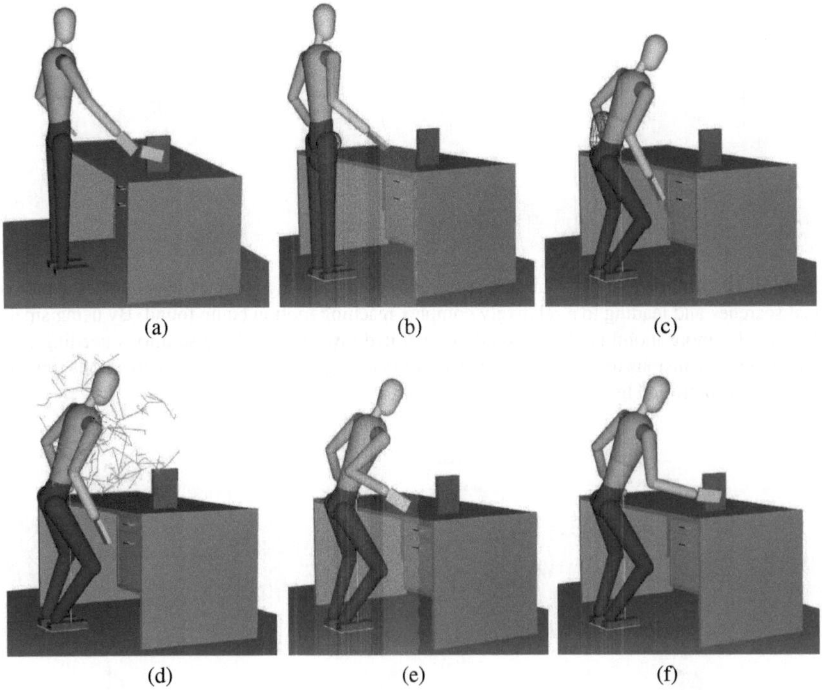

Figure 10.8 Example of sequencing of stepping, balance and reaching skills. Image (d) shows the generated exploration tree which was able to find the collision-free motion produced by the reaching skill

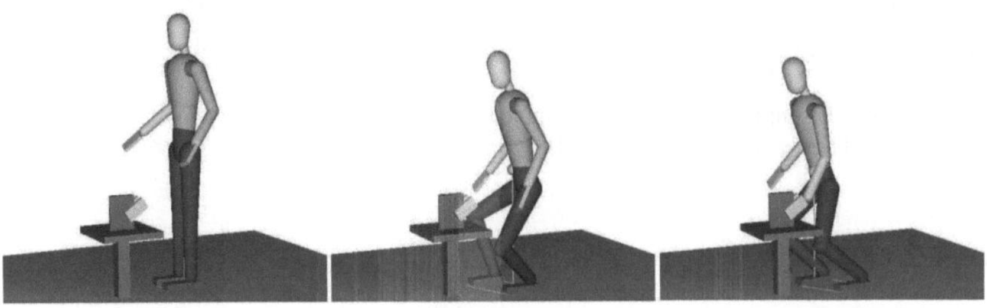

Figure 10.9 Example of a solution coordinating mobility and manipulation skills in order to reach for a low target

response to readings from the feet pressure sensors. The results presented demonstrate the capabilities of the multi-skill planning algorithm.

Figure 10.10 In this example, a large k_1 value was used allowing many expansions in the bidirectional searches and leading to a relatively complex reaching motion being found. By using smaller values for k_1, more mobility skills would be recruited favoring reaching solutions needing fewer bidirectional expansions to reach the goal. The colored trajectories illustrate the locations traversed by the joints controlled by each skill instantiated in T

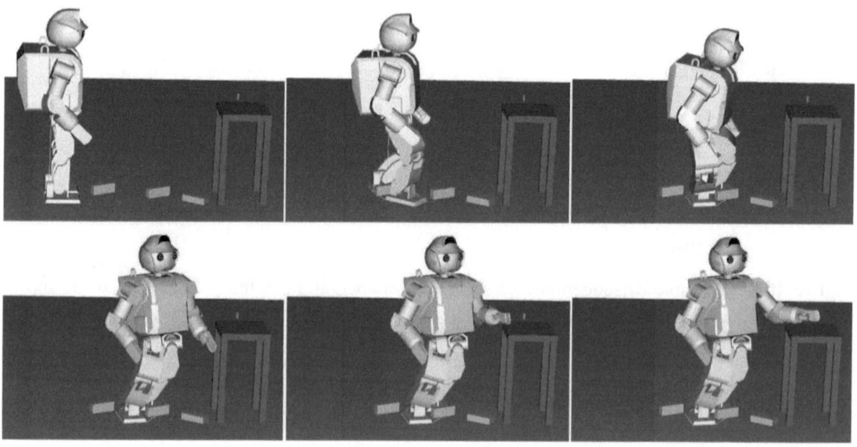

Figure 10.11 Example of a solution computed for the HOAP-3 model which requires several steps among obstacles until the target can be reached

10.4 Learning

Learning is essential for improving motion planning. In our framework (Figure 10.3) learning starts by analyzing executed tasks in order to extract useful information to improve individual skills and also to improve the overall planning process. Learned information may be directly sent to individual skills or may be stored in the task memory database, which will be accessible to each skill and as well to the multi-skill planner.

We describe now our first experiments in designing such learning algorithms with an overview of the learning strategies employed in our *Attractor-guided Planning* (AGP) approach for reaching tasks [29]. It includes a simple task analysis module for selecting previous tasks to be reused with attractor points which guide the planning of similar subsequent reaching motions.

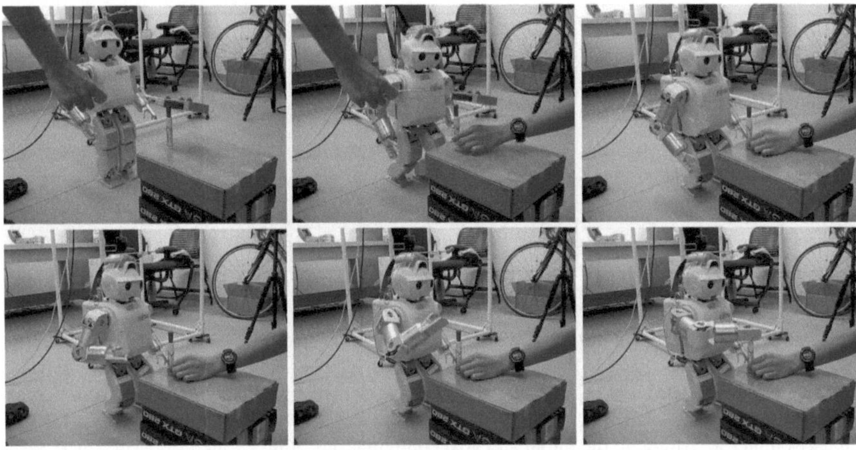

Figure 10.12 Example of applying a planned motion to the HOAP-3 humanoid. The target location to reach is specified by markers tracked with motion capture cameras (first two images). The upper three snapshots show the few steps planned for approaching the target, and the lower three snapshots show the arm motion for finally reaching the target

The task analysis metric and the extraction of attractor points are discussed in the following sections and provide a good example of features which can be learned specifically to improve the exploration search of a reaching skill.

Finally, we also present an approach for extracting motion constraints from motion capture examples. Constraints will, for instance, inform that a certain object should be relocated only with specific types of motions. As a result, an imitation-based approach for learning motion constraints is achieved.

10.4.1 A Similarity Metric for Reaching Tasks

A similarity metric is needed to identify previously planned reaching tasks which are similar to a new task to be solved. Consider the case of comparing the query task T with a previous task T' from the task memory database. Task T' is considered reusable only when its local environment and solution motion will share similar characteristics. The main question is what information should be taken into account. One possible approach is to define a metric based only on the distances between the initial and goal configurations of the tasks. Such a naive metric works well with dynamic queries in a static environment, where all the obstacles are not moving. When it comes to a dynamic environment, the change of obstacle positions may largely affect the structure of a solution path, as illustrated in Figure 10.13. This motivates us to include obstacle positions in the task comparison metric.

When considering obstacles, one important factor is how much influence an obstacle has on the solution path. In Figure 10.13, obstacle B shows a great change

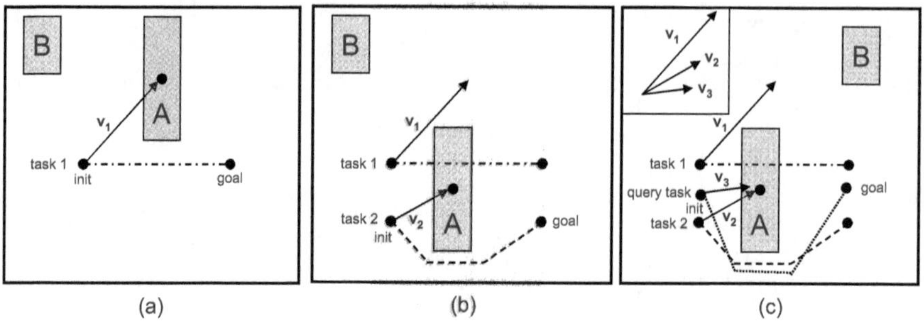

Figure 10.13 A planning environment with moving obstacles A and B. Task 1 and task 2 are previously solved (image (a) and (b)), and the query task is selecting an example from the database. v1, v2 and v3 denote the local coordinates of obstacle A with respect to the three initial configurations. Note that obstacle B is out of the local coordinate system range, thus is not included in the comparison metric. Although the initial and goal configurations of task 1 are closer, the query task chooses task 2 as example because v2 is closer to v3

of position from task 2 to the query task (images (b) and (c)), but since this change has limited impact on the solution path, this obstacle should not influence the comparison. Our comparison technique represents obstacles in local coordinates with respect to the initial and goal configurations.

For each task, two coordinate systems are built with origins located at the initial and goal configurations. Obstacles that are too far away from an origin are not included in the respective coordinate system. The comparison metric accumulates the distance between each obstacle o from the query task and o' from the example task that is closest to o, computed in their local coordinates. The metric is described as follows:

$$dist(T,T') = h_1 * dist(c_{init}, c'_{init}) + h_1 * dist(c_{goal}, c'_{goal}) \\ + h_2 * \textstyle\sum_{o'} \min_o dist(o,o').$$

The weights of the motion query distance and the obstacle distance are tuned by heuristics h_1 and h_2.

10.4.2 Learning Reaching Strategies

Once a previous similar task is identified from the task memory base, the stored motion is used to guide the planning of the new task at hand. Attractor points extracted from the previous motion are used. Since the focus here is on reaching tasks, we fit the path of the humanoid wrist approximately into a sequence of piecewise linear segments, and then define attractors to be the points where two segments connect. Given a set of 3D points denoting the path of the wrist joint in workspace, we employ a variation of a line tracking algorithm [61], which was further experimented and modified by Nguyen and colleagues [52].

The modified line tracking algorithm starts by constructing a window containing the first two points, and fits a line to them. Then it adds the next point to the current line model, and recomputes the new line parameters. If the new parameters satisfy the line condition up to a threshold, the window incrementally moves forward to include the next point. Otherwise, the new included data point is detected as an attractor point, and the window computation restarts from this attractor.

After a list of attractors is detected, a validation procedure is used to ensure that the attractor list defines a valid collision-free path. If a collision is detected between two consecutive attractors, the middle point on the original path is inserted in the attractor list and the process iterates until all of the straight segments are free of collisions. The goal is to detect valid attractors even in difficult regions. If the sequence of valid attractors is applied to guide the exact same task again, the solution path will be trivially solved without any search outside the guided path.

Once a set of attractors is decided to be reused, the sampling strategy in the motion planner is biased toward regions around the attractors. Taking advantage of the attractors is beneficial for two main reasons: first, it preserves the structure of a successful solution path. Second, whenever a new obstacle location collides with part of the solution path, a large portion of the path is still reusable. Note that since only similar tasks are used by the query, the environment does not differ much.

During the planning, dynamic Gaussian distributions are placed around each attractor point in the configuration space. The distributions are used as a replacement for the random sampling procedure of the exploration trees in the planner. The scale of the Gaussian distribution is initialized as zero, which means samples start at exactly the attractor point, guiding the search directly toward the current attractor. When samples are not able to attract the exploration tree due to variations in the new task, the scale of the Gaussian distribution is increased. In the limit the bias will gradually disappear and the sampling will return to a random sampling. When samples get very close to the attractor, the Gaussian centered on the next attractor point is then used. The process repeats for the whole list of attractors until a solution is found or the task is considered not solvable.

The same procedure can be applied to single or bidirectional search trees, in which case opposite attractor orders are taken in each tree (see Figure 10.14). Several experiments were conducted and they demonstrate a significant improvement in computation speed for dynamic environments which maintain some structure between queries [29].

10.4.3 Learning Constraints from Imitation

Different kinds of motion constraints are important in object manipulation, for example, changes in orientation should not be allowed when relocating a cup of water. Moreover, the pattern of a movement is also meaningful, for example, whether the hand of the character is moving along a straight line on a surface or just moving unconstrained in space.

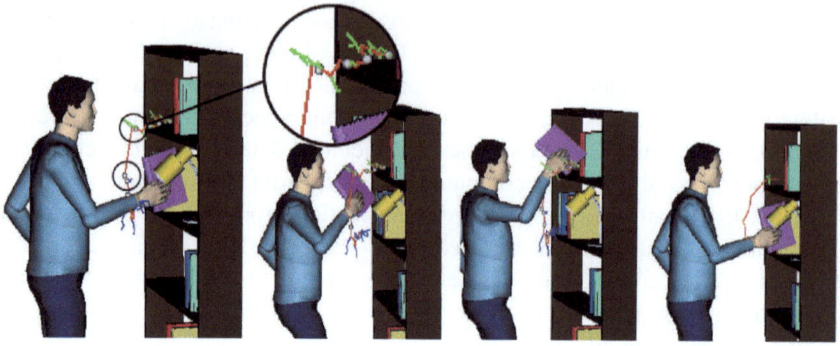

Figure 10.14 Attractors (gray spheres) are extracted from a path (red line) computed by a bidirectional RRT tree expansion (green and blue branches). The attractors specifically represent postures which are trying to avoid collisions, as shown in the second and third images. The attractor-guided planning results in less tree expansions and in a smoother solution path (last image) [29]

Note that it is assumed here that an object being manipulated (or relocated) has a fixed attachment to the hand of the humanoid, and therefore finger manipulations are not considered, meaning that the hand and the object can be considered to be one single end-effector being controlled.

Motion constraints of objects or tasks have to be somehow programmed in the humanoid. We explore the approach of learning constraints from direct demonstrations, i.e. from imitation. The idea is that users can demonstrate how tasks should be performed and the observed motions are then analyzed and learned. We consider that a demonstrated motion is captured and represented as the trajectory of the user's hand. The demonstrated trajectory can then be analyzed and classified. A classification of the types of motion constraints considered is given below:

- Stationary: when both the position and the orientation of the end-effector remains the same. This happens for instance when holding an object still (like a photograph camera) while the body may be moving. Note that constraints are detected in global coordinates.
- Translation-only: when only the orientation of the end-effector remains the same. This will result in a motion composed only of translations (for example to relocate a cup of water).
- Rotation-only: when only the translation of the end-effector remains the same. This constraint happens when no translations are present.
- Patterned-rotation: this constraint refers to cases where the end-effector is rotating around an axis or a point. This constraint appears in particular rotational manipulations (like when opening a door).
- Patterned-translation: for the cases where the end-effector is translating in a plane or along a straight line. This constraint appears in particular translational manipulations (like painting a straight line).

The first three types of constraints are basic constraints. The stationary constraint has to be detected first. The translation-only constraint can exist at the same time with a patterned-translation constraint, but not with a patterned-rotation constraint. Similarly, rotation-only constraints can exist at the same time with patterned-rotation constraints, but not with patterned-translation constraints. Based on these observations, Figure 10.15 shows a decision tree for detecting constraints based on a series of tests.

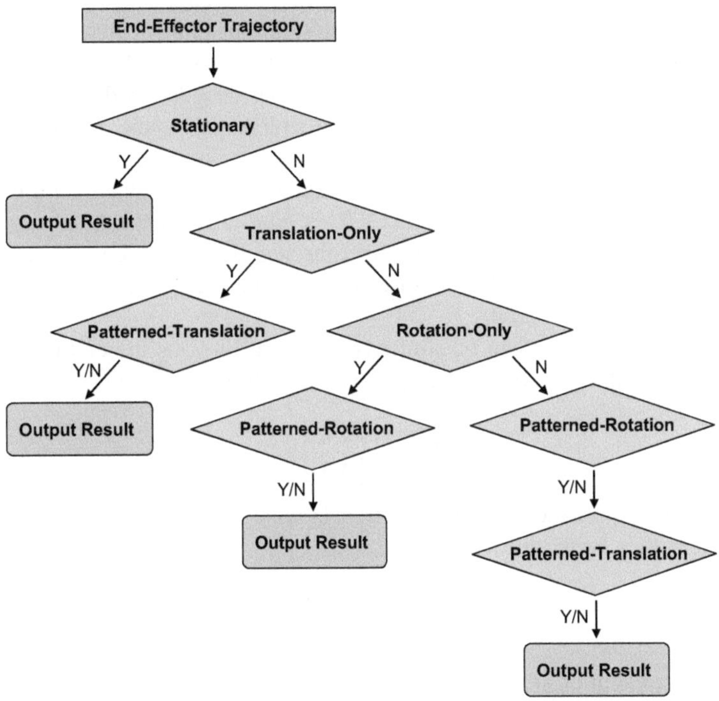

Figure 10.15 Decision stages for detecting constraints. Letters Y or N denote whether the constraints can be detected or not

A constraint detection algorithm can then be devised for analyzing the demonstrated motion. Given a specific time interval $[t_1, t_n]$, the demonstrated motion m of the end-effector is denoted as the following sequence of captured frames:

$$(f_1, f_2, \ldots, f_k, f_{k+1}, \ldots, f_n).$$

Each frame f_k contains geometric information (position and orientation) about the end effector at that time step, i.e., the position and orientation of the end-effector with respect to the global coordinate system. Instantaneous constraints are first detected between each pair of frames. Then, the sequence of instantaneous constraints will be merged so that motion m is segmented by constraint type.

10.4.3.1 Detection of Instantaneous Constraints

An instantaneous constraint C_k is detected between time k and $k+1$. Given a pair of frames f_k and f_{k+1}, the definition of C_k depends on the 3D transformation from time k to time $k+1$, which is represented here by the displacement matrix D_k. Therefore if x_k denotes the demonstrated hand position at time k, then $x_{k+1} = D_k x_k$.

If the motion is stationary from time k to $k+1$, then D_k will be the identity transformation and $(D_k - I_4)x_k = 0$. Since the solutions to this equation include all the observed points on the end-effector, the elements in matrix $(D_k - I_4)$ will contain very small values. Note that the translational values encoded in the homogeneous matrices have to be scaled to a compatible unit with respect to the rotational values, which are inside $[-1, 1]$.

The squared sum of all the elements in matrix $(D_k - I_4)$ is then computed and called the *stationary value* of constraint C_k. This value denotes how much movement the end-effector has performed during this time step. If this value is smaller than a stationary threshold τ_s, then the end-effector is considered stationary from time k to time $k+1$.

Rotational movements can be similarly detected. First, as the end-effector can only perform rigid transformations, D_k can be re-written as:

$$D_k = \begin{bmatrix} R_k & t_k \\ 0_3 & 1 \end{bmatrix},$$

where R_k and t_k are the rotational and translational components of D_k. The matrix $(D_k - I_4)$ can then be written as:

$$D_k - I_4 = \begin{bmatrix} R_k - I_3 & t_k \\ 0_3 & 0 \end{bmatrix}.$$

The squared sum of all the elements in matrix $(R_k - I_3)$ is then called the *rotational value* of C_k. If it is smaller than a rotational threshold τ_r, then the movement will be identified as translation-only, i.e. with no rotation involved.

Similarly, for a rotation-only movement during one time step, the constraint can be detected by comparing the squared sum of vector t_k (the *translational value*) with a translational threshold τ_t.

In summary, the first three types of constraints are detected as follows:

- For an instantaneous constraint C_k, if its stationary value minus the squared sum of all the elements in (D_k-I_4) is smaller than threshold τ_s, then the end-effector is stationary between time k and $k+1$.
- For an instantaneous constraint C_k, if its rotational value minus the squared sum of all the elements in (R_k-I_3) is smaller than threshold τ_r, then the end-effector is not rotating between time k and $k+1$.
- For an instantaneous constraint C_k, if its translational value minus the squared sum of all the elements in t_k is smaller than threshold τ_t, then the end-effector is not translating between time k and $k+1$.

To detect the other two constraints (patterned-rotation and patterned-translation), successive transformations have to be merged and tested if they have similar patterns.

10.4.3.2 Merging Transformations

Transformation R_k denotes the rotational component of the transformation between time k and $k+1$. The rotation axis d_k = (r_{21}-r_{12}, r_{02}-r_{20}, r_{10} - r_{01}) can then be determined for each frame where r_{ij} denotes the elements of matrix R_k. Given a sequence of frames (f_1, f_2, ..., f_k, ..., f_n), the rotational axes (d_1,d_2, ...,d_k,d_{k+1}, ...,d_n) can then be computed. If they can be approximately fitted by a straight line L, then the motion is a rotation around the line L, and the transformations between time 1 and n can be merged.

Similarly, if all the translational vectors can be approximately fitted to a straight line, then the end-effector will be translating along the line. Two new thresholds τ_{rfit} and τ_{tfit} are needed to decide the rotational and translational fitting limits. The fitting algorithm employed is the same as the one used to detect attractors for the AGP planner (Section 10.4.2). Note that the same procedures can also be applied to detect translations in a plane.

Since real motion capture data is being analyzed, the existence of noise or performance imprecisions will create outliers, i.e., one or more frames that divide the motion into two parts, which should have the same constraint. To handle outliers, a frame buffer is used to store aberrant frames with a different motion pattern than in previous frames. A new motion will be created only when the number of aberrant frames in the frame buffer reaches a limit. Two limits are needed:

- Rotation buffer size: this threshold decides if a patterned rotation has ended and it is denoted as τ_{rbuf}. When the number of inconsistent frames in a patterned rotation is greater than τ_{rbuf}, a new motion constraint will be created.
- Translation buffer size: this threshold decides if a patterned translation has ended and it is denoted as τ_{tbuf}. When the number of inconsistent frames in a patterned translation is greater than τ_{tbuf}, a new motion constraint will then be created.

After a set of constraints are detected, they need to go through a final filter in order to ensure that the motion does not contain unnecessary segments. Two constraints are then merged if they belong to the same constraint category and are close enough to each other.

10.4.3.3 Computing the Thresholds

A number of threshold parameters for detecting and merging constraints have been used: $\tau_s, \tau_r, \tau_t, \tau_{rfit}, \tau_{tfit}, \tau_{rbuf}, \tau_{tbuf}$. In order to perform a robust computation of the constraints, these values will be determined from template motions which can be provided for calibrating the constraint detection. Three types of calibration motions are needed: a stationary motion m_1, a rotation-only motion m_2 around an axis (and without translation), and a translation-only motion m_3 (without rotation) along a straight line.

From motion m_1 it is possible to determine the stationary threshold τ_s. First all stationary values for the frames of the motion are computed with:

$$V_s = (\sigma_1, \sigma_2, ..., \sigma_k, \sigma_{k+1}, ..., \sigma_n).$$

Due to the presence of noise and imprecisions from the human operator during the motion capture process it is not correct to simply take the highest σ value and assign it as the stationary threshold. There might be outliers in V_s, for instance, a subtle shake of the hand could lead to a very high σ value.

In order to reduce overestimation of the stationary threshold, outliers have to be detected and rejected. First, the median number σ_m from set V_s is computed. Then for each $\sigma_k \in V_s$, the squared distance $dist_k = (\sigma_k - \sigma_m)^2$ is computed. Let the smallest squared distance be $dist_{min}$. A σ_k is detected to be an outlier if $dist_k > h * d_{min}$, where h is a heuristic weight which is greater than 1. After the outliers are removed from the set of stationary values, the stationary threshold τ_s is set to be the highest stationary value from the filtered V_s set.

The rotational threshold and the translational threshold are similarly computed. Since motion m_2 is a rotation-only motion without translation, most of its instantaneous constraints should be lower than the translational threshold τ_t. First, the translation vectors $(t_1, t_2, ..., t_k, t_{k+1}, ..., t_{n-1})$ are computed from each of the frames. After removing the outliers, τ_t is assigned the highest translational value. The rotational threshold τ_r can then be obtained from motion m_3 in the same way.

The line fitting thresholds τ_{rfit} and τ_{tfit} can also be obtained from m_2 and m_3. Since motion m_2 is a rotation around a specific axis, all detected instantaneous rotational axes can be fitted to an approximate straight line. After removing outliers, τ_{rfit} is set as the largest distance between the instantaneous rotational axes and the fitted line. The computation of τ_{tfit} is similarly performed from motion m_3. Finally, the maximum number of continuous outliers can then be assigned to τ_{rbuf} and τ_{lbuf}.

10.4.3.4 Reusing Detected Constraints in New Tasks

Object manipulation tasks are often complicated and will typically be composed of multiple motion constraints. For example, a water pouring task will typically include a translation without rotation followed by a rotation without translation. The constraint segmentation procedures described will divide the motion into a sequence of constrained sub-motions according to their constraint types. The connections between these sub-motions can then be represented by constrained attractors which will guide the execution of similar tasks, for example for pouring water in similar environments or to different target locations.

The attractor-guided planning described in Section 10.4.2 is then employed with the extension that configurations will always be sampled respecting the constraints associated with attractors. In this way constrained attractors will be able to guide the planning of future tasks which will respect constraints to be imitated. Even in a changing environment the constraints of the original motion will still be maintained.

10.4.4 Results and Discussion

A FASTRAK magnetic motion capture system was used for capturing example hand motions with constraints. Although a high frame rate is not necessary, the system is able to capture motions at 60 Hz. First, template motions were processed in order to determine the thresholds needed and then several motion constraints could be successfully detected.

Figure 10.16 illustrates segmentation results of demonstrations containing different types of motions: (1) two translations; (2) translation and rotation; (3) translation, rotation and again translation; and (4) translation and random movements.

Each segment is shown as a pair of attractors (gray spheres). A segment represents a motion with a different constraint type than its neighbor segments. The constraint detection procedure described is capable of detecting all constraints. Stationary motions are also detected and capture a natural stop when the performer switches between translations and rotations.

Figure 10.16 The segmentations of four motions with different types of constraints. The attractors (denoted by gray spheres) represent a constraint type change and can then be used to guide the planning of future similar tasks

Figure 10.17 shows the AGP planner being used to solve the constrained manipulation task of relocating a cup without rotating it. The motion demonstrated includes only translation. After recording the demonstration, an obstacle is placed in the demonstrated path in order to force the imitation procedure to find an alternative path. The AGP approach is then able to find a suitable solution maintaining the main characteristic of the demonstrated motion: the cup is never allowed to rotate during the relocation.

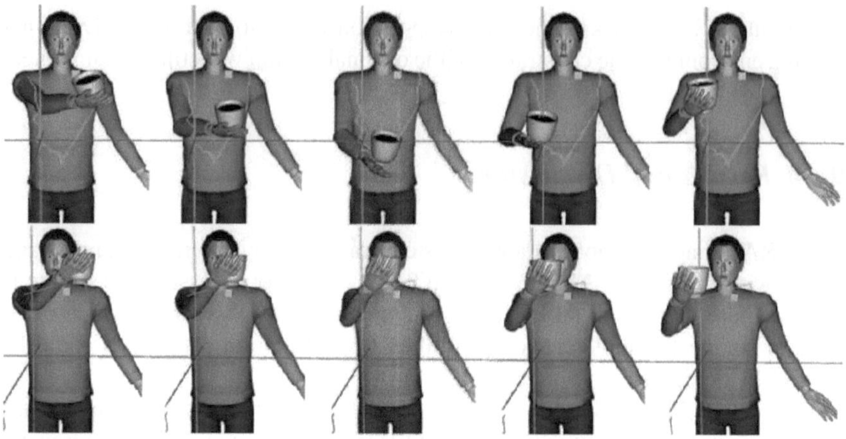

Figure 10.17 The upper sequence shows a solution path found by a non-constrained RRT planner. The lower sequence shows a solution path found by the AGP planner with attractors constrained to translational motion. The small cube in the environment represents an obstacle and demonstrates adaptation of the solution respecting the constraints. The image also illustrates that, due the much more constrained search procedure, AGP also favors finding shorter solutions

10.5 Conclusion

We have described several pieces of a framework integrating learning with motion planning in different ways. Several results were also presented. We believe that the proposed approach is able to address a wide range of humanoid applications requiring autonomous object manipulations.

The presented framework also naturally accounts for the possibility of including new motion skills which could be automatically learned from experiences and then added in the skill base in order to provide new specialized skills to be used by the multi-skill motion planner. The possibility of achieving an automatic open-ended learning structure for accumulating motion skills and object-related constraints has the potential to empower the multi-skill planner with the capability to solve generic and increasingly complex tasks. The proposed framework presents our first steps in this direction.

Acknowledgments This work has been partially supported by NSF Awards IIS-0915665 and BCS-0821766.

References

[1] M. A. Arbib. Perceptual structures and distributed motor control. In: V. B. Brooks, editor, Handbook of Physiology, Section 2: The Nervous System Vol. II, Motor Control, Part 1, pp 1449–1480. American Physiological Society, Bethesda, MD, 1981.

[2] P. Baerlocher. Inverse kinematics techniques for the interactive posture control of articulated figures. PhD thesis, Swiss Federal Institute of Technology, EPFL, 2001. Thesis number 2383.

[3] P. Baerlocher and R. Boulic. Parametrization and range of motion of the ball-and-socket joint. In: proceedings of the AVATARS conference, Lausanne, Switzerland, 2000.

[4] D. Bertram, J. Kuffner, R. Dillmann, and T. Asfour. An integrated approach to inverse kinematics and path planning for redundant manipulators. In: proceedings of the IEEE international conference on robotics and automation, pp 1874–1879. IEEE, May 2006.

[5] A. Billard and M. J. Matarić. Learning human arm movements by imitation: Evaluation of a biologically inspired connectionist architecture. Robotics Autonomous Syst, 37(2-3), pp 145–160, November, 30 2001.

[6] C. Breazeal, A. Brooks, D. Chilongo, J. Gray, G. Hoffman, C. Kidd, H. Lee, J. Lieberman, and A. Lockerd. Working collaboratively with humanoid robots. In: proceedings of humanoids, Los Angeles, CA, 2004.

[7] C. Breazeal, D. Buchsbaum, J. Gray, D. Gatenby, and D. Blumberg. Learning from and about others: Towards using imitation to bootstrap the social understanding of others by robots. Artif Life, 11(1–2), 2005.

[8] T. Bretl. Motion planning of multi-limbed robots subject to equilibrium constraints: The free-climbing robot problem. Int J Robotics Res, 25(4), pp 317–342, 2006. ISSN 0278-3649.

[9] B. Burns and O. Brock. Sampling-based motion planning using predictive models. In: proceedings of the IEEE international conference on robotics and automation (ICRA), pp 3120– 3125, Marina del Rey, CA, April 18-22 2005.

[10] S. R. Buss and J.-S. Kim. Selectively damped least squares for inverse kinematics. J Graphics Tools, 10(3), pp 37–49, 2005.

[11] J. Chestnutt, M. Lau, K. M. Cheung, J. Kuffner, J. K. Hodgins, and T. Kanade. Footstep planning for the honda asimo humanoid. In: proceedings of the IEEE international conference on robotics and automation, April 2005.

[12] J. Chestnutt, K. Nishiwaki, J. Kuffner, and S. Kagami. An adaptive action model for legged navigation planning. In: proceedings of the IEEE-RAS international conference on humanoid robotics (Humanoids), 2007.

[13] B. Dariush, M. Gienger, B. Jian, C. Goerick, and K. Fujimura. whole body humanoid control from human descriptors. In: proceedings of the international conference on robotics and Automation (ICRA), pp 3677–2684, May 19–23 2008.

[14] J. Decety. Do imagined and executed actions share the same neural substrate? Cognitive Brain Res, 3, pp 87–93, 1996.

[15] R. Diankov and J. Kuffner. randomized statistical path planning. In: proceedings of the international conference on robotics and automation (ICRA), pp 1–6, May 19-23 2008.

[16] M. A. Diftler and R. O. Ambrose. Robonaut, a robotic astronaut assistant. In: international symposium on Artificial Intelligence, robotics and automation in space (ISAIRAS), Montreal Canada, June, 18 2001.

[17] E. Drumwright and V. Ng-Thow-Hing. Toward interactive reaching in static environments for humanoid robots. In: proceedings of the IEEE international conference on intelligent robots and systems (IROS), Beijing, China, October 2006.

[18] C. Esteves, G. Arechavaleta, J. Pettré, and J.-P. Laumond. Animation planning for virtual characters cooperation. ACM Trans Graphics, 25(2), pp 319–339, 2006.

[19] P. Faloutsos, M. van de Panne, and D. Terzopoulos. Composable controllers for physics-based character animation. In: proceedings of SIGGRAPH, pp 251–260, New York, NY, USA, 2001. ACM Press. ISBN 1-58113-374-X.

[20] S. F. Giszter, F. A. Mussa-Ivaldi, and E. Bizzi. Convergent force fields organized in the frog's spinal cord. J Neurosci, 13(2), pp 467–491, 1993.

[21] A. Goldman and V. Gallese. Mirror neurons and the simulation theory of mind-reading. Trends Cognitive Sci, 2(12), pp 493–501, 1998.

[22] S. Grassia. Practical parameterization of rotations using the exponential map. J Graphics Tools, 3(3), pp 29–48, 1998.
[23] D. B. Grimes, D. R. Rashid, and R. P. N. Rao. Learning nonparametric models for probabilistic imitation. In: Neural Information Processing Systems (NIPS), pp 521–528.
[24] H. Hauser, G. Neumann, A. J. Ijspeert, and W. Maass. Biologically inspired kinematic synergies provide a new paradigm for balance control of humanoid robots. In: proceedings of the IEEE-RAS international conference on humanoid robotics (Humanoids), 2007.
[25] K. Hauser, T. Bretl, K. Harada, and J. Latombe. Using motion primitives in probabilistic sample-based planning for humanoid robots. In: workshop on algorithmic foundations of robotics (WAFR), pp 2641– 2648, July 2006.
[26] K. Hauser, T. Bretl, and J. Latombe. Non-gaited humanoid locomotion planning. In: proceedings of the IEEE-RAS international conference on humanoid robotics (Humanoids), pp 2641– 2648, December 2005.
[27] K. Hauser, V. Ng-Thowhing, Gonzalez-Baos, H. T. Mukai, and S. Kuriyama. Multi-modal motion planning for a humanoid robot manipulation task. In: international symposium on robotics research (ISRR), 2007.
[28] J. K. Hodgins, W. L. Wooten, D. C. Brogan, and J. F. O'Brien. Animating human athletics. In: proceedings of ACM SIGGRAPH, pp 71–78, New York, NY, USA, 1995. ACM Press.
[29] X. Jiang and M. Kallmann. Learning humanoid reaching tasks in dynamic environments. In: proceedings of the IEEE international conference on intelligent robots and systems (IROS), San Diego CA, 2007.
[30] S. Kagami, J. Kuffner, K. Nishiwaki, S. Kagami, M. Inaba, and H. Inoue. Humanoid arm motion planning using stereo vision and rrt search. J Robotics Mechatronics, April 2003.
[31] M. Kallmann. Scalable solutions for interactive virtual humans that can manipulate objects. In: proceedings of the artificial intelligence and interactive digital entertainment (AIIDE'05), pp 69–74, Marina del Rey, CA, June 1–3 2005.
[32] M. Kallmann. Analytical inverse kinematics with body posture control. Computer Animation and Virtual Worlds, 19(2), pp 79–91, 2008.
[33] M. Kallmann, A. Aubel, T. Abaci, and D. Thalmann. Planning collision-free reaching motions for interactive object manipulation and grasping. Computer graphics Forum (proceedings of eurographics'03), 22(3), pp 313–322, September 2003.
[34] M. Kallmann, R. Bargmann, and M. J. Matarić. Planning the sequencing of movement primitives. In: proceedings of the international conference on simulation of adaptive behavior (SAB), pp 193–200, Santa Monica, CA, July 2004.
[35] L. Kavraki, P. Svestka, J.-C. Latombe, and M. Overmars. Probabilistic roadmaps for fast path planning in high-dimensional configuration spaces. IEEE Trans Robotics Automat, 12, pp 566–580, 1996.
[36] C. A. Klein and C.-H. Huang. Review of pseudoinverse control for use with kinematically redundant manipulators. IEEE Trans Syst Man Cybern, SMC-13(3), pp 245–250, March-April 1983.
[37] S. Koenig. A comparison of fast search methods for real-time situated agents. In: proceedings of the international joint conference on autonomous agents and multiagent systems (AAMAS), pp 864–871, 2004.
[38] S. Koenig and M. Likhachev. Real-time adaptive a*. In: proceedings of the international joint conference on autonomous agents and multiagent systems (AAMAS), pp 281–288, 2006.
[39] Y. Koga, K. Kondo, J. J. Kuffner, and J.-C. Latombe. Planning motions with intentions. In: proceedings of SIGGRAPH, pp 395–408. ACM Press, 1994. ISBN 0-89791-667-0.
[40] J. Kuffner, K. Nishiwaki, S. Kagami, M. Inaba, and H. Inoue. Motion planning for humanoid robots. In: proceedings of the 11th international symposium of robotics research (ISRR), November 2003.
[41] J. J. Kuffner and S. M. LaValle. RRT-Connect: An efficient approach to single-query path planning. In: proceedings of IEEE international conference on robotics and automation (ICRA), San Francisco, CA, April 2000.

[42] K. Kurashige, T. Fukuda, and H. Hoshino. Motion planning based on hierarchical knowledge for a six legged locomotion robot. In: proceedings of IEEE international conference on systems, man, and cybernetics, vol. 6, pp 924–929, 1999.
[43] J.-C. Latombe. Robot Motion Planning. Kluwer Academic Publisher, December 1990.
[44] J.-P. P. Laumond. Robot Motion Planning and Control. Springer-Verlag New York, Inc., Secaucus, NJ, USA, 1998.
[45] S. LaValle. Rapidly-exploring random trees: A new tool for path planning. Technical Report 98-11, Iowa State University, Computer Science Department, October 1998.
[46] S. M. LaValle. Planning Algorithms. Cambridge University Press (available on-line), 2006. URL `msl.cs.uiuc.edu/planning/`.
[47] M. Likhachev, G. J. Gordon, and S. Thrun. Ara*: Anytime a* with provable bounds on sub-optimality. In S. Thrun, L. Saul, and B. Schölkopf, editors, Advances in Neural Information Processing Systems 16. MIT Press, Cambridge, MA, 2004.
[48] V. Manikonda, P. Krishnaprasad, and J. Hendler. A motion description language and hybrid architecure for motion planning with nonholonomic robots. In: proceedings of IEEE international conference on robotics and automation (ICRA), May 1995.
[49] M. J. Matarić. Socially assistive robotics. IEEE Intelligent Systems, August 2006.
[50] J. McCann and N. S. Pollard. Responsive characters from motion fragments. ACM Trans Graphics (SIGGRAPH 2007), 26(3), Aug. 2007.
[51] M. Morales, L. Tapia, R. Pearce, S. Rodriguez, and N. Amato. A machine learning approach for feature-sensitive motion planning. In: proceedings of the workshop on the algorithmic foundations of robotics, 2004.
[52] V. T. Nguyen, A. Martinelli, N. Tomatis, and R. Siegwart. A comparison of line extraction algorithms using 2d laser rangefinder for indoor mobile robotics. In: international conference on intelligent robots and systems (IROS05), Edmonton, Canada, 2005.
[53] M. N. Nicolescu and M. J. Matarić. Natural methods for robots task learning: Instructive demonstration, generalization and practice. In: proceedings of the 2nd international joint conference on autonomous agents and multi-agent systems (AAMAS), Melbourne, Australia, 2003.
[54] A. Olenderski, M. Nicolescu, and S. Louis. Robot learning by demonstration using forward models of schema-based behaviors. In: proceedings of international conference on informatics in control, automation and robotics, pp 14–17, Barcelona, Spain, September 2005.
[55] J. Pettré, M. Kallmann, M. C. Lin, J. Kuffner, M. Gleicher, C. Esteves, and J.-P. Laumond. Motion planning and autonomy for virtual humans. In: SIGGRAPH'08 Class Notes, 2008.
[56] A. Ramesh and M. J. Matarić. Learning movement sequences from demonstration. In: proceedings of the international conference on development and learning (ICDL), pp 302–306, MIT, Cambridge, MA, 2002.
[57] S. Schaal. Arm and hand movement control. In M. Arbib, editor, The handbook of brain theory and neural networks, pp 110–113. The MIT Press, second edition, 2002.
[58] S. Schaal, A. Ijspeert, and A. Billard. Computational approaches to motor learning by imitation. The Neuroscience of Social Interaction, 1431, pp 199–218, 2003.
[59] L. Sentis and O. Khatib. A whole-body control framework for humanoids operating in human environments. In: proceedings of the international conference on Robotics and automation (ICRA), pp 2641–2648, May 15–19 2006.
[60] A. Shapiro, M. Kallmann, and P. Faloutsos. Interactive motion correction and object manipulation. In: ACM SIGGRAPH symposium on interactive 3D graphics and games (I3D), Seattle, April 30 - May 2 2007.
[61] A. Siadat, A. Kaske, S. Klausmann, M. Dufaut, and R. Husson. An optimized segmentation method for a 2d laser-scanner applied to mobile robot navigation. In: proceedings of the 3rd IFAC symposium on intelligent components and instruments for control applications, 1997.
[62] W. Suleiman, E. Yoshida, F. Kanehiro, J.-P. Laumond, and A. Monin. on human motion imitation by humanoid robot. In: proceedings of the international conference on robotics and automation (ICRA), pp 2697–2704, May 19–23 2008.

[63] K. A. Thoroughman and R. Shadmehr. Learning of action through combination of motor primitives. Nature, 407, pp 742–747, 2000.
[64] A. Treuille, Y. Lee, and Z. Popović. Near-optimal character animation with continuous control. In: proceedings of ACM SIGGRAPH. ACM Press, 2007.
[65] L. C. T. Wang and C. C. Chen. A combined optimization method for solving the inverse kinematics problems of mechanical manipulators. IEEE Trans Robotics Automat, 7(4), pp 489–499, 1991.

MIX
Papier aus verantwortungsvollen Quellen
Paper from responsible sources
FSC® C105338

If you have any concerns about our products,
you can contact us on
ProductSafety@springernature.com

In case Publisher is established outside the EU,
the EU authorized representative is:
Springer Nature Customer Service Center GmbH
Europaplatz 3, 69115 Heidelberg, Germany

Printed by Libri Plureos GmbH
in Hamburg, Germany